Acquiring Better Seismic Data

Acquiring Better Seismic Data

William Carr Pritchett

London New York
Chapman and Hall

First published in 1990 by
Chapman and Hall Ltd
11 New Fetter Lane, London EC4P 4EE

Published in the USA by
Chapman and Hall
29 West 35th Street, New York NY 10001

Typeset in 10/12 Times New Roman by
Thomson Press (India) Ltd, New Delhi
Printed in Great Britain by
St Edmundsbury Press, Bury St Edmunds,
Suffolk

ISBN 0 412 35000 9

British Library Cataloguing in Publication Data

Pritchett, W. C. (William Carr), 1925–
Acquiring better seismic data
1. Seismic waves. Measurement. Instrumentation
I. Title
551.2'2
ISBN 0–412–35000–9

Library of Congress Cataloging in Publication Data

Pritchett, W. C., 1925–
Acquiring better seismic data/W. C. Pritchett.
p. cm.
Includes bibliographies and index.
ISBN 0–412–35000–9
1. Petroleum—Prospecting. 2. Seismic prospecting. I. Title.
TN271.P4P735 1989
622'.1828—dc20

This book is dedicated to the memory of John Waller Pritchett, Jr, the outstanding student in the petroleum engineering class of 1942 at the University of Texas. Rather than seeking a scholarship to pursue a graduate degree or pursuing his profession, he enlisted in the US Marine Corp. On 2 July 1943 while serving as company commander and leading an assault, he was mortally wounded. The world lost a highly promising engineer and I lost a brother and buddy. Together we played ball; together we hunted, fished, and explored our twin uncles' ranch; together we cut and burned cedar to clear land; but together we did not write books on both exploration and production.

Contents

Preface

I assume that most readers of this book will be at least moderately familiar with the basics of seismology. Readers less familiar with seismology may find it helpful to consult a more fundamental text such as Coffeen (1986). This book is based upon my own experience and is therefore somewhat limited in scope. Also, I have touched only lightly or skipped altogether certain techniques that I consider to be impractical or unimportant, so as to focus on what I consider to be practical and significant for those who are interested or involved in seismic data acquisition. My emphasis on the practical is not meant to disparage those whose interest is more academic. What is academic today may be quite practical tomorrow.

In seismic data acquisition we need: (1) broad bandwidth, usually of at least two octaves, so that our reflection wavelets are not ringy, (2) suppression of noise to give us an adequate signal-to-noise ratio without serious damage to reflections of interest, (3) enough source energy so that data from our deepest objectives can be recorded, (4) frequencies in our reflections that are high enough to provide the vertical resolution required, and (5) minimum costs consistent with acquiring data that meet the needs of the interpreting explorationist for defined exploration objectives. The purpose of this book is to help you meet these needs more fully and thereby acquire better seismic data.

During my thirty seven years with a major oil and gas company many people have contributed to the understanding and knowledge that I seek to document and explain in this book. They deserve much credit. But I alone am responsible for errors, oversights, and any lack of clarity. Much of my material would be subscribed to by most experts in the fields involved, but a few of the ideas, opinions, and recommendations are controversial. Free expression and free interchange of ideas are keys to progress, and I offer my ideas, opinions, and recommendations for the consideration of all in the hope that they will be helpful, stimulating, or possibly both.

There are many excellent geophysicists, geologists, data processors, interpreters, and 'landmen' who are women. I use masculine pronouns to avoid awkwardness, but they should be interpreted as including both men and women equally.

Unfortunately the American Revolutionists did not listen to the recommendations of a surveyor named George Washington who advocated the adoption of the metric system of measurements. (It is ironic that they revolted over revenue stamps and taxes on tea, but would not give up a crazy system of units based upon the length of a kingly foot.) This book is meant to be practical and to relate to the world as it is. The United States has been largely surveyed in feet, miles and acres and my experience is mainly with English units. Most of my examples are in English units with translation to metric. Please forgive the odd numbers that result.

In this book I felt the need to cover a broad range of knowledge and I have sought help from expert colleagues where I have felt the need to do so. They have been quite generous with their time and patience.

Most of the seismic data used in illustrations were supplied by the Atlantic Richfield Company who were very generous in granting permission to use their data and in supplying the data to me. Some data were supplied by Grant–Norpac, Inc. and by Northern Michigan Exploration Company. There are many people to whom I owe much, and I cannot possibly give credit to all. Colleagues that deserve special acknowledgement are Dr Henry F. Dunlap, John B. Etnyre, Virgil R. Pyatt, and James D. Yancey. I greatly appreciate the encouragement given to me by Nigel A. Anstey.

REFERENCES

Coffeen, J. A. (1986) *Seismic Exploration Fundamentals.* PennWell, Tulsa, OK
Sheriff, R. E., *Encyclopedic Dictionary of Exploration Geophysics*, SEG

Glossary

Acoustic impedance

Acoustic impedance is analogous to electric impedance. For a compressional wave acoustic impedance is equal to the ratio of the amplitude of the pressure wave to the amplitude of the displacement velocity. In an infinite, isotropic, homogeneous material the acoustic impedance is equal to the product of the density of the material and the velocity of sound in the material. Near the boundary between two semi-infinite, isotropic, homogeneous materials the acoustic impedance has a value that is intermediate between the density and velocity products of the two materials.

Air guns

The most popular seismic source for marine work utilizes groups of what are called air guns. These 'guns' are steel chambers which are filled with high-pressure air that is released into the water to generate the seismic signal. The seismic signal from an air gun includes the initial pulse plus subsequent pulses resulting from the expansion and contraction of the air bubble. A tuned air gun array is one which has been designed such that the noise from the expansion and contraction of the air bubbles is minimized so that the dominant seismic signal is from the initial pulse when the high-pressure air is injected into the water.

Alias

An ambiguity in the frequency of sampled data. If a signal is sampled at discrete times with less than two samples per period for some frequency component, then for that component the sampling is ambiguous due to *temporal aliasing.* If a signal is sampled at various positions with less than two samples per wavelength for some frequency component, then for that component the sampling is ambiguous due to *spatial aliasing. Alias filters* are filters designed to suppress frequency components that would be aliased due to temporal aliasing—and hence avoid temporal aliasing. (Section 2.8.)

Angle of incidence

The angle between an incident ray and the line that is perpendicular to the interface at the point of incidence.

Angle of reflection

The angle between a reflected ray and the line that is perpendicular to the interface at the point of incidence.

Angle of transmission

The angle between a transmitted ray and the line that is perpendicular to the interface at the point of incidence.

Apparent velocity

The velocity indicated by a measurement. For example, a refraction from a subsurface layer may appear to move along the surface of the earth when in actuality the refraction arrives at the surface at an oblique angle and is reflected back downward and there is no wave traveling along the surface. The apparent velocity is faster than the true velocity and is equal to the true velocity divided by the sine of the angle between the true direction of propagation and the apparent direction of propagation for the measurement. (Section 4.3 and Fig. 4.3.)

Apparent wavelength

The wavelength indicated by a measurement and equal to the *apparent velocity* divided by the frequency. For waves that arrive at the surface at a small angle of incidence (nearly perpendicular to the surface) the apparent wavelength and apparent velocity are much greater than the *true wavelength* and *true velocity*. (Section 4.3 and Fig. 4.3.)

Array

A group of seismic detectors designed to be responsive to a desired seismic signal but to be less responsive to some seismic noise. In marine work the word 'array' often refers to a group of seismic sources rather than seismic detectors.

Array length

The length of a linear array. For linear arrays with a uniform spacing between elements the effective length of the array is equal to the number of elements n times the spacing d which is a length d longer than the physical length of the array.

Attenuation

A loss of amplitude and energy of a signal. In this book the word 'attenuation' will be used primarily for loss of amplitude and energy as a result of conversion of sonic (elastic wave) energy to heat. (Chapter 22.)

Automatic gain control

Automatic means for reducing the gain of the recording and display system as signal level increases and for increasing the gain as signal level decreases, so as to keep the recorded or displayed signal level nearly constant.

Beam steering

A *source beam* from multiple sources is 'steered' or directed at a desired angle by adjustment in the excitation times between the source elements or by differences in the depths of the source elements. A *receiver beam* is 'steered' by separately recording signals at a number of positions and then shifting the relative times before summing the individual signals. Alternatively the recording positions can be at different depths in which case the signals would not have to be recorded separately and then summed. (Chapters 8 and 11.)

Causal wavelet

A wavelet that is caused by some excitation. A causal wavelet is zero until a time after an excitation that is equal to the propagation time from the point of excitation to the point of observation.

Coherence

In addition to the general meaning of the word, coherence has a special meaning relating to the degree to which two wave trains are in phase. Coherence in the frequency domain is analogous to correlation in the time domain.

Common depth point

The approximate common center of the reflection areas of multiple measurements of reflections from a subsurface interface. Before migration a common depth point is usually assumed to exist and to lie directly below the *common mid-point* of traces in a *common mid-point gather*. (Section 2.2 and Fig. 11.5.)

Common mid-point

Traces that have a common, surface mid-point between their source points and receiver points. Such traces are usually assumed to have common reflection points at depth. See *common depth point*.

Converted shear wave

When a *compressional wave* impinges at an oblique angle upon a boundary between layers with contrasting acoustic impedance, the particle motion at the wavefront has a component which is parallel to the interface. As a result, part of the compressional wave energy is converted to *shear wave* energy. The resulting shear waves are called *converted shear waves* or PS waves. (Chapters 12 and 15.)

Converted compressional wave

When a shear wave impinges at an oblique angle upon a boundary between layers with contrasting acoustic impedance, there may be a component of particle motion at the wavefront which is perpendicular to the interface. If so, part of the shear wave energy is converted to compressional wave energy. The resulting compressional waves are called *converted compressional waves*, or SP waves.

Convolution

A change in the wave shape of a signal as a result of some linear filter such as the earth filter, an electrical circuit filter, or a digital filter in a computer program.

Correlation

A process (frequently digital) by which two signals are compared for the degree of correspondence between them. If the two signals are identical it is called an *autocorrelation*. If the two are not identical it is called a *crosscorrelation*. Equipment used to correlate signals is called a *correlator*. (Appendix H and Figs. H–1 and H–2.)

Crest

A maximum point of a waveform. If the polarity were reversed a crest would become a trough.

Critical angle

The angle of incidence at an interface between two layers which results in an *angle of transmission* which parallels the interface and sets up a refraction in the second layer. (Appendix D and Fig. D-1.)

Cross-dip

The dip of a bed in a vertical plane that is perpendicular to a seismic line.

Deconvolution

The inverse of convolution, i.e. the undoing of the effect of a previous filter on a signal (to the extent that this previous effect can be undone). (Chapter 19.)

Degree

Besides the common usage, degree is used in a very specific sense in regard to *zero-phase wavelets*. A zero-phase wavelet of zero degree can be represented by a summation of cosine waves with phase angles of zero degrees. A zero phase wavelet of ninety degree can be represented by a summation of cosine waves with phase angles of ninety degrees. (Appendix G.)

Depth point

The depth point is the reflection point on an interface for a ray from the source to the receiver. Usually we think in terms of an earth with horizontal layers and with a horizontal surface. For this idealized earth model the depth points for all reflecting interfaces lie directly below a surface point that is half-way between the source and the receiver. For dipping interfaces and for complex earths the reflection points at depth will not be half-way between the source and the receiver positions. But we often handle the data as if the depth points were still half-way between. Frequently we ultimately migrate the data to consolidate events and to put them in their proper locations.

For converted waves the angles of incidence and reflection are not equal at the conversion interface. As a result the depth point is closer to the receiver and farther than the source for PS waves and is closer to the source and farther from the receiver for SP waves. For PS and SP waves in a horizontally layered earth, the depth points for different interfaces will not project vertically to the same surface point. This adds considerable complexity to common depth point processing of converted wave seismic data.

Dispersion

A change in the phase velocity of a wave as a function of the frequency. (Chapter 22)

Downsweep

A vibrator *sweep* which starts at the highest frequency of the sweep and the frequency is reduced progressively during the sweep.

Dynamic range

The amplitude range of signals that can be recorded with very little distortion. In a practical sense with modern seismic instruments, the lower limit of the dynamic range is usually restricted by the level of 'outside', background noise rather than noise generated within the amplifiers and other electrical circuits. This background noise includes noise generated by animals, people, trucks, boats, other crews, wind, waves, power line pick-up, vehicles running over cables, etc. The upper limit of the dynamic range is usually restricted by the detector or a by some circuit element such as an amplifier that precedes the analog-to-digital converter.

In most modern digital recording systems the effective dynamic range is not limited by the number of bits recorded to represent each signal sample. Both the lower and upper limits of the dynamic range may vary considerably with frequency. The dynamic range is also affected by the measurement technique and by data processing. The compositing of multiple measurements can improve the dynamic range. The Vibroseis technique provides significant opportunity to improve the dynamic range by editing before correlation and by correlation itself. But correlation also creates side lobes which are a type of noise in themselves which limit dynamic range.

End-on cable
A cable in which all of the traces to be recorded extend down the seismic line in only one direction from the source.

Fold
The number of separate measurements of a single reflection (with different source points) that are recorded for each *mid-point* with the *common mid-point* technique.

Gather
A display of seismic traces with some common element. A *source or shot gather* is a display of traces from the same source or shot point. A *receiver gather* is a display of traces recorded with the same receivers. A *CDP gather* is a display of traces with the same reflection depth point. A *CMP gather* is a *common mid-point gather* which is generally used as an approximation to a CDP gather.

Ghosts
Multiple events on seismic sections that involve downward reflections from above the depth of the source, from above the depth of the receiver, or both. (Chapters 8 and 9 and Fig. 9.4.) See *harmonic ghost* for a different type of ghost.

Group
All of the receivers of an *array* that feed the same data channel.

Group interval
The distance between the centers of adjacent groups along a seismic line.

Group velocity
The velocity with which energy is propagated. It is the velocity of the envelope of a wavelet. (Chapter 22.)

Harmonic ghost
In the Vibroseis technique, when downsweeps are used, events called harmonic ghosts may appear on correlated records. These ghosts result from correlation of the sweep with harmonics of the intended sweep frequencies.

Lateral
In a horizontal direction perpendicular to the seismic line.

Linear sweep
A vibrator sweep where the instantaneous frequency is a linear function of time during the sweep.

Listening time

That part of the recording time after the end of a vibrator sweep. (Actually the listening time begins with the initiation of the sweep.)

Live

A geophone or other detector is live if it is operative and connected to a data channel that is to be or is being recorded.

Longitudinal

In a direction parallel to the seismic line.

Marker bed

A bed or sequence of beds that is relatively planar which has little dip and yields a measurable and consistent seismic reflection over a relatively large area.

Minimum-phase filter

All physically realizable filters are minimum-phase. The phase delay as a function of frequency is the minimum that will preserve *causality* for an input wavelet that is minimum-phase. (Appendix G.)

Minimum-phase wavelet

A *causal wavelet* that has the minimum values for the phase angles of its phase spectra. (Appendix G.)

Mixed-phase wavelet

A wavelet that is neither *minimum-phase* nor *zero-phase*. (Appendix G.)

Multiple reflection

A seismic reflection from depth that has undergone at least one downward reflection and therefore includes more than one, two-way passage though at least one interval.

Normal moveout

The delay in the arrival time of a reflection with increasing *offset* distance between the source and receiver due to the increase in path length. (Chapter 4.)

Normal moveout velocity

The velocity that best fits the measured *normal moveout* of a reflection from an interface that is parallel to the surface based upon the slant path distance of the hypotenuse where the one side is twice the depth and the other side is the offset distance between the source and receiver positions. (Chapter 4.)

Notch filter

A narrow band, rejection filter designed to reject a strong noise with a narrow bandwidth (often used to reject noise picked up from power lines).

Notch

A minimum response. In particular a very low output response of a filter near some frequency, wavelength, or velocity for which the theoretical response may be zero.

Nyquist frequency

The frequency with a period that is twice the sample interval. Frequencies above the Nyquist frequency are aliased. (The Nyquist frequency is equal to the reciprocal of twice the sample interval.)

Offset

The horizontal distance from the center of a *source pattern* to the center of a *detector array.*

One-way time

Half of the reflection time after correction for *normal moveout.*

Patch

A large area over which many geophone *arrays* are laid out. The arrays are recorded separately, but are summed, after suitable corrections, to produce a single trace.

Pattern

A group of source positions that are designed to effectively produce a desired seismic signal and to minimize some *source-generated noise.* The positions are frequently energized simultaneously to record a single seismic record. The excitation times may be slightly different in order to deflect the beam angle from the vertical to a more favorable angle (*beam steering*). In this book, 'pattern' will refer to a configuration of sources but not to a configuration of detectors.

Peak

A maximum response. In particular a high-output response of a filter near some frequency, wavelength, or velocity.

Phase velocity

The velocity with which a given crest or trough of a wavelet travels. It is the apparent velocity with which a continuous, sinusoidal wave of a single frequency travels. In non-dispersive media it is the same as the group velocity. In dispersive media it differs from the group velocity. (Chapter 22.)

Podded geophones

Geophones that are planted very close together.

Predictive deconvolution

Deconvolution of the latter part of a seismic trace by the use of information from the early part of the same trace.

Primary reflection

A seismic event that results from only one reflection and that reflection is from an interface that is deeper than the source.

Q

A measure of the efficiency for storing and releasing energy. It is defined as 2π times the peak energy in a cycle divided by the energy dissipated in the cycle.

Reverberation

Multiple reflections in a layer bounded by interfaces with high coefficients of reflection that result in an oscillatory noise which interferes with the recording of primary reflections. Reverberation is most severe in marine seismology where there is a hard water bottom and the multiple reflections are at the top and bottom of the water layer. The problem occurs frequently in land data, but it is usually much less severe. (Chapters 8 and 19.)

Ringy

Data characterized by many cycles of nearly constant period and nearly constant amplitude.

Semblance

A measure of the coherence between traces. It is equal to the zero lag value of the autocorrelation of the sum of the selected traces divided by the average of the zero lag values of the autocorrelations of all of the selected traces.

Signal-to-noise ratio

The ratio of the energy of the desired signal to the rest of the energy during a given time interval.

Single-ended spread

A seismic recording with the live receiver positions extending in only one direction from the source. See *end-on cable.*

Source-generated noise

Noise that is generated by the source. For example, in a reflection survey refractions as well as air waves and ground roll are source-generated noises. (Chapter 2.)

Split spread

A seismic recording with live receiver positions extending along the seismic line in both directions from the source. In a symmetrical split spread the live offsets are the same in both directions. In an unsymmetrical split spread the offsets are not the same in both directions and usually the number of traces differs in the two directions.

Spread

The geophone *arrays* and connecting cable used to record a single seismic record.

Stack

A record made by summing traces from different recordings where the desired signals should reinforce and improve the *signal-to-noise ratio.*

Stacking velocity

The velocity used to make a *normal moveout* correction before stacking (summing) the traces of a *common mid-point gather*. The stacking velocity is the same as the moveout velocity.

Statics

Variations in the arrival times of reflections due to variations in surface elevation, shot depth, weathering thickness, weathering velocity, etc. (also corrections made to compensate for these variations).

Sweep

A Vibroseis sweep is the sinusoidal wave of slowly varying frequency that is applied to the earth by one or more vibrators.

Trace

The seismic response from a single geophone array and from a single source position.

Traveltime

The time between the initiation of a seismic source and the arrival of a response at a detector.

Trough

A minimum point of a waveform. If the polarity were reversed a trough would become a crest.

Upsweep

A vibrator *sweep* that begins at the lowest frequency of the sweep and the frequency is progressively increased during the sweep.

Vibroseis

A system for seismic measurements in which the input signal is not a short duration pulse; instead, it is a sinusoidal signal with a slowly varying frequency. The input signal is generated by vibrators which use hydraulic force to accelerate a reaction mass.

Wavenumber

The number of wave cycles per unit distance. ($1/\lambda$ but in Chapter 22 it is defined as $2\pi/\lambda$.)

Zero-phase filter

A filter which changes the amplitude spectra of the input wavelet, but does not change the phase spectra. Zero phase filters are not physically realizable filters.

Zero-phase wavelet

A wavelet which can be represented by a series of cosine waves of different frequencies with the same phase angles. (Appendix G.)

References

Sheriff, R. E. (1973) *Encyclopedic Dictionary of Exploration Geophysics*, SEG.

Nomenclature

ABBREVIATIONS

AGC	automatic gain control
CDP	common depth point
CMP	common mid-point
cm	centimeter or centimeters
dB	decibel or decibels
ft/s	feet per second
ft	foot or feet
Hz	Hertz (cycles per second)
km	kilometer or kilometers
ln	natural logarithm (to the base e)
m	meter or meters
mi	mile or miles
mm	millimeter or millimeters
m/s	meters per second
μs	microseconds
NMO	normal moveout
s	second or seconds
S/N	Signal-to-noise or signal-to-noise ratio
SP	shot point or source point
VP	vibrator point

GREEK LETTER SYMBOLS

α	attenuation exponent in nepers per meter or foot
ϕ	phase angle in radians or degree, or the porosity of a formation
ϕ_g	phase delay of a ghost due to the gun depth
ϕ_c	phase delay of a ghost due to the cable depth
κ	wavenumber in feet or meters ($1/\lambda$ or $2\pi/\lambda$ in Ch. 22)
λ	wavelength in feet or meters
λ_a	apparent wavelength in feet or meters

μ	micro (10^{-6})
π	3.14159...
θ	angle (frequently measured from vertical)
θ_{i1}	angle incidence of a ray in layer 1 at the interface between layers 1 and 2
θ_{r12}	angle of refraction of a ray in layer 1 at the interface with layer 2
θ_{t2}	angle of transmission of a ray in layer 2 from layer 1
ω	angular velocity in radians per second or degrees per second

ENGLISH LETTER SYMBOLS AND COMBINATION SYMBOLS

A	Amplitude of a wave or wave component
a_{bp}	acceleration of the baseplate of a vibrator
a_{rm}	acceleration of the reaction mass of a vibrator
c	earth constant in the attenuation exponent
d	spacing between adjacent elements in a geophone array or source pattern, or travel distance of a ray through the earth
d_g	depth of a gun array
d_c	depth of the cable
D	depth in feet or meters
f	frequency in Hertz (or megahertz)
$f_{0.64}$	frequency for which the response is 64 per cent of maximum
$f_{0.3}$	frequency for which the response is 30 per cent of maximum
f_0	frequency for which the response is zero
f_a	apparent frequency
f_{am}	apparent mean frequency
f_e, F_F, F_2	ending or final frequency of a sweep
f_s, F_I, F_1	starting frequency or initial frequency of a sweep
f_{a1}	apparent lower frequency of wavelets
f_u	apparent upper effective frequency allowed by the earth filter
j	square root of -1
k(f)	sensitivity of a geophone as a function of frequency
L	longitudinal offset distance from source to receiver
L_s, SL	length of a vibrator sweep in seconds
m_{bp}	baseplate mass of a vibrator
m_{rm}	reaction mass of a vibrator
n	number of elements in a geophone array or source pattern
P	path length of a ray in feet or meters (or centimeters)
R_{wa}	reflection coefficient at the water–air interface
s(t, x)	sinusoidal wave in time and in space
SP	Shot point or source point
t	time in seconds or milliseconds (or microseconds)

t_0	time zero from which other times are referenced
T	period of a wave (reciprocal of frequency)
U	group velocity in feet/second or meters/second
v	velocity in feet/second or in meters/second
v_a	apparent velocity in feet/second or in meters/second
v_p	phase velocity in feet/second or meters/second
VP	vibrator point
w	weighting factor
w(t)	wavelet as a function of time
$\Delta\phi$	a change in formation porosity
Δt	an increment of time
Δx	an increment of distance in the x direction
ΔL	an increment of longitudinal offset
ΔP	an increment of path length of a ray

REFERENCES

Sheriff, R. E. (1973) *Encyclopedic Dictionary of Exploration Geophysics*, SEG.

1

Introduction

You can acquire *better seismic data.* Emphasis in this book is upon concepts and techniques to help you select equipment, techniques and recording parameters to acquire better seismic data. Better seismic data means higher quality data to solve precisely defined exploration problems. But it also means minimum cost consistent with quality data. In many areas of the world previous exploration has already found most of the better reserves except where the traps are subtle or where previous seismic data are poor. Successful exploration today often requires better seismic data.

This book is written for explorationists in general and as a reference for those involved in seismic data aquisition who we will refer to as field geophysicists (even if they are in an office much more than they are in the field). In small organizations the field geophysicist may also be the interpreter or the data processor. In large organizations, geophysicists and geologists who are mainly interpreters sometimes become involved in seismic data acquisition. Perhaps more explorationists should and would be involved if they were more knowledgeable. They and seismic data processors need a better understanding of the acquisition of better seismic data and the problems, choices and opportunities faced by field geophysicists. The perspective is petroleum oriented, but much of the material should apply to mining and engineering work as well. When we recorded long records, we often saw crustal reflections from within the basement. I am confident that techniques discussed in this book could be applied successfully to the acquisition of better seismic data for crustal studies. Hopefully this book will be helpful to many explorationists, seismic contractors, and earth science students and teachers.

Instead of a deliberate, pedalogical approach this book is organized so as to quickly present practical information and interesting and sometimes controversial ideas and conclusions. Subjects discussed in this book, including special problems and solutions, are indicated in the Contents pages and are summarized in this chapter. Some mathematical derivations, equations, numerical examples, and tables are given in appendices so that the main text can present concepts and conclusions quickly for easier reading and reference. A glossary and lists of abbreviations and symbols are provided. In Chapter 2 we plunge into a wide range of interesting and useful concepts, conclusions and observations. The chapter discusses noises and how noises can be suppressed by geophone arrays, source patterns, and common mid-point stacks. The advantages of Nigel Anstey's stack array (Anstey, 1986) are pointed out, but with emphasis on advantages of array lengths that are twice the group interval rather than equal to the group interval. Other items discussed include multiples, signal sampling and aliasing, and nonlinear effects and dynamic range.

Chapters 3 and 4 develop ideas and concepts required before we proceed with more advanced material. Chapter 3 and Appendix D deal with the design of practical geophone arrays and source patterns. Chapter 4 discusses moveout velocity, differential moveout across a geophone array or source pattern, and the loss of high-frequency components of reflected signals as a result of differential moveout. The chapter also deals with apparent wavelength and apparent velocity as a function of the vertical angle of arrival. Techniques and equations are given for extracting from field data information including moveout velocities, zero offset reflection times, and reflection times as a function of offset. Techniques for estimating the frequency bandwidth of reflections are discussed. Chapter 5 discusses the Vibroseis technique, its advantages, and the practical operation of Vibroseis crews. Chapter 6 discusses the need for high resolution seismic data and how to improve resolution with either impulsive sources (such as explosives) or vibrator sources with nonlinear sweeps. Chapter 7 illustrates in detail the practical design of field tests with example cases for a Vibroseis crew, a shot hole crew, and a portable crew using mini-hole shot patterns. Chapter 8 discusses marine seismology including the limitations of a wide gun array for the suppression of scattered noise. This chapter includes a discussion of beam steering and its application at sea. Chapter 9 deals with shot hole crews including the effect of ghosts on reflected wavelets and means for suppressing ghosts. The discussion also includes the use of shallow, mini-hole patterns. Chapter 10 discusses off-road seismic coverage and three-dimensional surveys including the use of techniques to record simultaneously from two or more sets of sources.

The next five chapters deal with special exploration problems and practical solutions to these problems. Chapter 11 deals with steep dips and complex geology and the possible advantages of beam steering and long offsets. Chapter 12 addresses problems presented by high-velocity surfaces such as karst and volcanic surfaces and the use of star geophone arrays and source patterns as well as large detector 'patches'. The possible use of shear waves for this problem is included. Chapter 13 discusses poor record areas in general as well as problems with a low-velocity surface underlain by a much higher velocity section. Chapter 14 deals with shallow objectives and the effective use of simultaneous recording to reduce costs. Chapter 15 deals with the detection and mapping of the development of porosity in objective layers and the detection and mapping of the presence of hydrocarbons. The discussion includes the use of comparative compressional and shear-wave data. The use of compressional-wave refraction data is also discussed.

The next three chapters deal with practical field operations. Chapter 16 discusses the selection and operation of equipment. Chapter 17 covers practical line layout, and Chapter 18 summarizes much of the previous discussion and deals with miscellaneous problems such as crew management and public relations.

Chapter 19 briefly discusses seismic data processing from the standpoint of what the field acquisition geophysicist needs to know to do his job well. Chapter 20 deals with the important subject of effective communication between the field geophysicists, the processing geophysicists, and the interpreting explorationists. Chapters 21 and 22 deal with theory. Chapter 21 handles the concepts of superposition, reciprocity, and wavefront healing along with theory of the suppression of random noise. Chapter 22 deals with seismic wave propagation and the alteration of seismic wavelets during propagation. Effects of both body-wave attenuation and body-wave dispersion are included along with a discussion of their effects on measured reflection times (and measured arrival times in well velocity surveys).

Chapter 23 deals with economics, seismic crew efficiency, future trends in seismology, and final conclusions.

Getting good, interpretable seismic data starts in the field with data acquisition and no amount of computer processing can overcome poor choices of acquisition equipment, techniques, or parameters, or poor execution. Since some seismic data problems can best be handled in processing, the raw field data may or may not look good. The criterion for data quality is how well the processed data meets the needs of the interpreter in solving the exploration problems. The field geophysicist must understand data processing techniques well enough to know which problems should be attacked in the field and which should be left for processing to handle. Without reasonably good processing capabilties in the field even experienced geophysicists may err in selecting recording techniques and parameters. Consequently, field data may be unnecessarily degraded by filters or arrays or, conversely, it may be contaminated by noise that could best be suppressed by better field techniques or better recording parameters. The many advantages for processing data in the field should be carefully considered. But if field processing is not feasible, a good field geophysicist can usually develop a near optimum technique with near optimum recording parameters.

In relatively good record areas some simple method such as Nigel Anstey's stack-array often yields satisfactory results (Anstey, 1986). (Chapter 2). But those results are not likely to be optimum and in problem areas there is no substitute for a more in-depth approach. Even in areas where good data have previously been recorded, some initial testing may be in order to see if you can expand the bandwidth of reflections or if you can better suppress multiples or other noise. Meaningful tests can frequently be brief if the tests are carefully selected and organized. Critical tests should be carried out early in the test program so that some subsequent tests may be modified or aborted if that becomes appropriate. The primary purpose of a test may be to speed up recording to save money (but still get quite satisfactory data). Every part of every test should have a clearly defined purpose that is pertinent to exploration objectives. If tests are planned to try all possible combinations of a number of parameters, then they become long and expensive. The key is to test a near minimum number of the more promising combinations and to evaluate as you proceed to see what can be eliminated or if any other combinations should be tested. The recording time required for a test program may range from an hour to a couple of days. Tests should rarely require more time unless a significant amount of high fold, common mid-point data (CMP) is recorded—in which case the test should provide data which can be incorporated into the survey itself. A good test program must provide information that is worth more than the cost of the tests. Then the information must be used effectively to record better seismic data.

There are other very short tests which may be carried out periodically throughout a survey. For example, with a vibrator crew that has been summing twelve sweeps per vibrator point (VP), you might sum and dump to tape after four, eight, twelve, and sixteen sweeps for the first two VPs every morning. This would allow you to see if fewer sweeps would be satisfactory (and save money) or if more sweeps would significantly improve the data quality (by overwhelming noise that is not generated by the vibrators).

There is no way that the field geophysicist can do an excellent job unless he knows the exploration objectives and as much as is reasonably possible about the geology from the surface to below the objective depths. I recommend the use of a form called a prospect form (Appendix A) which is filled out and signed by an interpreting explorationist. This form asks for key information including the objectives and the geology. It also asks for seismic information such as velocities and for anticipated problems such as severe ground roll. It asks for line location and for numbers of the closest seismic lines available for study. The explorationist fills in whatever pertinent information he can in less than half an hour. (Certainly he can spend more time finding additional information if he wishes, but the filling in of the prospcct form should not become a burden.) The field geophysicist should then review the prospect form and add whatever information he can. He should consult with the explorationist for clarification or for help in deciding what other information is needed and where it can be obtained.

If other seismic data are available nearby or in similar geologic circumstances, they should be studied to see what problems exist and what band of frequencies are present in reflections from objective depths. At the ends of lines and in the vicinity of skips the geophysicists can gain some idea of the effectiveness of short and long offsets for recording reflections from various depths. If time permits, the geophysicist may want to display some pre-stack data from previous lines. Displays may be shot gathers (with and without automatic gain control), CMP gathers (with and without mute or normal moveout corrections), etc. He may wish to have various filters applied to a short portion of a line. For example, he might want to see if there is coherent energy above 50 Hz at times corresponding to reflections from objective depths (or from deeper depths). He could apply a sharp cut-off, low-cut filter of 50 Hz. If the section contained reflections with frequency components in the 50–60 Hz range, then he should observe coherent events near these reflection times. Such events would be extremely ringy because of the limited bandwidth.

In problem areas where data quality varies, the interpreting explorationist should plot all available seismic lines on an overlay to a geologic map. He should then color-code the seismic lines to show the relative data quality such as red for bad, orange for poor, yellow for fair and green for good. He should then look for any correlation between data quality and geologic outcrops, elevation, surface condition (swampy ground, cap rock), etc. If any correlations are found he may find it feasible to avoid locating future lines where conditions are similar to those where data quality was not satisfactory. Where lines must cross areas similar to those where previous data quality was poor, the field geophysicist should be so advised. He can then guess more effectively what problems may be encountered. He may want to conduct some special tests in a tough area to better define the problem and to seek a solution. He may want to increase source energy just in that area or to increase CMP fold just in that area (if the line was not planned initially for full CMP fold recording). In a tough area he may find that a different combination of source pattern and geophone array would improve data quality. To be successful the field geophysicist needs as much pertinent information as possible and to get it he requires excellent communication with understanding and co-operative explorationists.

The field geophysicist also needs excellent communication and co-operation with the processing geophysicist. There is a great deal that the processor can determine from the data that would be helpful to both the field geophysicist and the interpreting explorationist. This subject of communication between the field geophysicist, the interpreting explorationist and the data processor is so important that it is covered more fully in Chapter 20.

Some less technical subjects are very important. To be successful the field geophysicist must establish excellent, two-way communication and an excellent working relation with the seismic crew. He should communicate frequently with the party manager and work with and through the party manager to develop and maintain an effective crew. He must have technical support to help him see that equipment is up to specifications and that crew members are competent and conscientious in the use and maintenance of their equipment.

The field geophysicist must understand crew problems and should work with the party manager to accommodate personal problems in so far as it is feasible to do so. Crew morale is important. Good morale is more likely where the crew has the right equipment, the equipment is in good condition, the crew understands exactly what you want them to do, and they appreciate that it is important for each of them to do their job well. Good morale is more likely if you challenge the crew to suggest ways to improve operations and if you accept appropriate ideas. Good morale is more likely if you set high standards, criticize the crew constructively when they do poorly, but give them recognition and praise when they perform at a high level. Everyone likes to see evidence that his hard work yields worthwhile results. Showing a crew an appropriately censored piece of a superior seismic section can help each crew member feel good about his efforts. A jug hustler may not understand much about the section, but if you compare it with some previously acquired data of lower quality you can point out differences and explain why the differences are significant.

The field geophysicist must work closely with the crew to maximize continuity of data and yet avoid expensive damage claims and unfavorable reaction of the public and of public officials. The field geophysicist (and his boss) should accept some moderate risks to minimize loss of data, but he should seek ways to minimize these risks. Particularly when seismic sources are located close to houses, wells, water lines, etc., the crew should be alert and should immediately inform the observer if there is any evidence of damage or of unusual concern by the public. The observer should immediately investigate to see what has happened. If nothing seems amiss he should make a notation on his observer's sheets and then proceed with the work. If he is in doubt, if concern is justified, or if damage has occurred, he should report to the party manager and then both of them should call the field geophysicist.

The party manager, permit agent, and surveyor should all be encouraged to note feasible alternative line routes that might not be known to you. They should bring to your attention routes which might be advantageous. If severe problems arise for initially planned line routes, they should immediately inform you of the problems and be prepared to suggest possible changes in line routes.

The field geophysicist should assist the crew in obtaining difficult permits from landowners, mineral owners, and local, state and federal officials. If a problem of skips due to culture or withheld permits is severe, the field geophysicist should discuss the problem with the appropriate explorationist to see if line location might be changed to reduce the skip problem and still meet the exploration objectives.

REFERENCES

Anstey, Nigel A. (1986) Whatever happened to ground roll? *The Leading Edge*, **5**, No. 3 (March), 40–5.

2

General concepts, conclusions and observations

Whereas the discussion in this and other chapters is related to land acquisition, the general concepts also apply to marine seismology. This chapter deals with a wide range of concepts, conclusions and observations. We start with noises and how noises can be suppressed by the combination of geophone arrays, source patterns and common mid-point stacks. Included is a discussion of Nigel Anstey's stack array (Anstey, 1986) but with a preference for arrays that are twice as long as the group interval where source-generated noise is severe. Multiples, signal sampling and aliasing, and nonlinear effects and dynamic range are also discussed.

2.1 COHERENT NOISE MEASUREMENTS

Often we record noise (any unwanted response by our instruments) on purpose using individually recorded, point receivers and point sources in what we call a noise test. In such a test the receivers are located along a line with a short spacing between adjacent positions (Appendix E, Fig. E-1). Figure 2.1 shows the results of such a noise test where the spacing between receivers was 50 ft (15.24 m) and the offsets between source and receiver ranged from 50 ft to 7200 ft (15.24 m to 2195 m). The first breaks are refractions from one or more shallow earth interfaces. Low-frequency noise is rather strong only at the

Fig. 2.1 Noise spread measurements with offsets from 50 to 7200 ft (15.24 to 2195 m). Forty-eight receivers positions were located at 50 ft (15.24 m) intervals along a line and data were recorded from vibrator points off one end at offsets of 50, 2450, and 4850 ft (15.24, 746.8, and 1478.3 m). At each receiver position six geophones were planted very close together in a 'pod'. Figure E-1 (Appendix E) gives a simple illustration of a similar layout with only six receiver positions.

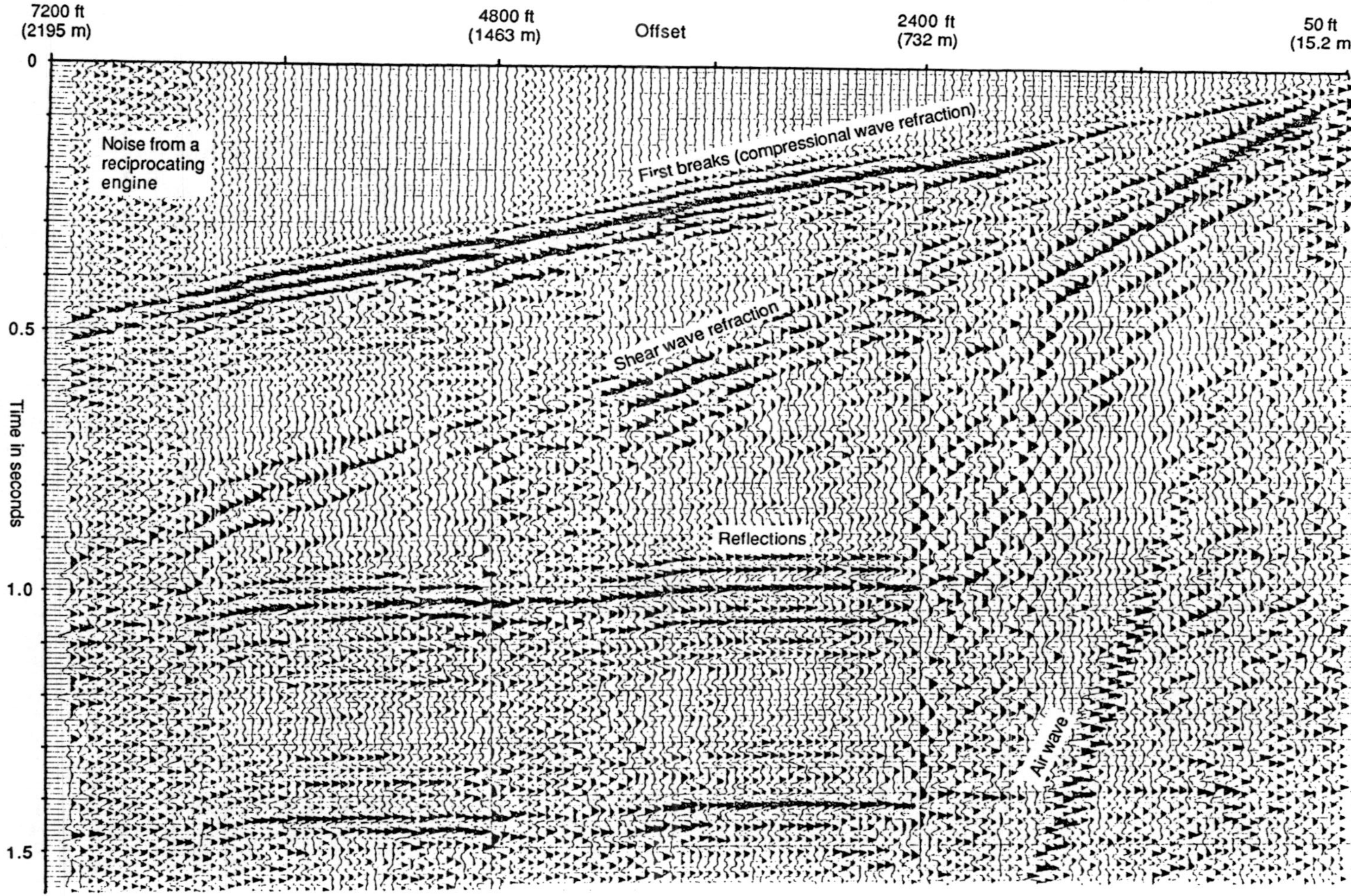
7200 ft
(2195 m)
4800 ft
(1463 m)
Offset
2400 ft
(732 m)
50 ft
(15.2 m)
0
0.5
1.0
1.5
Time in seconds
Noise from a
reciprocating
engine
First breaks (compressional wave refraction)
Shear wave refraction
Reflections
Air wave

shorter offsets and primarily at the early record times. The air wave is aliased by the 50 ft (15.24 m) spacing between adjacent receiver positions, but you can still judge its relative strength. The reciprocating engine that produced the indicated noise was the engine in the instrument truck that drove the alternator for the instrument power supply. The observer should have noticed this noise and should have reduced it (if necessary, by moving the instrument truck laterally off line).

Figure 2.2 is an enlargement of a section of this noise test. Trend lines have been drawn through troughs for two coherent noises on this enlargement. Drawing in trend lines helps to average out variations due to statics and overlapping noises. The apparent velocity is equal to the change in distance for a given change in time for a given event. For noise A the trend line at 0.18 s is at an offset of 2500 ft (762 m). One-tenth of a second later at 0.28 s the trend line is 31 traces farther out at an offset of 4050 ft (1234 m). The apparent velocity of noise A is:

$$V_a = (4050 - 2500)/0.1 = 15\,500\ \text{ft/s}$$

$$= [(1234 - 762)/0.1 = 4724\ \text{m/s}]$$

The time between the crests of noise A varies from about 0.02 to about 0.025 s showing a dominant frequency range of about 50–40 Hz. Since wavelength is equal to velocity divided by frequency (or velocity times period), the apparent wavelength of noise A is about 310–388 ft (94–118 m).

For noise B the trend line at 0.45 s is at an offset of 2500 ft (762 m). One-tenth of a second later at 0.55 s the trend line for noise B is 17 traces farther out at an offset of 3350 ft (1021 m). The apparent velocity of noise B is:

$$V_a = (3350 - 2500)/0.1 = 8500\ \text{ft/s}$$

$$= [(1021 - 762)/0.1 = 2590\ \text{m/s}$$

For noise B the time between crests varies from about 0.035 to about 0.043 s. The frequency and apparent wavelength range respectively from about 29–23 Hz and 298–366 ft (91–112 m).

Fig. 2.2 Measurement of noise characteristics. This figure is an enlarged portion of Fig. 2.1. Trend lines have been drawn through the troughs of two sets of coherent noise labeled A and B. For Noise A the trend line shows a change of 0.1 s in arrival time for an offset change of 1550 ft (472.4 m) which is an apparent velocity of 15 500 ft/s (4724 m/s). The period of noise A ranges from about 0.02 to about 0.025 s which corresponds to a dominant frequency range of 50–40 Hz. The corresponding apparent wavelength is 310–388 ft (94–118 m). The noise at B has an apparent velocity of 8500 ft/s (2590 m/s), a period of 0.035–0.043 s, a dominant frequency of 29–23 Hz, and an apparent wavelength of 298–366 ft (91–112 m).

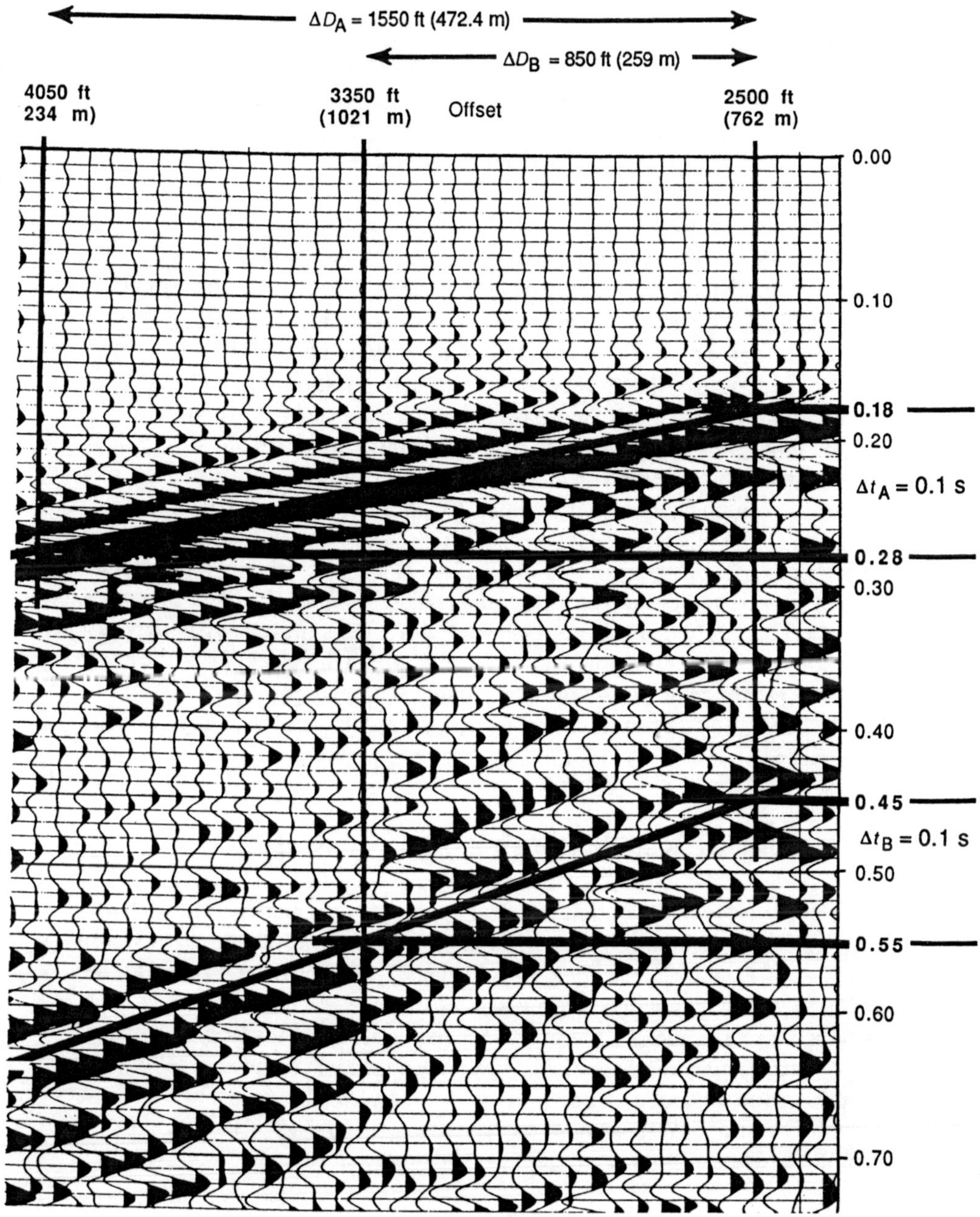

$(\Delta D_A)/(\Delta t_A) = 15\,500$ ft/s (4724 m/s)

$(\Delta D_B)/(\Delta t_B) = 8500$ ft/s (2590 m/s)

Look again at Fig. 2.1 and the aliased air wave. We could draw a trend line through the air wave and back to zero time at zero offset. We could then calculate the air wave velocity of about 1150 ft/s (350 m/s), measure its period of about 0.025–0.028 s, and calculate the frequency and wavelength ranges respectively of 40–36 Hz and 29–32 ft (8.8–9.8 m). There is nothing more we need to know, and in this case spatial aliasing was of no consequence.

In a similar manner the velocities, periods, frequencies and wavelengths of other coherent noises could be determined. Likewise measurements of the characteristics of unsuppressed noise can be made on routine field data. This allows us to determine the apparent velocities, frequencies and apparent wavelengths of noises that are not well suppressed in the field by the geophone arrays and source patterns. We can then consider whether or not the processed data will be seriously affected at objective times by these noises and whether or not we should change our field parameters or even our field technique.

Note in Fig. 2.1 that reflections from horizontal interfaces show very little increase in arrival time with an increase in offset. This is because the reflections arrive at a very small angle to the vertical for offsets that are small compared to the depth to the reflector. From measurements along the surface, the arriving reflections appear to move along the surface at a very high apparent velocity whereas in actual fact they arrive nearly vertically from below at a moderate velocity. The calculated apparent velocity along the surface is extremely fast and the apparent wavelengths extremely long. At much longer offsets the angle of arrival for the reflections would be greater, the apparent velocity would be much slower, and the apparent wavelengths much shorter (Chapter 4 and Fig. 4.3).

2.2 SUPPRESSION OF COHERENT NOISE

Coherent noises will have some range of apparent wavelengths. If the distance between two geophone positions in an array (or between two source positions in a pattern) is half of some apparent wavelength, then noise and signal at that apparent wavelength will be suppressed. With more geophones, more sources, or both, we can suppress coherent noises over a significant range of apparent wavelengths. Appendix D explains how geophone arrays and source patterns suppress source-generated noises and how they can be designed to fit our needs. The practical design of arrays and patterns is also discussed in Chapter 3. These design techniques are illustrated in Chapter 7 and throughout the book.

Appendix D also deals with the concept of sinusoidal waves in space and in time and the representation of these waves by vectors. Readers with limited background in array theory or in the representation of waves as vectors may want to read Appendix D before proceeding with this chapter.

The noise suppression capabilities of arrays and of source patterns are limited by variables that cannot be controlled or for which control is impractical. Some examples are variations in near-surface statics, variations in earth coupling, and variations in precise location of individual geophones. Because of these variables, complex arrays and complex patterns fail to provide theoretical promises of superior noise suppression. Therefore, relatively simple arrays and patterns should normally (but not always) be used to record seismic data. Simple arrays and patterns are easier and quicker to lay out and errors are less likely to occur. This does not mean, however, that we cannot achieve high levels of noise suppression over a reasonable band of noise wavelengths. For example, we may achieve about 18 dB of suppression with a practical geophone array, another 18 dB with a practical source pattern, and 20 dB more with the common mid-point (CMP) stack, yielding a total of 56 dB of suppression over a significant band of noise wavelengths!

Can we really suppress noises this well? Yes, if care is used in the design of the overall acquisition plan. In the wavelength rejection band of any array or pattern there are peaks where the noise suppression is relatively poor (or sometimes nil). There are also notches in the rejection bands where noise suppression is very good. With care we can adjust the source pattern to provide notches at some wavelengths where the geophone array has peaks. By this technique we can increase the width of the band of wavelengths for which the composite response yields good noise suppression (Chapter 3 and Appendix D).

Whereas the reader is presumed to be familiar with common mid-point (CMP) recording, this concept is so basic to modern seismic exploration that a short review may be appropriate. A number of traces with different longitudinal offsets are recorded for a sequence of shot points along a seismic line. From this data set, shot point and receiver position pairs are selected which have a common mid-point on the surface. It is assumed that traces with a common surface mid-point will also have a common reflection point at depth. For steeply dipping reflectors this is only approximately true (Chapter 11 and Fig. 11.5). A set of common mid-point traces is called a CMP gather. Corrections are made for differences in reflection time due to differences in the path lengths within each CMP gather. Then the traces are summed together (stacked). The reflected signals are largely additive, but the phase of the noise is variable. Consequently the S/N ratio is improved. An illustrative example that is used in several figures is as follows:

Number of traces	12
Group interval	200 (arbitrary units)
Source interval	200
Gap	4 positions (2 each side of source)
Cable	Symmetrical split

1500, 1300, 1100, 900, 700, 500–0–500, 700, 900, 1100, 1300, 1500
(Back Cable) (Forward Cable)

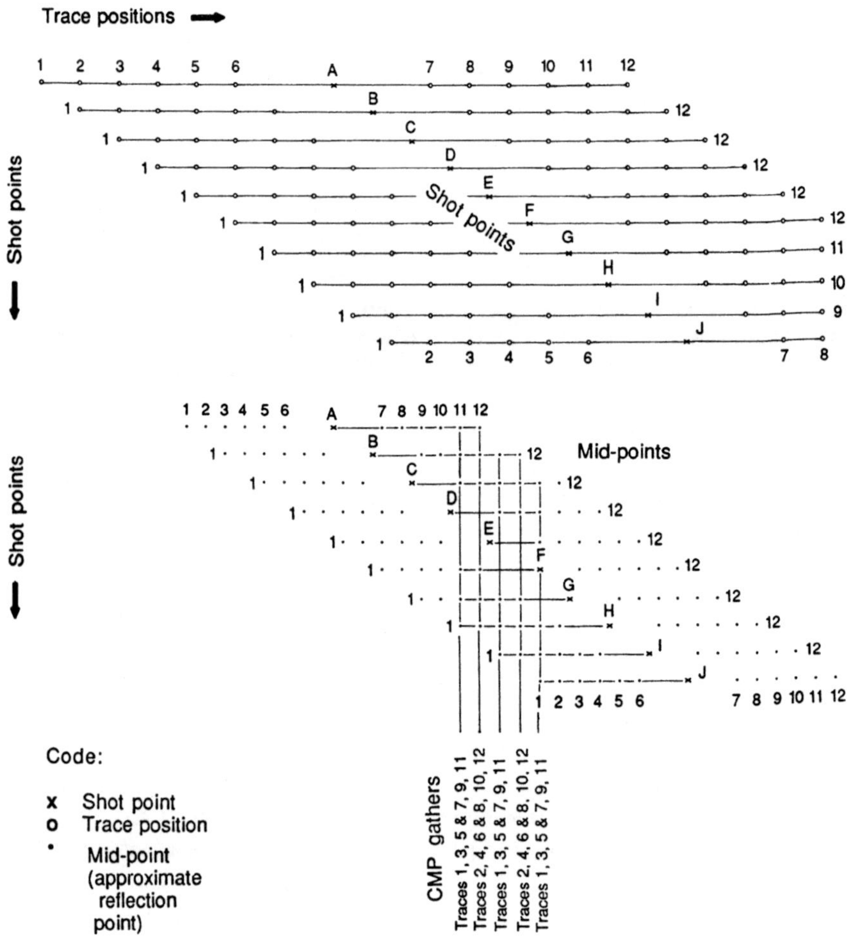

Fig. 2.3 Example of the use of a symmetrical split cable with an even trace gap. Shot points and cables are not laterally displaced, but are shown that way to illustrate the longitudinal movement of shot points and live traces. For horizontally layered earths the reflection points at depth are directly below a point half-way between the shot point and each receiver position. In this example the geophones at the two closest positions on each side of the shot point are not recorded and are said to be in the gap.

Figure 2.3 illustrates this example for shot points A through J with twelve traces recorded for each shot. The shots and traces are not laterally displaced but are shown that way in the figure for clarity. As illustrated, the first two trace positions each side of the shot point are not recorded. Geophones at these four positions are said to be 'in the gap'. This is an 'even gap' since four is an even number. In a typical case there would be a geophone array centered at each trace position. In the middle of Fig. 2.3 we show the locations of the reflection points at depth for horizontal reflectors. These illustrated reflection points at depth are halfway between the shot point and each trace position. Consider the reflection point just beyond shot point D. We will record six different traces for this reflection point as follows: A11, B9, C7, F5, G3 and H1.

For the next reflection point beyond we will also record six different traces as follows: A12, B10, C8, F6, G4 and H2. Figure 2.4 illustrates this second set of ray paths with this second common mid-point for the reflections. The ray paths A12, B10, and C8 for the front cable interlace with the ray paths F6, G4, and H2 for the back cable. This interlacing of ray paths is advantageous. First, there are no duplicate ray paths in each gather, which results in less dependence between the samples of both the reflected signals and the samples of the source-generated noise. Second, when the shot point interval is equal to

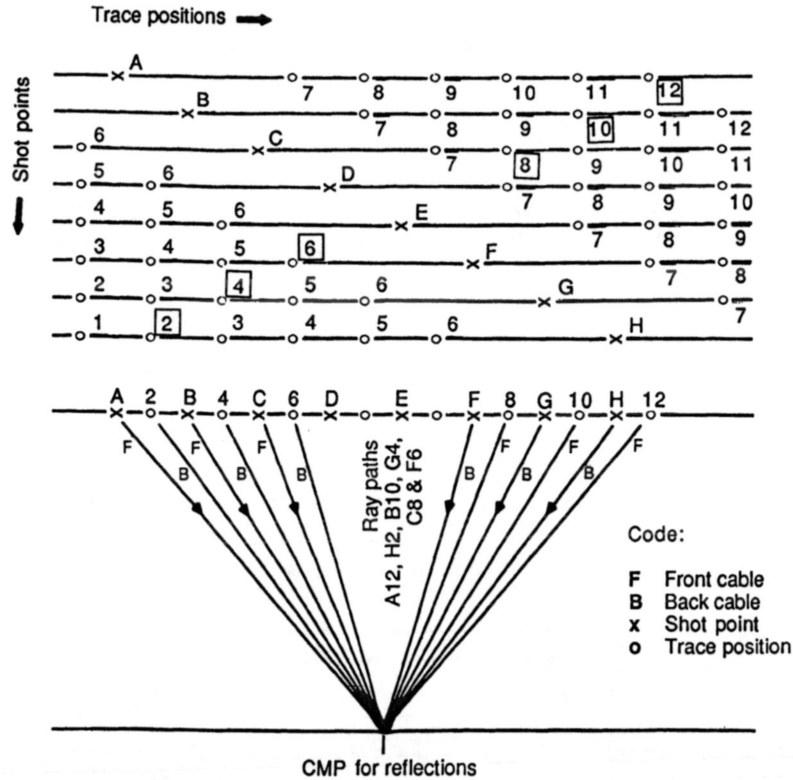

Fig. 2.4 12-trace symmetrical split cable with a 4-trace gap showing the interlacing of offsets from the front and back cables in a CMP gather. Within CMP gathers, rays from the forward cable (A-12, B-10, and C-8) interlace with rays from the back cable (H-2, G-4, and F-6). The difference between increasing offset distances in each CMP gather is one group interval with this interlacing but would be two group intervals without interlacing.

the group interval (trace interval), interlacing results in a minimum differential between the source to detector offset distances of the traces in each CMP gather. Minimizing the differential offset helps in the suppression of source-generated noise.

If we had located shot points (or shot pattern centers) at trace positions, the traces from the front and back cables would yield duplicate ray paths and duplicate offsets. The traces would not interlace, and we would have an odd trace gap. To get the desired interlacing we must locate shot points halfway between the trace positions which will result in an even trace gap. The CMP stack can be quite effective in suppressing very long wavelength noises (without simultaneously suppressing the desired reflections from depth). Each CMP gather contains information from offsets ranging from the minimum offset (traces 5 or 6, in Fig. 2.3) to the maximum offset (traces 1 or 12 in Fig. 2.3). With recording systems with many more than twelve traces, each stacked trace represents the equivalent of the signal from a giant array with the advantage of normal moveout corrections (NMO) to preserve the reflected signals. For land work, the highest practical CMP fold is equal to half the number of traces and is achieved with a source interval equal to the group interval. We will refer to this as full-fold recording.

Consider the case where the spacing between adjacent geophones is constant in each array, the shot point interval is equal to the group interval, and the gap is even. Figure 2.5 illustrates the sampling of a sinusoidal wave along the surface for arrays with effective lengths equal to half a group interval, a group interval, two group intervals, and three group intervals. For half group interval arrays there are gaps in the sampling of the sinusoidal wave. But for group interval arrays the interlaced samples from the front and back cables provide a continuous sampling of the sinusoidal wave. There are no gaps other than the relatively short spacing between adjacent geophones in each array. For double group interval arrays the sampling is continuous for the front cable considered alone and for the back cable considered alone. If there were some consistent differences in the sinusoidal waves between the front and back

---

Fig. 2.5 Sampling within CMP gathers of a sinusoidal noise component by geophone arrays. With half group interval arrays there are gaps in the sampling. With group interval arrays there are no gaps and the combined sampling for all the traces represents a single, giant array with uniform spacing and uniform weighting. With double group interval arrays the front and back cables can be considered separately and there are still no gaps in the sampling. The combined sampling for all the traces represents the sum of two, giant arrays with uniform spacing. This sum represents a trapezoidally weighted array with weights of 2 and 1. With triple group interval arrays the arrays overlap in offset coverage even for the front and back cables taken separately. The combined response is a trapezoidally weighted array with weights of 3, 2 and 1.

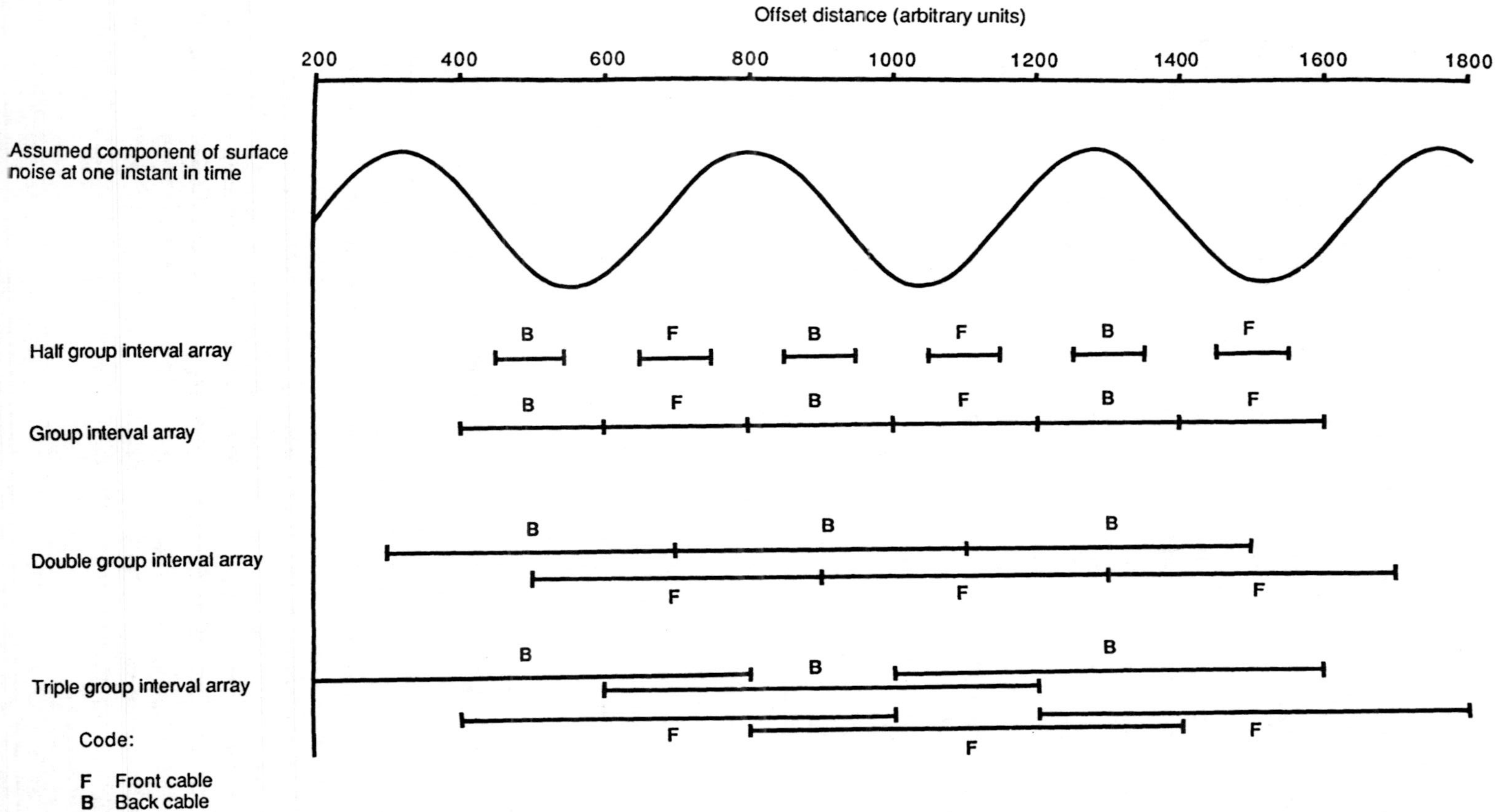
Offset distance (arbitrary units)
200
400
600
800
1000
1200
1400
1600
1800
Assumed component of surface noise at one instant in time
Half group interval array
Group interval array
Double group interval array
Triple group interval array
B
F
Code:
F Front cable
B Back cable

cables, with double group interval arrays we would have a continuous sampling of both of the sinusoidal waves. With triple group interval arrays the sampling is continuous and triple fold with a tapering off at each end to double fold and then to single fold.

If the length of each geophone array is equal to an integral number of group intervals, then the CMP stack (sum of each CMP gather) represents a uniform and continuous sampling of signals from all offset distances with a fine sample interval that is equal to the spacing between adjacent geophones in each geophone array. Under ideal conditions this would result in effective suppression of noise wavelengths ranging from as short as twice the spacing between adjacent geophones up to as long as the distance between the minimum and maximum offsets. But for real earths, noise suppression by the CMP stack (in combination with geophone arrays) is more limited.

By careful design we can provide good overlap of the band of wavelengths that are more effectively suppressed by the CMP stack with those that are effectively suppressed by the geophone array and source pattern. There are compromises involved and the best design depends upon many considerations. But the CMP stack only works to the extent that static corrections and velocity functions for moveout correction can be determined from the data. In areas where noise is very strong it may be difficult or impossible to extract this necessary information unless noise is more effectively suppressed in the field with a better combination of geophone arrays and source patterns. Particularly if we cannot see some reflections on our field records, we should analyze any coherent noises in the field records to determine the wavelengths, dominant frequencies, and velocities of these noises (Fig. 2.2). If some of the stronger noises are dominated by low frequency components, we should try a suitable low-cut filter on our field records to see if this will allow us to see reflections. If a low-cut filter does allow us to see reflections, then we should feel reassured that the processor will also be able to find reflections. But we should not use the filter in the field while recording production data. If filtering does not allow us to see some reflections, we should consider changing our geophone arrays and source patterns to suppress some of the stronger noises. If longer geophone arrays or source patterns are required, check to see if longer arrays or patterns would severely limit the high frequency response (Chapter 4).

If practical, we should record land surveys with symmetrical split spreads and with full CMP fold (source interval equal to the receiver group interval) with the sources centered half way between the centers of the geophone arrays. We will call this the "full-fold, even-gap, symmetrical split spread'. With this recording system use linear geophone arrays with equal spacing between the geophones. The effective length of the arrays should usually be an integral number of group intervals. We will refer to such linear arrays with equally spaced geophones as 'group interval' arrays, 'double group interval' arrays, 'triple group interval' arrays, etc. For uniformly sampled, composite arrays and with ideal conditions, the noise response (for wavelengths greater than twice the geophone spacing) approaches the quantity $(\sin \theta)/\theta$ where the argument (θ) in radians is (π) (length of the array in wavelengths) (see Appendix D). For long, composite arrays this quantity becomes quite small. With the recommended recording system and with arrays that are an integral number of group intervals, the sum of the traces in each CMP gather approximates a giant array and the angle θ becomes very large resulting in excellent noise suppression over a wide range of wavelengths when conditions are ideal.

When reflected waves arrive at the surface at a rather large angle from the vertical, the arrival time will be significantly different between the two ends of a long geophone array (Fig. 4.3). This difference in reflection arrival time (differential NMO) is most serious for the higher frequency components of the reflected signal. We need to limit the length of the geophone arrays sufficiently to avoid excessive loss of the higher frequencies. Even where vertical resolution is critical (and hence higher frequencies are critical), we should look at the loss of high frequencies for the CMP gather as a whole to see if the loss is excessive. In Chapter 4 equations are given for calculating the high-frequency loss for long arrays and a numerical example is included.

Double group interval arrays are often the best choice for land surveys, unless they result in too much loss of needed higher frequency components. With double group interval arrays, and with the full-fold, even-gap symmetrical split spread, there is good overlap in the wavelengths of noises that are well suppressed by the geophone arrays and the CMP stack. Also with double group interval arrays, for each CMP gather, offsets from just the back cable and from just the forward cable result in the desired continuous and uniform sampling of offsets with a fine sample interval equal to the spacing between geophones (Fig. 2.5). This desired sampling for suppressing noise is not then dependent upon the interlacing of the offsets from the forward and back cables. Why this is advantageous is discussed next.

For his stack-array method in land surveys, Anstey (Anstey, 1986) favors group interval arrays with the full-fold, even-gap, symmetrical split spread. Under relatively ideal conditions these group interval geophone arrays do prevent any gap in the wavelengths that are suppressed and provide opportunity for good suppression of noises with very long wavelengths (Fig. 2.5). But they are less effective under less ideal circumstances. Suppose that the signal injected into the earth were to change in amplitude or in its amplitude spectra. Such a change might result from a change in the lithology at the shot depth or a change in the outcrop for a surface source. A change in outcrop would also change the response of the geophones, but the effects on the sources and on the receivers would not likely be offsetting. (Reciprocity of sources and receivers is not perfect because geophone arrays and source patterns are not identical and because there are nonlinear effects near high-energy sources.)

If the change were abrupt and located at the CMP point, then the double group interval array would still be effective since the separate sums for the back and front cables should each effectively suppress the noise (Figs. 2.6 and 2.5). But for group interval arrays the difference in the injected signals at the sources would result in differences in the noises between the front and back cables which would upset the balance and could cause the noise sum to be relatively large. If the change in lithology at shot depth or the change in outcrop were about half the maximum offset in front of the CMP point, then the double group interval array would still be effective. The sum of the noises from the back cable would be affected in that the earth at the shot point would change. But the sum of the noises from the back cable could be considered as two sums. The first sum would be for offsets up to where the source changed and the second sum for offsets beyond where the source changed. Both sums for the back cable would be small for noise wavelengths that were short

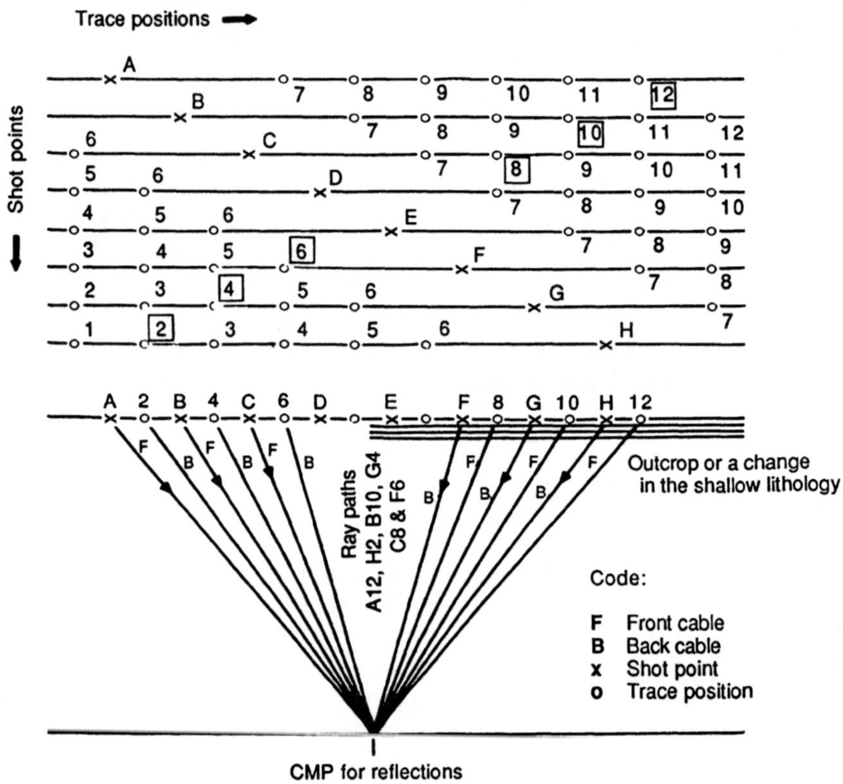

Fig. 2.6 Effect of an outcrop or a change of shallow lithology on CMP gathers. If there is a change in outcrop or shallow lithology at the center of a CMP gather, then data from consecutive offset distances in the CMP gather are from shots in different outcrops or different shallow lithologies.

compared to these still large, composite arrays. The sum of the noises from the front cable would be affected in that the earth at the geophones would change. But again, the sum of the noises from the front cable could be considered as two sums. The first sum would be for offsets up to where the earth at the geophones changed and the second sum for offsets beyond where the earth changed. Both sums for the front cable would also be small for noise wavelengths that were short compared to these still large, composite arrays. Therefore double group interval arrays would still suppress noises effectively even with an abrupt change in lithology or outcrop somewhere near the middle of the front or back offset range.

The last section of Appendix D also discusses the suppression of noise by the CMP stack considered both by itself and in combination with the geophone array. The conclusions drawn in Appendix D are consistent with the stack-array concept where that concept is applicable (i.e. where the effective length of the geophone array is equal to an integer times the incremental offset distance between a set of CMP samples of signal and noise that are consistent).

Often we want to record refractions which are the 'first breaks' for use in making static corrections*, but we want to suppress later refractions (which often include shear wave as well as compressional wave refractions). Figure 2.7 illustrates refractions from a dipping refractor that would be included in a CMP gather. Consider these refractions as noise to be suppressed. The angle of arrival of the refractions for the updip cable is steeper than for the downdip cable therefore the apparent wavelength of a refraction will be different for the two cables (Fig. 4.3). A given geophone array updip might suppress a component of the refraction quite well but the same array downdip might be relatively ineffective. With group interval arrays the interlacing of samples from the front and back cables in the CMP gathers may not yield a small, composite response to a dipping refractor. This complex case is considered in more detail in Appendix C.

Double group interval arrays are generally preferred, but group interval arrays are frequently effective and should be used in those cases where double interval arrays are too long to allow adequate recording of higher frequency components that are important to exploration objectives. In good record areas where source-generated noise is only a minor problem, group interval arrays may be optimum. For marine work group interval arrays as proposed by Anstey (Anstey, 1986) are effective.

Where strong noises with very long wavelengths are present, it may be

*The static correction program we used routinely sought to align reflections in one or more specified time windows. Our program proved to be so successful that we rarely found it beneficial to use shallow refraction data to aid in static correction even when static corrections were quite large. Therefore we did not usually consider it to be important to record shallow refraction data and we would readily sacrifice shallow refraction data if reverberating first breaks were a serious noise problem.

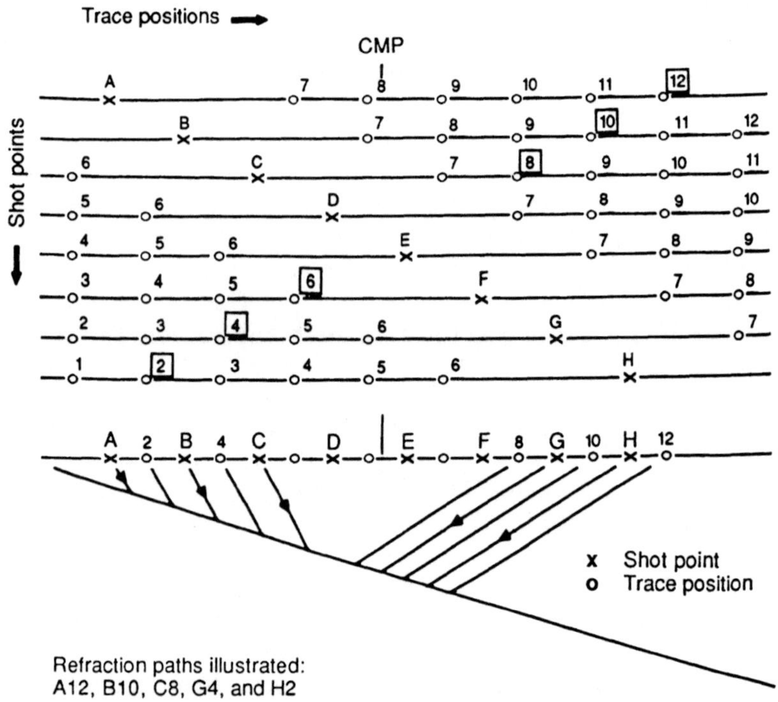

Fig. 2.7 Response of a CMP gather to noise from a dipping refractor. The angle of arrival of refractions from a dipping refractor in a CMP gather is different for updip and downdip receivers. The apparent wavelength of the refraction and the suppression of the refraction by the arrays will be different updip from downdip (Fig. 4.3).

feasible to use triple group interval or even quadruple group interval geophone arrays to suppress these noises more adequately and provide more suppression before stacking. Some argue that you lose horizontal resolution with long arrays. But seismic waves do not consist of pencil-like rays that reflect from points. Instead they are waves that reflect from rather large areas of subsurface interfaces and these areas increase in size with depth which limits horizontal resolution (Appendix F, program B). In most cases arrays have to be extremely long to seriously reduce horizontal resolution.

For sources such as vibrators which are quite mobile, it is recommended that data be recorded with maximum CMP fold by making the source point interval equal to the group interval. When the seismic source is an explosion in a deep hole, it is often uneconomic to record maximum CMP fold (with the shot point interval equal to the group interval). In Chapter 9 we show that with a half maximum fold (shot point interval equal to twice the group interval), you cannot achieve the uniform sampling of the noise advocated by Anstey unless you can use quadruple group interval arrays with an odd trace gap. You can, however, achieve the desired uniform sampling in processing by properly summing pairs of traces in processing, but you must use geophone arrays that are an integral number of group intervals (Chapter 9).

2.3 DIPPING REFLECTORS

When a reflecting interface has significant but moderate dip and the geophone arrays are downdip from the source, reflected waves will arrive at a substantial angle to the vertical (Fig. 2.8 and Chapter 11). Difference in the arrival times across a long geophone array would reduce the response at the higher frequencies. The result could be a significant loss of higher frequency components. But if you were recording with a symmetical split cable the geophones would be updip from the source for the other half of the cable and the difference in arrival times across an array would be reduced as a result of the dip, if the source pattern were short (Fig. 2.8 and Chapter 11). So the response at the higher frequencies would be reduced for half the traces in each CMP gather, but would actually be somewhat better for the other half of the traces in each gather. If the dip is too steep, the effect of the dip would much more than offset the effect of moveout and the angles of arrival within a CMP gather would become large for most of the traces. In this case we would lose significant reflection energy at the higher frequencies. For each gather the composite responses of the higher-frequency components of a reflection from a

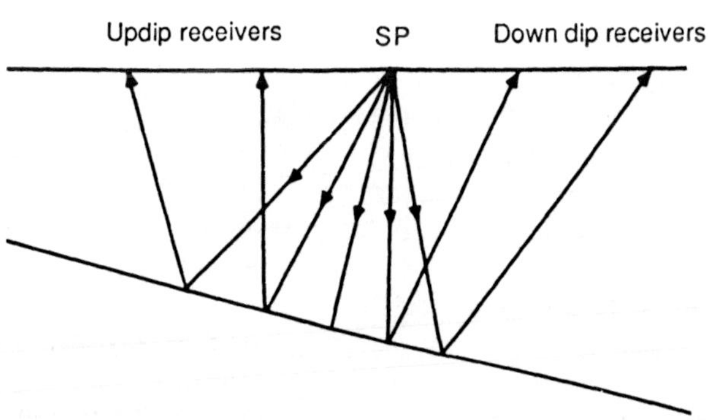

Fig. 2.8 Refractions from a dipping reflector. Note that the angles of arrivals at updip receivers are steeper than angles of arrival at downdip receivers at equal offsets.

moderately dipping reflector would not be less than half as a result of the dip (with a symmetrical split cable and a relatively short source pattern).

2.4 AIR WAVES AND OTHER SHORT-WAVELENGTH NOISES

Air waves can be a serious source of noise and particularly of noise with short wavelengths. Wavelengths for air waves are short because the velocity of sound in air is only about 1100 ft/s (335 m/s). A short spacing between geophones and/or a short spacing between sources in a composited source pattern can be effective in suppressing air waves. Ideally you would prefer a spacing of half a wavelength of the highest frequency component of the air wave. If you limit the upper frequencies you record to 55 Hz, then you would prefer a spacing of no more than 10 ft (3 m). This may or may not be practical. With vibrators working 'inline' on or along a road it is not feasible to get the vibrator pads so close together because the vibrator trucks are too long. But you can record and move the whole set of vibrators forward a short distance such as 10 ft (3 m) and record again. This procedure is repeated a number of times and then all of the data for each trace position is summed up. For the composited records, noise wavelengths longer than twice the short move-up distance will be suppressed. For a high resolution survey designed to record reflections with frequency components up to 110 Hz, the desired spacing or move up distance would decrease to only 5 ft (1.5 m).

The combination of a short spacing between geophones combined with a possible need for long geophone arrays would lead to the need for many geophones per trace. Therefore, with long geophone arrays (used with mobile, surface sources) we may choose to rely on short spacings in the source patterns and not have closely spaced geophones in the arrays.

The CMP stack (by itself alone) has limited capability to suppress short-wavelength noises. The relative phase of the signal samples becomes random for short-wavelength noises (particularly at long offsets) and the relative response is equal to the square root of the CMP fold divided by the CMP fold (which is the reciprocal of the square root of the CMP fold).

Another way to handle air wave noise is simply to use a minimum offset long enough to record the desired reflection data before the air wave arrives. This approach may be the most practical to use for portable crew operations in very rugged terrain (where use of many geophones or closely spaced sources would be costly).

Air waves cause some strange noises on seismic records. If your source point is near the top of a bare hill and your cable descends into a wooded area and then crosses another bare hill, you are likely to see little or no air wave in the low area but the air wave may be quite strong for traces near the top of both hills. Another strange effect can occur with long offsets if you record data from multiple source excitations that are separated in time by a few seconds (which is common with the Dynaseis* technique). What happens is that on the longer offsets you have air waves that arrive very early yet exhibit the normal low air wave velocity. The problem is noise from the air wave from an earlier excitation of the source. The same thing can occur with vibrators if you record from a series of very short sweeps.

2.5 NOISE FROM THE ALTERNATOR IN THE RECORDING TRUCK

Particularly with the Vibroseis technique we have a big alternator in operation in the recording truck while data are being recorded. Typically this alternator is driven by a reciprocating engine which puts unwanted noise into the ground through two different mechanisms. The first is by vibrations of the engine coupled to the chassis of the recording truck and then to the ground. The other mechanism is from an air wave from the exhaust system and this is usually the main source of the problem. The noise from the air wave can be nearly sinusoidal and often at a frequency of about 50 Hz. The frequency depends upon the number of cylinders of the engine and the revolutions per minute of the engine. If some manufacturer would use a turbine engine, the exhaust air wave problem would be moved to a frequency well above the normal seismic band. Until this is done, however, you may need to move the recording vehicle laterally off the line a distance of a few hundred feet (more than 100 meters). If you are working along a road, look for a side road or trail that will get you off the line. Sometimes you can reduce the exhaust air wave considerably by directing the exhaust away from the seismic line. A rotatable 90 degrees bend at the end of the exhaust pipe is easy to adjust (much easier than having to turn the recording truck at some odd angle and perhaps block the road).

The 60 Hz power generated to supply the instruments and other equipment in the recording truck can be a nuisance. Aircraft use 400 Hz systems to save on the weight of transformers. It would be great if seismic instrument trucks also used 400 Hz power. Normally the 60 Hz electrical energy is successfully blocked from the low-level circuits, but it may cause problems indirectly. For example, there are means for suppressing 60 Hz energy (in seismic data) that is picked up by the geophones or cables from commercial power systems (Chapter 16). But these methods depend upon a single, stable source of this unwanted 60 Hz energy. The commercial power systems are extremely stable,

*Trademark of the Atlantic Richfield Company.

but the 60 Hz power systems that supply our seismic instruments are very unstable. The special methods used to reject power line pick-up must avoid being misled by the unstable 60 Hz energy from the recording truck power supply. This is not a big problem, but it is one more problem to handle and one more possible source of trouble.

2.6 SOURCE-GENERATED NOISES VERSUS OTHER NOISES

A distinction needs to be made between noise generated by the seismic source and noise from other sources such as wind, atmospheric disturbances, rain, animals, cars, trains, pumps, pipelines, and people. More source power can help to overwhelm noises which are not source-generated. But for source-generated noises, more source power cannot help. But more source elements and more positions for these elements in a well-chosen, composited source pattern can effectively suppress source-generated noises.

Suppose that we ran a seismic line through a field with pumping wells. Noise that is not source-generated would impinge upon the geophone arrays from many directions. If we used areal geophone arrays such as stars, they would tend to suppress the pump noise no matter what the azimuthal angle of arrival of that noise. But 15 hole star source patterns would not aid in the same way. The arrangement of the hole patterns would not be critical yet the arrangement of the geophone arrays would be critical. Conversely, for source-generated noise that is scattered laterally and consequently arrives at large azimuthal angles, both the arrangement of the star geophone arrays and the arrangement of the star source patterns would be critical to the suppression of the noise. If a star source pattern does not emit much energy at the azimuth to a scatterer, the scattered noise will be small. If a star geophone array is relatively insensitive at the azimuth from a scatterer, then the response to the scattered noise will also be small.

If we lower vibrators and move them up immediately after each sweep, we generate some noise that may or may not cause difficulty. (In my own experience this has not been a serious problem and to speed work and save money we often operated this way.) Even though this noise is generated by the sources, it would not be considered to be source-generated noise because this noise would not be related to the use of the sources in generating the desired reflections. (In this case if we changed the source energy by increasing or decreasing the drive level of the vibrators, we would change the amplitude of our reflections as compared with the subject noise.) Conversely, noise from mud and rocks ejected from the shot holes should be classified as source-generated noise since increasing the charge size would likely also increase the subject noise.

2.7 MULTIPLES

Multiples can be a serious problem particularly if they are prominent at or near objective times. Multiples can be suppressed in processing if the moveout of multiples is significantly different from the moveout of primary reflections. We need to record data that will yield a substantial difference in the moveouts. To do this we need long offsets and an adequate range of offsets in the CMP gathers. (But there is no value in using offsets that are too long to be effective for recording primary reflections from objective interfaces.) The maximum effective offset is normally about equal to the depth of the reflector, but this will vary. In some cases at the same record time a multiple velocity is only slightly different from the primary velocity. In this case, at short offsets there may be almost no difference in the moveout of the primary and multiple between adjacent offsets in the CMP gather. Therefore the summation of traces from a number of relatively short offsets may enhance the multiple nearly as well as it does the primary. It may therefore be advisable in some cases to use a rather long minimum offset. Ideally you would use a short minimum offset but use a longer group interval for the shorter offsets. But this would not be a practical field operation.

2.8 SIGNAL SAMPLING AND ALIASING

In modern seismology we use digital recording which means that we do not record continuously but rather we sample waves at closely spaced but discrete times (Appendix D). To define the waves from such samples we need at least two samples per period of the highest frequency components we wish to utilize, otherwise we will encounter what is called 'temporal aliasing'. This means that our ability to define the waves from our digital data is limited and frequency components above the alias frequency will be misconstrued as being frequency components below the alias frequency. Components at frequencies 35 Hz above the alias frequency will be interpreted as components at a frequency 35 Hz less than the alias frequency. The frequencies above fold back and appear to be frequencies as far below the alias frequency as the actual frequencies were above. On this basis we can use a 4 ms sample rate (with an alias frequency of 125 Hz) for frequencies up to 90 Hz with an effective 90 Hz alias filter. Then noise above 160 Hz (125 + 35) will appear to be noise below 90 Hz. Unless noise above 160 Hz is unusually strong, it should be suppressed adequately by the 90 Hz alias filter.

Yet there are explorationists who insist that they can see significant differences between data recorded at a sample rate of 2 ms and that recorded at 4 ms with a 90 Hz alias filter for both. Under normal circumstances any significant differences that exist must be the result of something that affected the data after it passed through the alias filter. Then any differences between the two data sets are not the result of any signal from the earth and do not contain information that we seek about the earth. We should not waste company money by recording and processing data at 2 ms that should be recorded at 4 ms. Any difference in costs is too much to pay unless there is a valid reason. For example, it would make sense to use a 2 ms sample rate if we plan to use beam steering and 2 ms shifts might fit our need. See Chapter 11.

Similarly we do not sample seismic waves with traces at all surface locations, but instead samples are separated by the group interval. Both reflections we want to extract and noises we want to suppress have apparent wavelengths that can be measured at the surface. To prevent spatial aliasing we need two samples per apparent wavelength along the seismic line. Since wavelength is inversely proportional to frequency, higher frequency components have shorter wavelengths and are more likely to be spatially aliased.

We must be concerned about aliasing whether it is temporal or spatial. But sometimes geophysicists and equipment manufacturers become over-concerned. In their zeal to prevent aliasing at all costs they impose severe restrictions without analyzing to see just what harm might result if some aliasing did occur. A case in point occurred several years ago with equipment that is now obsolete but the lesson remains valid. We had a 48-channel DFS IV recording system and normally recorded at a 4 ms sample rate using an alias filter that began cutting off for frequencies above 64 Hz in order to have excellent rejection above the alias frequency of 125 Hz. (The period for a frequency of 125 Hz is 8 ms and 4 ms sampling would just provide the needed two samples per period.) We wanted to record frequencies up to about 90 Hz so we needed to use an alias filter which would pass these frequencies. The next highest filter made by the manufacturer passed frequencies below 128 Hz so as to reject very well frequencies above 250 Hz. This alias filter was intended to be used with a sample rate of 2 ms. The manufacturer, in his zeal to protect the user against aliasing, had set up checks in the recording system to prevent use of the higher frequency filter with a 4 ms sample rate. We would have gladly used the 2 ms sample rate except that this would have required us to use only 24 channels for recording rather than all 48.

The observer found a way to trick the system into letting us record at a 4 ms sample rate and use the higher frequency alias filter. But was it dangerous to subject ourselves to the horrors of aliased noises contaminating our records? In this case we sought frequencies only up to 90 Hz and aliased noises that would fold back below 90 Hz would have to be above 160 Hz ($125-90=35$ and $125+35=160$). The higher alias filter which passed signals below 128 Hz would provide only moderate suppression of frequencies above 160 Hz. But how much seismic noise would you expect to encounter above 160 Hz? Such high frequencies would be severely attenuated if they propagated through the earth from much distance. We recorded 24 channels of data at a 2 ms sample rate with the higher alias filter but with no source input. Sure enough, background noise level above 160 Hz was quite low. We conducted a very successful survey with a 4 ms sample rate with no adverse effects due to aliasing. The point of this example is that possible alias problems may not translate to significant alias problems, and before accepting a serious limitation you should evaluate the probable severity of any adverse effects from aliasing.

2.9 NONLINEAR EFFECTS AND DYNAMIC RANGE

We know that there are nonlinear effects close to a dynamite explosion. But it is important that beyond the immediate vicinity of the source both the earth response and the response of our instruments are linear. Except near powerful sources, we can be quite comfortable in the expectation that the earth is linear. Geophones can be overdriven and become very nonlinear, but I have been surprised how close you can put a geophone to a big vibrator and measure a response that is not badly distorted. When a geophone is overdriven, we have certainly exceeded its dynamic range. We should avoid recording from geophones that are so close to a powerful source that they could be overdriven. Even at lower excitation levels the geophone response will still not be perfectly linear. Amplifiers and filters can also be overdriven, but even if not overdriven they are not perfectly linear. What is the effective dynamic range for good, modern seismic equipment? It depends upon the background noise levels at seismic frequencies in the earth and in our instruments. Where the earth is very quiet, the effective dynamic range of modern instruments is surprisingly high. Nevertheless, we need to be careful not to waste it and to look for ways to increase it further.

Generally we do not try to record traces from geophones that are very close to our source because usually the data are not good. What if we used accelerometers instead of geophones? Would this increase our effective dynamic range? Relative to normal velocity geophones, accelerometers have less sensitivity at low frequencies, but their sensitivity increases with frequency. Since strong noise is frequently at rather low frequencies, it is possible that the use of accelerometers could increase our dynamic range, at least as it pertains to recovering high-frequency components of our reflections. Also, accelerometers might have higher limits for maximum excitation level which might allow measurements closer to our sources. Special geophones with electronic feedback have been built that yield a response similar to that of accelerometers. Both accelerometers and the feedback geophones have a very linear phase response throughout the seismic spectrum including the very lowest frequencies. This is advantageous. But cost and probably a higher noise level for the special geophones and the accelerometers have been impediments to acceptance to date. Perhaps in the future these problems will be overcome and the velocity sensitive geophone will at last become obsolete. But digital-grade, velocity sensitive geophones available today are amazingly good in terms of linearity, sensitivity, dependability, dynamic range, ruggedness, and cost.

Why do geophone arrays that are very close to the source yield poor data? Is it because the geophones are overdriven? Perhaps. But it may also be because the noise level so close to the source is so high that reflected signals are overwhelmed. Also geophone arrays and source patterns become relatively ineffective at short offsets because divergence of energy near the source causes the amplitude of noises to decrease rapidly with distance, whereas our design of arrays and patterns assumes that noise to be suppressed is constant in amplitude over the length of arrays and patterns. Therefore, strong noises on short offset traces will not be suppressed effectively by arrays and patterns. Frequently the nearest effective offset will be one where the distance from the nearest geophone to the closest source position is a few hundred feet (about 100–200 m).

2.10 MISCELLANEOUS

I recommend that geophones in the gaps (between the source and active traces) be disconnected from the cable to prevent cross-talk in the cable. With telemetry systems this is not required. Disconnecting geophones and then reconnecting them when using the preferred split cable does add to crew manpower requirements. However, you can gain some advantages with these two people. We required our contractors to provide these people (unless we were using a telemetry system). These two people should have radios and should communicate information to the observer. They should read the number on each stake as they either connect or disconnect the geophones from the cable. This allows the observer to check to see that the geometry is correct. They should do a tap test on one of the geophones of the array that has just been hooked up to be sure that polarity is correct. The two people can see that the arrays are correctly laid out and can report to the observer any deviations from the norm such as an array that is shortened in the forward direction due to a bridge. Such information should be written on the observer's sheets. They should immediately and emphatically report any evidence of damage such as any indenture of soft pavement by vibrators. If nearby citizens seem troubled this should be reported. They should see that geophone plants are good and that geophones have not been loosened by exposure to high energy levels from nearby source excitations. Geophones are more likely to be loosened if the soil is very loose or if the geophones were planted in snow, ice, or frozen soil and subsequent melting has occurred.

2.11 SUMMARY

The full-fold, even-gap, symmetrical split spread used with double group interval geophone arrays is often an effective choice for land surveys. It even works quite well for recording reflections from beds with significant dip if the source patterns are not too long (Chapter 11). With a short source pattern, reflections from interfaces with significant dip will be poor for the downdip half of the cable, but will be good for the updip half of the cable. At worst, the effective CMP fold is halved for reflections from beds with significant dip. This recommended technique should not be used universally. In some cases double group interval arrays may be too long to record needed higher-frequency components of reflections from the longer offsets or from dipping interfaces. In tough areas you could consider doubling the number of traces, doubling the CMP fold, thereby halving the group interval, and using double group interval geophone arrays that would no longer severely limit the high-frequency response of the CMP gathers as a whole. There is no one method or any two methods that guarantee better seismic data.

REFERENCES

Anstey, Nigel A. (1986) Whatever happened to ground roll? *The Leading Edge*, **5**, No. 3 (March), 40–5.

Kestle, A. N. (1983) A Practical High Resolution Geophone. Presented at the 53rd annual meeting of the SEG, September 1983. (Mr Kestle was with Terra Dynamics, a subsidiary of De Regt Special Cable of Holland.)

3

Practical geophone arrays and source patterns

The theory and mathematics of geophone arrays and source patterns are developed in Appendix D. In this chapter we discuss practical means for utilizing the theory to design effective combinations of geophone arrays and source patterns. We illustrate the techniques with a numerical example.

Geophone arrays can be designed to suppress coherent noise based solely upon the apparent wavelength of the noise along the seismic line. The arrays are filters which reject specific wavelengths. What we actually do is create an array that has a null or a very weak response at the angle of arrival of any noise or signal with a specified apparent wavelength. Fortunately, we do not have to know the velocity of the shallow earth or the angle of arrival of the noise. One complication is that noises usually have significant frequency components over a band of frequencies which means that they have a band of wavelengths. For example, we may have a noise with an apparent velocity of 5000 units per second and a period of 40 ms (frequency 25 Hz). The apparent wavelength is 200 units. But the noise wavelet may have significant frequency components from 10–40 Hz with wavelengths ranging from 125–500 units. Therefore, we have to design arrays to suppress a band of wavelengths to suppress any given noise.

Source patterns can also be designed to suppress coherent noise based solely upon the apparent wavelength of the noise along the seismic line. What we actually do is to create a pattern that has a null at the angle of emission from the sources which generate the noise. Again, we do not have to know the angle of emission or the velocity of the shallow earth.

Elaborate designs of geophone arrays and source patterns with complex weightings can theoretically provide excellent noise suppression over wide bands of wavelengths. But the effects of variations in the surface and near surface, and variations in the geophone and source coupling to the earth upset these precisely weighted designs and field results are disappointing. (Also imperfect layout of complex designs contributes to their disappointing results.) But with rather simple designs we can control the approximate wavelengths for nulls (notches) in the array and pattern responses and set these nulls at or near known noise wavelengths. Also to a reasonable degree, we can provide notches in the source pattern for wavelengths for which the geophone array is relatively less effective, and vice versa. We can do this with very simple calculations that are exact in terms of the location of notches. We only approximate the suppression, but from a practical standpoint an approximation of the theoretical suppression is quite satisfactory. The precise response will be altered anyway because of variations in the shallow earth and because we do not want to spend an inordinate amount of crew time accurately locating every geophone position.

3.1 PRACTICAL DESIGN OF GEOPHONE ARRAYS AND SOURCE PATTERNS

We can use a numerical example from an actual field case to illustrate the very simple calculations and the practical method recommended to design geophone arrays and source patterns.

We began a survey in an area where we had never been. Commercially available data in the area were not good enough in terms of quality and vertical resolution to meet our exploration objectives. We conducted a noise test to see what the major noise problems were. The first breaks died away quickly and were not considered to represent a serious problem in recording reflections from our objectives. However rather strong shear wave refractions and ground roll were significant problems. These noises varied with time and with offset. A rather strong air wave was a problem for the shorter offsets. A summary of some of the noises found is shown in Table 3.1.

We had a 48-trace crew with five vibrators. At the time this work was done we did not have a nonlinear sweep generator. We had no means for suppressing power line pick-up (60 Hz) other than with notch filters in the recording truck. Use of these notch filters essentially eliminated frequencies between 56 and 64 Hz in our data.

Our data acquisition objective was to record better data from objectives at estimated depths of 4000 and 8000 ft (1220 and 2440 m). For the shallow objective we needed frequencies of at least 70 Hz. From sweep and array tests we determined that the record time for the shallow objective was about 0.6 s and the moveout velocity was about 15 000 ft/s (4570 m/s). If we used geophone arrays and source patterns long enough to suppress most of the noise, we would also suppress needed high-frequency components of our reflection from the shallow objective for offsets longer than about 3600 ft (1100 m). As a compromise between our two objectives we chose a receiver group interval of 165 ft (50 m) with a symmetrical split cable with offsets of 4537.5–742.5–0–742.5–4537.5 ft (1383–226–0–226–1383 m).

Table 3.1 Significant noises from the noise test

Apparent velocity (ft/s)	(m/s)	*Frequency (from period)* (Hz)	*Wavelength* (ft)	(m)
1150	350	40	29	8.8
1150	350	36	32	9.8
4250	1295	26	163	50
4250	1295	20	212	65
5000	1524	21	235	72
5143	1568	23	221	67
5143	1568	19	267	81
5143	1568	18	283	86
8500	2591	29	298	91
8500	2591	23	366	112
9167	2794	20	458	140
9167	2794	17.5	523	159

We chose a 16–80 Hz upsweep despite the problem with long wavelength noise below 20 Hz. We did this because of concern that we might not get back frequencies above the power line notch filter for our deeper objective. We wanted a two-octave frequency range in our reflection wavelets, but were willing to settle for an effective 16–56 Hz range for the deep objective. We softened slightly the effect of low-frequency noise in our field data by using an 18 Hz low-cut filter with an 18 dB per octave slope. All filters were included in the filtered sweep used to correlate the data. Correlation then doubled the dB rejection, but canceled the phase effects (Chapter 5).

Figure 3.1 shows sketches of geophone array and source pattern suppression of wavelengths using the very simple formulas at the bottom of the preprinted form to calculate notch wavelengths where the suppression is theoretically infinite and to calculate peak wavelengths where suppression is relatively poor, (Appendix D has more details). Vertical lines are drawn in Fig. 3.1 at notch and peak wavelengths; downward lines for notches and upward lines for peaks. Also shown is the composite response of the geophone array and the source pattern (with a different scale for the suppression in dB). Near all notches the suppression exceeds 24 dB. At the peak with the longest

Fig 3.1 Design of a geophone array and a source pattern to suppress noises from a real field case. The wavelengths for notches and peaks in the responses are indicated by vertical lines. The combined, composite response in decibels is the sum of the two responses in decibels. (Note the change in *d***B** scale for the composite response).

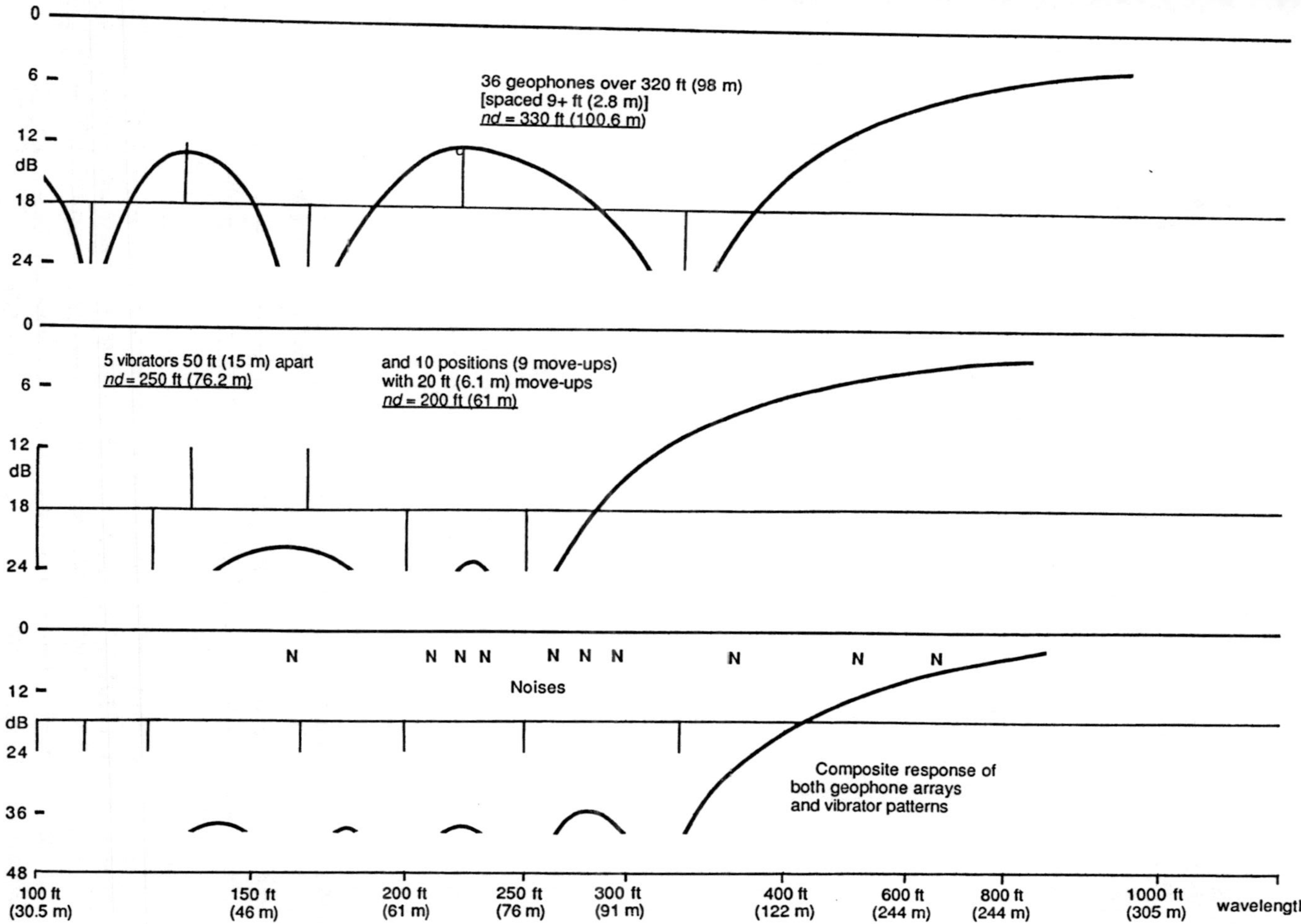

n = number of elements. *d* = spacing between adjacent elements. Notches at *nd*, *nd*/2, *nd*/3,...*nd*/(*n*-1)

Peaks at infinity, **2*nd*/3, 2*nd*/5, 2*nd*/7,...*d*** Peak at ***d*** has no attenuation

Table 3.2 Notches and peaks for a geophone array and source pattern

Geophone array *36 over 320 ft (98 m)* *nd = 330 ft (101 m)*		*Source pattern with move-up* *5 spaced 50 ft (15 m)* *nd = 250 ft (76 m)*		*9 moves of 20 ft (6.1 m)* *nd = 200 ft (61 m)*	
Notch	*Peak*	*Notch*	*Peak*	*Notch*	*Peak*
330 ft (101 m)					
		250 ft (76 m)			
	220 ft (67 m)				
				200 ft (61 m)	
165 ft (50 m)			167 ft (51 m)		
	132 ft (40 m)				133 ft (41 m)
		125 ft (38 m)			
110 ft (34 m)					
			100 ft (30 m)	100 ft (30 m)	
	94 ft (29 m)				
83 ft (25 m)		83 ft (25 m)			80 ft (24 m)
Notches at:	330, 250, 200, 165, 125, 110, 100, 83,... ft				
	(101, 76, 61, 50, 38, 34, 30, 25,... m)				

wavelength the suppression may be no better than about 12 dB, but improves about 2 dB for each succeeding peak at shorter wavelengths if the number of elements (n), is large.

The designs were actually made from Table 3.2 and then plotted on the form. First, a 330 ft (100 m) array was chosen [actually a 321 ft (98 m) array with 36 geophones spaced 9+ ft (2.8 m) apart] to suppress many of the stronger noises including the air wave. The 9 ft (2.8 m) spacing was less than half of the wavelength of the main airwave components. This double group interval array is compatible with the chosen 165 ft (50 m) group interval for easy field layout and for effective combination with the CMP stack in suppressing noise over a wide range of wavelengths (Chapter 2).

The source pattern for five vibrators was designed to provide notches near the wavelength of 220 ft (67 m) where the suppression from the geophone array was relatively poor and where there were strong noises to be suppressed. This was done with both the vibrator spacing and the move-up pattern. By spacing five vibrators 50 ft (15 m) apart we put a notch at a wavelength of 250 ft (76 m). By choosing a move-up of 20 ft (6.1 m) between each of the ten sweeps we put another notch at a wavelength of 200 ft (61 m). The composite result provided excellent suppression of all but the longest wavelength noises. These longest wavelength noises had to be suppressed by the CMP stack. The array was long enough to limit the higher frequency response for the shallower objective at the longer offsets, but the overall results for both objectives were considered to be excellent. In the event that one vibrator was temporarily out of service, work could proceed with four vibrators spaced 60 ft (18 m) apart making a number times spacing product of 240 with notches at 240, 120, and 80. The number of sweeps and sweep positions could be increased to fourteen with a 14 ft move-up between sweeps to yield a number times spacing product of 196 with notches at 196 and 98.

The source pattern with move-ups is a trapezoidally weighted pattern, see Fig. D-6. The response at very long wavelengths is determined mostly by the larger of the two *nd* products. Appendix D gives an example of a precise determination of the response at long wavelengths, in particular the section on a parallelogram array.

A few years later in the same general area we used a nonlinear, logarithmic sweep of 10–100 Hz to emphasize high frequencies. With the nonlinear sweep the output at the lower frequencies was relatively low despite sweeping as low as 10 Hz. Consequently the low-frequency noise did not overwhelm the data even though shorter arrays were used (so as to avoid loss of the higher-frequency components of reflections from the shallower objective). The same approach to geophone array and source pattern design was used but was based upon the objective of still higher resolution and upon the noises in response to the nonlinear sweep. The stacked section was superb with a high S/N ratio and wide bandwidth including frequency components of up to 100 Hz for reflections from the shallower objective (and from still deeper reflectors). This time the data were recorded with a power-line noise cancellation technique and without a power-line notch filter, so there was no gap in the spectra near 60 Hz. (Chapter 6 and Figs 6.1–6.3 give examples of other cases of data acquired with linear and nonlinear sweeps.)

3.2 USE OF *F-K* PLOTS IN ARRAY AND PATTERN DESIGN

We ordinarily display seismic data as waves as a function of time at a number of surface locations closely spaced along the surface. The data can also be displayed in what are called f-k plots (Embree, 1963) where the ordinate is frequency f and the abscissa is $+k$ and $-k$ where k is the wave number (the reciprocal of the wavelength). Waves that appear to move along the surface away from the shot point will plot in the first quadrant whereas waves that are back-scattered and move inward will plot in the second quadrant. The amplitude can be indicated by contours or by various codes including colors. Both reflections and nondispersive noises will plot as rays through the origin since $f/k = V_a$, where V_a is the apparent velocity. Waves with near-vertical

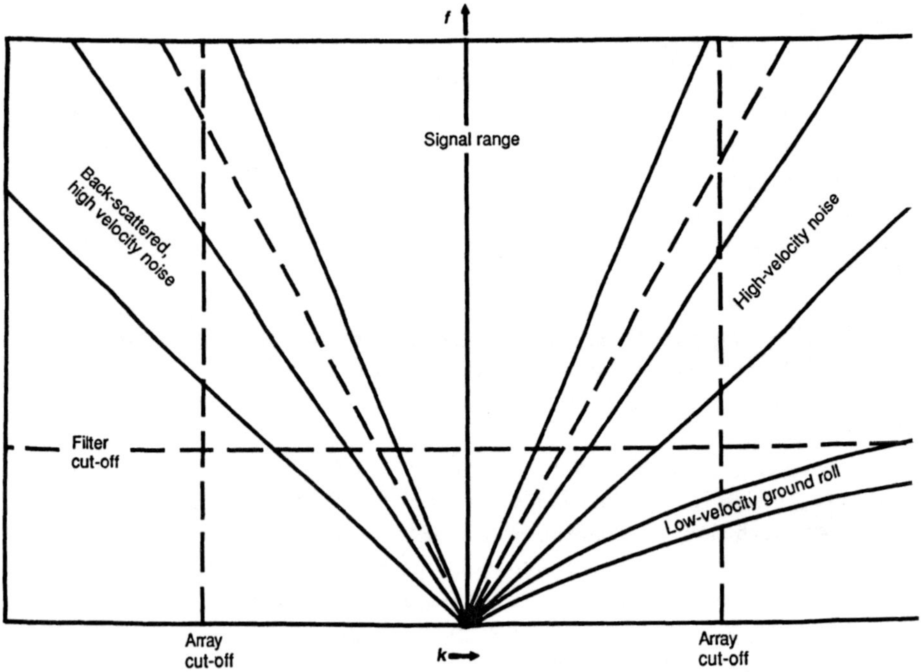

Fig. 3.2 Frequency versus wave number (f-k) plots of noise and signal. Constant-velocity events plot along radial lines through the origin. Noises with high velocities plot along lines with high slopes and low-velocity noises plot along lines with low slopes Reprinted from *Geophysics* by permission of the SEG, from Embree (1963).

angles of arrival (such as reflections from horizontal interfaces at short offsets) will plot near the ordinate because of their high apparent velocity.

Figure 3.2 illustrates an f-k plot and how such plots separate reflections from noise and one type of noise from another. If we use a low-cut filter then we will suppress both signal and noise that plots below a horizontal line illustrated by the dashed line labeled 'filter cut-off'. If we design an array or source pattern (or combination of both) that suppresses wavelengths below some value, then it will suppress signal and noise that plots beyond vertical dashed lines at a value of $+k$ and $-k$, where k is the reciprocal of the cut-off wavelength. With wavelength filters such as geophone arrays and source patterns and with frequency filters we may or may not be able to separate needed reflection data from noise, depending upon where the signals and noises appear on f-k plots.

The author personally has not made much use of f-k plots, primarily because they were not generally available for newly acquired test data and other methods were available to guide his decisions. Field displays of f-k

plots are now practical and may become a widely used tool for analyzing field data (including test data) and for designing geophone arrays and source patterns. But f-k plots have limitations and disadvantages, which include:

1. We have no information about the record time of events on f-k plots unless we make a number of plots for different ranges of record times. Therefore we do not know which noises occur at record times of our objective reflections.
2. It is difficult to display much dynamic range on f-k plots.
3. There is a problem in aliasing which makes some noises plot at erroneous locations.

When f-k plots are available or can be made available at modest cost, they will prove helpful and should be used by field geophysicists (Rigdon, 1987). But they are not necessary and will not take the place of conventional displays of data for use in designing arrays and patterns and deciding on methods and recording parameters to acquire seismic data. They are quite useful to data processors (Chapter 19).

3.3 DESIGN OF AREAL ARRAYS AND PATTERNS

There are situations where linear geophone arrays and/or linear source patterns are not satisfactory. One example of the need for areal arrays and patterns is where scattering in the near surface is severe (Chapter 12). Another case is where we have either severe, source-generated or background noise that arrives at steep azimuths relative to the seismic line. In areal or three-dimensional surveys we have geophones that are laterally displaced from the seismic line and some type of areal array pattern may or may not be most appropriate. Parallelograms made up of laterally displaced lines of elements can be designed like weighted, in-line arrays with regard to the suppression of noise that travels parallel to the seismic line (Appendix D and Fig. D-8). In fact, this is handled just as we handled the source pattern with move-up in the preceding numerical example. To a limited degree one can handle noise with other azimuths of arrival by treating each linear line of elements as if it were shortened by multiplying by the cosine of the azimuthal angle to the seismic line as follows:

$$\{nd' = (nd)[\cos(\text{azimuthal angle})]\}$$

We would then need also to treat the sum of the laterally displaced lines as an array with the spacing being the lateral separation times the sine of the azimuthal angle.

For complex areal arrays and source patterns, we have to work with computer-generated responses for various azimuthal angles of arrival for noises. Chapter 12 gives ideas for designing areal arrays. Star arrays and patterns consist of at least three linear arms extending from a center point with an azimuthal separation between arms of 360 degrees divided by the number of arms (Chapter 12). Preferably stars should have an odd number of arms so we cannot have a wave arriving from some azimuthal direction such that it will hit all the geophones of more than one arm simultaneously. Also, a source pattern with an odd number of arms will not have the sources from more than one arm precisely in phase in any one azimuthal direction.

Star geophone arrays and star source patterns can be effective where the need for areal arrays and patterns justifies their consideration and possible use. Stars are quite poor compared with linear designs for in-line noise, but they certainly have their place in the field geophysicists tool kit to handle special problems. Sometimes star geophone arrays in combination with appropriate inline source patterns are quite effective, or vice versa. In other cases, both geophone arrays and source patterns should be stars, or stars can be used with parallelograms or other areal arrays or patterns. In any case effort should be made to place notches in the responses where they are needed to handle identified, coherent noise or to broaden the band of wavelengths that are well suppressed by the composite responses plus the CMP stack. In bad record areas where reflections may be difficult to extract from the data, try to handle as much noise as is feasible before stack.

Star arrays and patterns are only approximately omnidirectional; therefore there is a best orientation for stars for the noise that travels along the seismic line. This best orientation of odd arm stars is with one arm perpendicular to the seismic line.

One use for star geophone arrays is to suppress wind noise where it is a severe problem. Another use is where a lot of cultural noise comes from many directions such as when a line passes though an oil field where many pumps are in operation. In these situations the star geophone array may help, but there is no advantage in using a star source pattern as opposed to a linear source pattern. The star source pattern is only effective where source-generated noise arrives at the detectors from steep azimuthal angles.

REFERENCES

Embree, P., Burg, J. P. and Backus, M. M. (1963). 'Wide-band velocity filtering – the pie-slice process'. *Geophysics*, **28**, 948–74.

Rigdon, H. and Hoover, G. (1987). 'Quantitative selection of seismic acquisition parameters'. *Geophysics: The Leading Edge of Exploration*, January 1987, 18–25.

4

Handy tools for field work

Sometimes in the field you will feel the pressure of needing to make some key decisions so that an expensive crew can start or continue their work. Under these circumstances it is good to have handy tools for little jobs so you can concentrate on your main tasks. Tools include exact and approximate equations and a few small programs and an inexpensive, programmable calculator. Appendix F contains a few programs for a Texas Instruments TI-66 calculator (available in Europe). These programs can be adapted to other calculators (including the TI-65). In this chapter we discuss moveout velocity, differential moveout across a geophone array or source pattern, and the loss of high-frequency components of reflected signals as a result of differential moveout. We also deal with apparent wavelength and apparent velocity as a function of the vertical angle of arrival. We deal with extracting from field data information including: moveout velocities, zero offset reflection times, reflection times as a function of offsets, and estimates of the frequency bandwidth of reflections.

4.1 MOVEOUT OR STACKING VELOCITY

The arrival times of a reflection from a horizontal interface to points on a horizontal surface can be related to the offset distance by a relationship that is frequently referred to as $X^2 - T^2$ (Sheriff 1973, p. 245). This relationship is:

$$(vt)^2 = (2D)^2 + L^2$$

Where

- t is the arrival time of the reflection at the surface;
- D is the depth of the reflection interface;
- L is the horizontal offset distance from the seismic source to the point of arrival of the reflection on the surface;
- v is a velocity that is called the moveout or stacking velocity.

This relationship is the equation of a hyperbola and we refer to moveout which satisfies this equation as being hyperbolic moveout. From Fig. 4.1 we can see

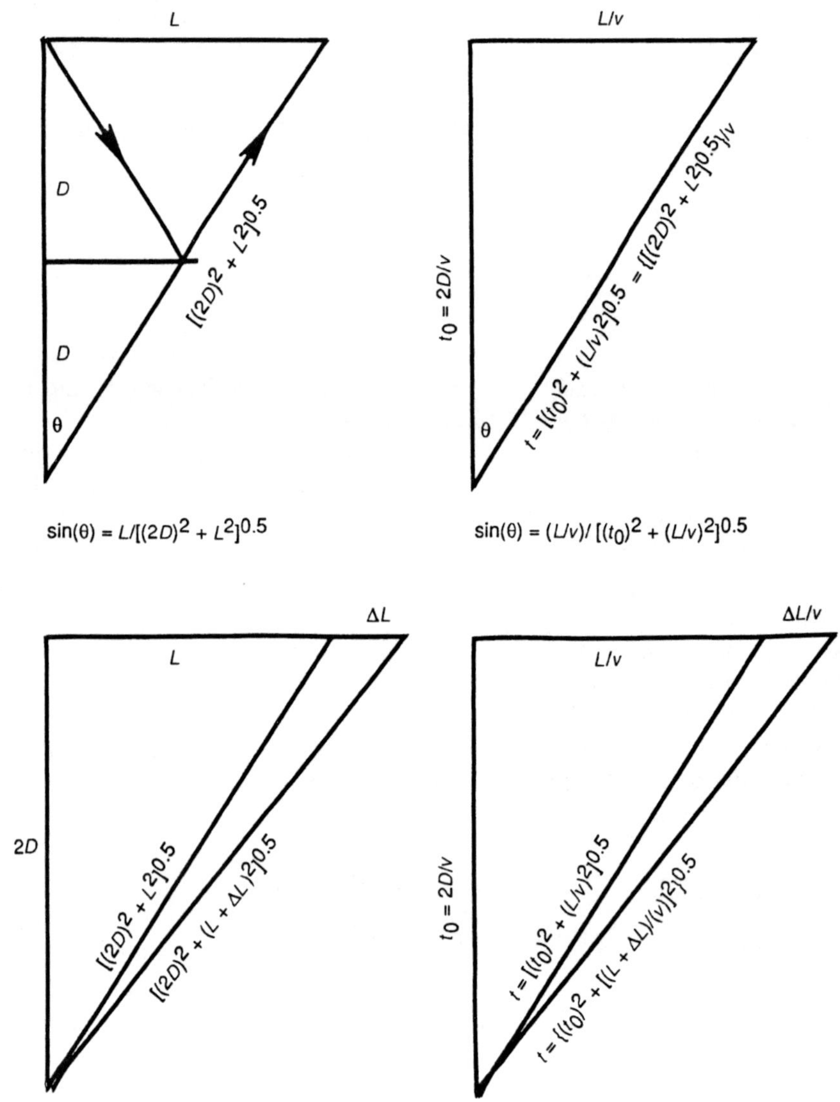

Fig. 4.1 Offset distance, depth, reflection time, and velocity. The total path length of a reflected ray is equal to the straight line distance from the mirror image of the source to the receiver position. For reflections from planes that are parallel to the surface we can solve right triangles to determine path length, travel time, differential path length, and differential travel time.

that this equation is consistent with the concept that the arrival time is the distance along the hypotenuse of a right triangle divided by the moveout velocity. One leg of this right triangle is twice the depth and the other leg is the horizontal offset distance.

The above relationship is approximately valid over a reasonable range of offsets even for earths that consist of many horizontal layers of different interval velocities. For such multi-layer earths the depth D in these relationships is an apparent depth which is only approximately equal to the true depth. The true depth is equal to half the zero offset reflection time divided by the average velocity (Sheriff, 1973) rather than divided by the moveout velocity. In these cases the interval velocity of a layer is approximately related to the moveout velocities at the top and bottom of the layer by the Dix formula (Sheriff, 1973). Figure 4.1 illustrates relationships between offset distance, reflector depth, reflection time, and moveout velocity. It will be helpful in visualizing and understanding much of the following text. It assumes hyperbolic moveout.

4.2 DIFFERENTIAL MOVEOUT AND ARRAY RESPONSE

A frequently encountered little job is to calculate the differential moveout for a reflection across an array at some large offset and then to calculate the signal loss for the higher-frequency components due to this differential moveout. If we speak of a group interval geophone array with a group interval of 220 ft (67 m), most geophysicists would probably envisage an array with an effective length of 220 ft (67 m) which would have an actual length of one geophone spacing less. (The actual length of an array is the distance from the nearest geophone of the array to the farthest geophone of the array.) Unless otherwise stated we will think in terms of effective lengths for both geophone arrays and source patterns.

A much-used program to calculate differential moveout and the loss of higher-frequency components is included in Appendix F, Program A′. But if your calculator slipped out of your shirt pocket while you were bending over a shot hole, all is not lost.

Let

t_0 = two-way reflection time at zero offset.

L = offset distance from a source to the middle of a geophone array minus half the effective length of the geophone array (or for a source pattern, from the middle of a source pattern to a receiver position minus half of the effective length of the source pattern).

ΔL = the effective length of an array (or the effective length of a source pattern). The effective length of a geophone array or a source pattern is the number of elements times the spacing which is one spacing more than the actual length.

V = moveout or stacking velocity for the reflection.

Δt = differential moveout for the reflection across the effective length of a geophone array (or the difference in propagation time to a receiver position over the effective length of the source pattern).

$f_{0.64}$ = frequency for which the response drops to 64 per cent because of the differential moveout across the array (or pattern).

$f_{0.3}$ = frequency for which the response drops to 30 per cent because of the differential moveout across the array (or pattern).

f_0 = frequency for which the response is zero because of the differential moveout across the array (or pattern).

An exact equation (assuming hyperbolic moveout) for the differential moveout is:

$$\Delta t = \sqrt{t_0^2 + \left(\frac{L + \Delta L}{V}\right)^2} - \sqrt{t_0^2 + \left(\frac{L}{V}\right)^2}$$

When $(L + \Delta L)/Vt_0$ is small (offset much less than twice the reflector depth), then an approximate equation is:

$$\Delta t \approx \frac{L(\Delta L)}{t_0 V^2}$$

We will use the differential moveout across a geophone array or source pattern to calculate the relative response to an objective reflection. The frequencies where the response drops to 64%, 30% and 0% are those where the effective length of the array or pattern is half a wavelength, three-quarters wavelength, and a full wavelength respectively. These are the same frequencies for which the differential moveout over the effective length is half a period, three-quarters of a period, and a full period. Equations for these frequencies are:

$$f_{0.64} = 1/(2\Delta t)$$
$$f_{0.3} = 3/(4\Delta t)$$
$$f_0 = 1/\Delta t$$

To get the array response for any frequency f:

Response $= \sin\phi/\phi$ (accurate for wavelengths much longer than the spacing between elements and for arrays with at least four elements)

$\phi = \pi(\Delta t/T) = \pi f \Delta t$ (ϕ in radians for half of the array)

Table 4.1 Differential arrival time across an array and the consequent loss of high frequency components

Offset distance (*ft*)	(*m*)	$\frac{L+\Delta L}{Vt_0}$	*Approx.* Δt (*ms*)	*Accurate* Δt (*ms*)	*Relative response* $f = 50$ Hz	*Relative response* $f = 55$ Hz
10890	3319	0.46	16.3	15.15	0.290	0.191
9790	2984	0.42	14.6	13.8	0.378	0.284
8690	2649	0.37	12.9	12.5	0.472	0.386
7590	2313	0.33	11.3	11.1	0.568	0.494
6490	1978	0.28	9.6	9.6	0.664	0.603
5390	1643	0.23	7.9	8.0	0.755	0.708
4290	1308	0.19	6.2	6.5	0.837	0.805
3190	972	0.14	4.5	4.8	0.907	0.888
2090	637	0.10	2.9	3.2	0.959	0.950
990	302	0.05	1.2	1.5	0.991	0.989

Let us illustrate the use of these equations with a numerical example. Assume the following:

Number of traces	96
Group and SP interval	220 ft (67 m)
Cable	10 890–550–0–550–10 890 ft (3319–168–0–168–3319 m)
Geophone array	Double group interval with 48 geophones over 440 ft* (134 m)
Reflection time	2 s
Moveout velocity	12 000 ft/s (3658 m/s)

The distance L is the offset minus half the effective length of the geophone array. The distance ΔL is 440 ft (134 m). Calculated results are given in Table 4.1 and are also plotted in Fig. 4.2.

For an offset of 10 890 ft (3319 m) with $\Delta t = 15.15$ ms, we calculate the following:

$$f_{0.64} = 1/(2\Delta t) = 33 \text{ Hz}$$
$$f_{0.3} = 3/(4\Delta t) = 50 \text{ Hz}$$
$$f_0 = 1/\Delta t = 66 \text{ Hz}$$

*Effective length of 440 ft (134 m) with an actual length of 431 ft (131 m) with a spacing between adjacent geophones of 9.17 ft (2.8 m), since the effective length is (nd) but the actual length is $(n-1)(d)$.

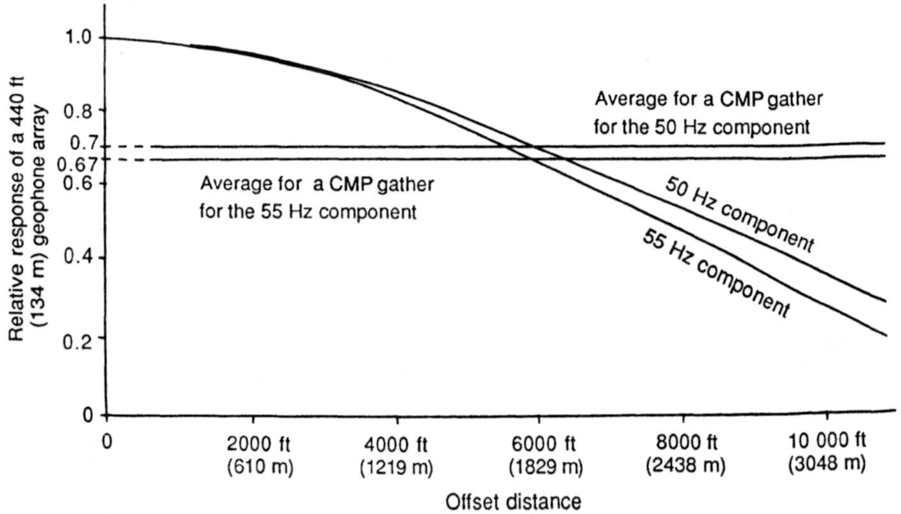

Fig. 4.2 Response of a geophone array as a function of offset distance. This example is for a 440 ft (134m) geophone array for a reflection time of 2 s with a movement velocity of 12 000 ft/s (3658 m/s). Whereas the normalized response at the longest offset is quite low for both 50 and 55 Hz components, for the CMP gather as a whole the composite, normalized responses are 70 and 67 per cent respectively.

Suppose in this numerical example you needed frequencies of up to 50 Hz in your final, stacked section for the reflection at 2 s. Would this 440 ft geophone array be too long? Actually it is not too long. You hate to lose any high-frequency energy, but the relative, composite response for the CMP gathers as a whole is 0.70 at 50 Hz and 0.67 at 55 Hz (Fig. 4.2). Any seismic noise at these frequencies is also going to be attenuated by the arrays so the signal-to-noise ratio may not be reduced. Data processors could easily compensate for the small loss in amplitude at 50 and 55 Hz. With some reluctance the author would accept this much suppression of higher-frequency components if it appeared to be necessary to handle strong noise.

4.3 ANGLE OF ARRIVAL, APPARENT VELOCITY, AND APPARENT WAVELENGTH

Figure 4.3 illustrates the arrival of parallel rays at the surface at an angle of θ to the vertical. If we consider a differential surface distance ΔL then the difference in path length for the rays arriving at the two ends is ΔP.

$$\sin\theta = \Delta P/\Delta L \quad \text{and} \quad \theta = \sin^{-1}\Delta P/\Delta L$$

$$\Delta t = \Delta P/V = (\Delta L \sin\theta)/V$$

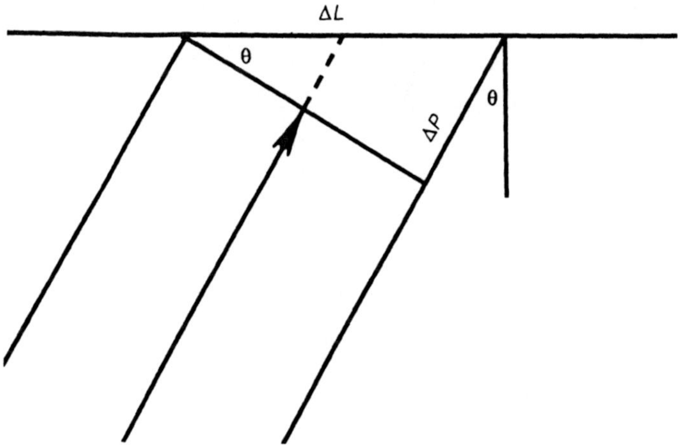

θ = The angle of arrival of parallel rays
ΔL = The differential distance along the surface
ΔP = The differential path length
v = The actual velocity along the oblique path
v_a = The apparent velocity along the surface
Δt = The differential time of arrival
λ = The true wavelength along the oblique path

$\sin(\theta) = \Delta P/\Delta L$
$\Delta t = \Delta P/v$
$v_a = \Delta L/\Delta t = \Delta L/(\Delta P/v) = v/[\sin(\theta)]$
$\lambda_a = v_a/f = v/[(f)\sin(\theta)]$

Fig. 4.3 Angle of arrival, apparent velocity, and apparent wavelength. If the angle of arrival is nearly perpendicular to the surface, then the apparent velocity along the surface is high and the apparent wavelength is long.

$$V_a = \Delta L/\Delta t = \Delta L/[\Delta P/V) = V/\sin\theta$$
$$\lambda_a = V_a/f = V/(f\sin\theta)$$

This approach can be used for any nearly parallel rays arriving at the surface at any angle. It can be used to obtain the apparent velocity and apparent wavelength for waves reflected from dipping reflectors, but we must know the angle of arrival θ and an estimate of the appropriate velocity.

For the ideal case of a horizontal surface and horizontal interfaces we can use Fig. 4.1 and write the previous equations as follows:

$$\theta = \sin^{-1}\frac{L}{\sqrt{(2D^2)+L^2}} = \sin^{-1}\frac{L/V}{\sqrt{t_0^2+\left(\frac{L}{V}\right)^2}}$$

$$= \sin^{-1}\frac{L}{\sqrt{(Vt_0)^2+L^2}}$$

$$\Delta t = \frac{L(\Delta L)}{V^2\sqrt{t_0^2+\left(\frac{L}{V}\right)^2}} = \frac{L(\Delta L)}{V\sqrt{(Vt_0)^2+L^2}}$$

$$= \frac{L(\Delta L)}{V\sqrt{(2D)^2+L^2}}$$

$$V_a = \frac{V^2\sqrt{t_0^2+\left(\frac{L}{V}\right)^2}}{L} = \frac{V\sqrt{(Vt_0)^2+L^2}}{L}$$

$$= \frac{V\sqrt{(2D)^2+L^2}}{L}$$

$$\lambda_a = \frac{V^2\sqrt{t_0^2+\left(\frac{L}{V}\right)^2}}{fL}$$

$$= \frac{V\sqrt{(Vt_0)^2+L^2}}{fL}$$

$$= \frac{V\sqrt{(2D)^2+L^2}}{fL}$$

How do results with this approach compare with the previously calculated case in the preceding table? Take an array centered at an offset distance of 10 890 ft (3319 m). The calculated value for Δt is:

$$\Delta t = (440)(10\,890)/\{(12\,000^2)[2^2+(10\,890/12\,000)^2]^{0.5}\}$$

$$= 0.01515\ \text{s}$$

This parallel ray solution agrees with the non-parallel ray solution to four significant figures for this example. Compare with Table 4.1.

4.4 MOVEOUT VELOCITY AND ZERO OFFSET REFLECTION TIME

Often we pick some reflection in our field data (shot gathers) and want to calculate the zero offset time and the moveout velocity. If we want maximum precision, and have not lost our calculator, we can use the very simple programs C and C′ of Appendix F. These programs utilize an internal linear regression feature for pairs of times and corresponding offset distances. We could resort to plots of squares of offsets and squares of record times (X^2-T^2 plots), but that is tedious.

If a programmable calculator is not handy then the following equations can be used:

$$V = \sqrt{\frac{L_2^2 - L_1^2}{t_2^2 - t_1^2}}$$

and

$$t_0 = \sqrt{t_1^2 + \left(\frac{L_1}{V}\right)^2}$$

where

L_1 = the shorter offset distance;
L_2 = the longer offset distance;
t_1 = the reflection time at the shorter offset;
t_2 = the reflection time at the longer offset;
t_0 = the reflection time at zero offset.

We should select two offsets in the same direction from the source (front cable or back cable). Preferably one offset should be rather short and the other rather long. If we have a split cable, we should also select two offsets from the other cable and repeat the calculations. If the results are different, then the reflector may be dipping or faulted. Of course, static variations may cause the results to be different.

4.5 REFLECTION TIMES FOR VARIOUS OFFSETS

We may want to calculate, and possibly prepare overlays, for anticipated reflections from objective interfaces. This is particularly helpful where record quality is poor or when reflections from the objectives are weak. If you have never recorded data in a really poor record area you may not appreciate what an overlay can do to help you find reflected signals obscured by strong noise.

Likewise an overlay can help you decide that an event you hoped was a reflection cannot be a reflection because the moveout is not reasonable. The programs C and C′ in Appendix F can be used to calculate reflection times for any desired offsets given the NMO velocity and the zero offset reflection time.

If you do not have a programmable calculator handy, Fig. 4.1 may prove helpful to you in solving whatever problems you encounter relating offset distance, reflector depth, reflection time, and moveout velocity.

4.6 MEASUREMENT OF NOISE CHARACTERISTICS

Chapter 2 briefly describes a technique for measuring coherent noise properties including period, frequency, velocity, and wavelength. For a given set-up on the field display (or for a stacked section), we can make overlays to allow us to quickly estimate velocity. An overlay would consist of lines with various slopes which are labeled in terms of apparent velocity. If a coherent noise line-up has a slope between two of our labeled slopes we can estimate the noise velocity by interpolation. The apparent noise wavelength is simply the apparent velocity times the period of the noise.

A pair of transparent drafting triangles can be used to transfer the slope of a coherent noise over to a location where the slope can be measured easily. It is convenient if we count the number of traces for a time change of either 1 s or 0.1 s. Then the velocity is equal to the offset change if the time change is 1 s or it is ten times the offset change if the time change is 0.1 s. A transparent timing line scale may be helpful in measuring the periods of wavelets.

4.7 ESTIMATING THE FREQUENCY BANDWIDTH OF REFLECTION WAVELETS

Suppose we plan to record data in an area where we have some previously recorded data. We need to estimate the frequency bandwidth of the earlier data at our current objective times. This will allow us to better estimate the frequency bandwidth we are likely to be able to achieve. When we record some data in the area we will need to estimate the effective upper frequency of the earth before selecting the sample rate, the filters, and the geophone arrays and source patterns. If we use vibrators as the source we will have to select the sweep which is another important reason that we need to estimate the upper effective frequency for our objectives.

The bandwidth of a wavelet refers to the range of frequency components in the wavelet. Relative bandwidth refers to the bandwidth divided by the center frequency of the bandwidth. If a wavelet has a very narrow relative bandwidth it will be ringy with a lot of lobes and with one or more side lobes nearly as large as the central lobe or lobe pair. As the relative bandwidth increases, a main peak or main trough or a main peak-trough pair will develop and side lobes will become relatively weak (Appendix H and Fig. H-3).

For Vibroseis wavelets the apparent frequency is approximately the mean of the lowest and highest significant frequency components. This is a reasonable assumption for wavelets from impulsive sources as well. Suppose we recorded some data that had significant frequency components from 10 Hz up to say 56 Hz. The mean frequency would be 33 Hz. Now if we recorded the same data but with a low cut filter with a cut-off at 20 Hz (or if we used damped geophones with a 20 Hz resonant frequency), the mean frequency would increase to 38 Hz. The reflection wavelets would have higher apparent frequencies yet the upper significant frequency component would still be 56 Hz. The apparent gain in higher frequencies would be an illusion. A record with apparently higher frequency wavelets may not really have a higher upper frequency limit. We should select reflection wavelets that are not ringy and which appear to have good relative bandwidth and measure the period of these wavelets to estimate the apparent mean frequency of our data.

We can generally estimate the lower frequency limit of the bandwidth. It is usually set by the low-cut filter, by the resonant frequency of the damped geophones, or with vibrators by the lowest frequency of the sweep. If we used an impulsive source with 12 Hz geophones and a 16 Hz low-cut filter, the low-frequency limit would be about 16 Hz, but if we changed to 24 Hz geophones it would rise to 24 Hz, assuming the geophones are damped at about 60 to 70% of critical (Sheriff, 1973, Fig. 43). With vibrators sweeping from 20 to 80 Hz and 12 Hz geophones with a 16 Hz low-cut filter the lower-frequency limit will be very close to 20 Hz. (We might want to correct for the effect of the sweep taper particularly with nonlinear sweeps, Chapter 6.)

If we estimate the lower-frequency limit and measure the apparent mean frequency of some reflected wavelets, we can then estimate the upper frequency limit as well.

$$f_u = f_{am} + (f_{am} - f_l) = 2f_{am} - f_l$$

where f_{am}, f_u, and f_1 are respectively the apparent mean frequency, the upper-frequency, and the lower frequency.

We should test our estimates of the effective upper-frequency limit of our data. If the wavelets tend to be somewhat ringy with no dominant peaks or troughs, then the relative bandwidth must be quite limited.

A different way to estimate the effective upper-frequency limit is to look for ringy, short-period reflection events with some trace to trace coherence. These short-period events may be reflections from a single, very thin bed or they may be the result of overlapping reflections from a sequence of rather thin beds. These events may be rather low in amplitude. The reciprocal of the period of

Table 4.2A Group interval, number of traces, and offset range for symmetrical split cables with even number of traces in the gap

		48 traces			*60 traces*			*96 traces*			*120 traces*		
Groups per mile	*Group Interval*	*Near offset*	*Far offset*	*Traces in gap*	*Near offset*	*Far offset*	*Traces in gap*	*Near offset*	*Far offset*	*Traces in gap*	*Near offset*	*Far offset*	*Traces in gap*
60	88	308	2332	6	308	2860	6	308	4444	6	308	5500	6
		396	2420	8	396	2948	8	396	4532	8	396	5588	8
		484	2508	10	484	3036	10	484	4620	10	484	5676	10
		572	2596	12	572	3124	12	572	4708	12	572	5764	12
48	110	385	2915	6	385	3575	6	385	5555	6	385	6875	6
		495	3025	8	495	3685	8	495	5665	8	495	6985	8
		605	3135	10	605	3795	10	605	5775	10	605	7095	10
		715	3245	12	715	3905	12	715	5885	12	715	7205	12
44	120	420	3180	6	420	3900	6	420	6060	6	420	7500	6
		540	3300	8	540	4120	8	540	6180	8	540	7620	8
		660	3420	10	660	4240	10	660	6300	10	660	7740	10
		780	3540	12	780	4360	12	780	6420	12	780	7860	12
40	132	462	3498	6	462	4290	6	462	6666	6	462	8250	6
		594	3630	8	594	4422	8	594	6798	8	594	8382	8
		726	3762	10	726	4554	10	726	6930	10	726	8514	10
		858	3894	12	858	4684	12	858	7062	12	858	8646	12
33	160	560	4240	6	560	5200	6	560	8080	6	560	10000	6
		720	4400	8	720	5360	8	720	8240	8	720	10160	8
		880	4560	10	880	5520	10	880	8400	10	880	10320	10
		1040	4720	12	1040	5680	12	1040	8560	12	1040	10480	12
32	165	$557\frac{1}{2}$	$4372\frac{1}{2}$	6	$557\frac{1}{2}$	$5342\frac{1}{2}$	6	$557\frac{1}{2}$	$8312\frac{1}{2}$	6	$557\frac{1}{2}$	$10292\frac{1}{2}$	6
		$722\frac{1}{2}$	$4537\frac{1}{2}$	8	$722\frac{1}{2}$	$5507\frac{1}{2}$	8	$722\frac{1}{2}$	$8477\frac{1}{2}$	8	$722\frac{1}{2}$	$10457\frac{1}{2}$	8
		$887\frac{1}{2}$	$4702\frac{1}{2}$	10	$887\frac{1}{2}$	$5672\frac{1}{2}$	10	$887\frac{1}{2}$	$8642\frac{1}{2}$	10	$887\frac{1}{2}$	$10622\frac{1}{2}$	10
		$1052\frac{1}{2}$	$4867\frac{1}{2}$	12	$1052\frac{1}{2}$	$5837\frac{1}{2}$	12	$1052\frac{1}{2}$	$8807\frac{1}{2}$	12	$1052\frac{1}{2}$	$10787\frac{1}{2}$	12

Table 4.2B Group interval, number of traces, and offset range for symmetrical split cables with even number of traces in the gap

		48 traces			*60 traces*			*96 traces*			*120 traces*		
Groups per mile	*Group Interval*	*Near offset*	*Far offset*	*Traces in gap*	*Near offset*	*Far offset*	*Traces in gap*	*Near offset*	*Far offset*	*Traces in gap*	*Near offset*	*Far offset*	*Traces in gap*
30	176	616	4664	6	616	5720	6	616	8888	6	616	11000	6
		792	4840	8	792	5896	8	792	9064	8	792	11176	8
		968	5016	10	968	6072	10	968	9240	10	968	11352	10
		1144	5192	12	1144	6248	12	1144	9416	12	1144	11528	12
24	220	550	5610	4	550	6930	4	550	10890	4	550	13530	4
		770	5830	6	770	7150	6	770	11110	6	770	13750	6
		990	6050	8	990	7370	8	990	11330	8	990	13970	8
		1210	6270	10	1210	7590	10	1210	11550	10	1210	14190	10
20	264	660	6732	4	660	8316	4	660	13068	4	660	16236	4
		924	6996	6	924	8580	6	924	13332	6	924	16500	6
		1188	7260	8	1188	8844	8	1188	13596	8	1188	16764	8
16	330	495	8085	2	495	10065	2	495	16005	2	495	19965	2
		825	8415	4	825	10395	4	825	16335	4	825	20295	4
		1155	8735	6	1155	10725	6	1155	16665	6	1155	20625	6
12	440	660	10780	2	660	13200	2	660	21340	2	660	26620	2
		1100	11220	4	1100	13640	4	1100	21780	4	1100	27060	4
52.8	100	350	2650	6	350	3250	6	350	5050	6	350	6250	6
		450	2750	8	450	3350	8	450	5150	8	450	6350	8
		550	2850	10	550	3450	10	550	5250	10	550	6450	10
		650	2950	12	650	3550	12	650	5350	12	650	6550	12
26.4	200	500	5100	4	500	6300	4	500	9900	4	500	12300	4
		700	5300	6	700	6500	6	700	10100	6	700	12500	6
		900	5500	8	900	6700	8	900	10300	8	900	12700	8
		1100	5700	10	1100	6900	10	1100	10500	10	1100	12900	10

such events is another estimate of the effective upper frequency limit of our data. Of course, the effective upper frequency limit will tend to decrease for later record times and it will tend to decrease for long offsets.

The preceding estimates of the effective upper frequency limit of our data may be too pessimistic in terms of the potential for the area. Earth attenuation of the higher-frequency components may well cause these components to be hidden, yet good data processing may bring them out (Chapter 19). Also we may need to alter our field techniques to record the higher-frequency components with an adequate signal-to-noise ratio (S/N ratio). The use of nonlinear sweeps with vibrators can be extremely effective in improving the signal-to-noise ratio of higher-frequency components and in allowing these components to be seen in the field data (Chapter 6).

With vibrators we can measure the effective upper-frequency limit by sweeping several narrow frequency bands near the expected limit (Chapter 5, advantage 4). For crews using impulsive sources, it would be desirable to run narrow band filter panels on our data to determine the effective upper-frequency limit if we have the capability of doing this.

4.8 TABLES OF SOME PRACTICAL CABLE CONFIGURATIONS

Tables are included here which we found to be helpful in planning cable layouts. They are of no use for metric layouts except that they illustrate how you could prepare similar tables for your own use. Tables 4.2 A and 4.2 B will save some time and some repetitive calculations. They may also prevent some error when you are tired and under stress. But even in the office and when writing this book, the author often turned to these tables and found them to be handy tools. They are particularly helpful when weighing choices of group interval, offset range, and number of traces. In Chapter 7 you may gain some appreciation for these tables as we search for effective combinations to meet specific exploration objectives.

REFERENCES

Sheriff, R. E. (1973) Encyclopedic Dictionary of Exploration Geophysics SEG.

5 Vibroseis crews

The subject of this chapter is the Vibroseis technique including its advantages and disadvantages and the practical operation of Vibroseis crews.

5.1 VIBROSEIS ADVANTAGES

For seismic surveying on land the Vibroseis technique provides opportunity for significant advantages over other techniques. Usually it is the preferred land technique wherever it is practical to operate vibrators along continuous lines. Advantages include both operational and technical.

In the Vibroseis technique the energy injected into the earth from the source consists of many cycles of quasi-sinusoidal waves that are slowly changed in frequency over a period of time called the sweep time. Whereas sweep times can be as short as 2 s (or even less), they are usually in the range 10–20 s and may be longer with modern equipment. We sweep through just the band of frequencies that we want to have, and can reasonably hope to see, in our seismic reflections. We record the earth response over the entire sweep time plus a time period called the listening time which is equal to the length of time we want for our seismic sections. The resulting recordings contain reflections, refractions, ground roll, etc. but they all overlap in time because each event is spread out over a time period approximately equal to the sweep length. The records are long trains of sinusoidal wavelets that reveal little or no evidence of the subsurface layering. With special computer equipment called a correlator we perform the mathematical process of crosscorrelation of the earth signals with the input sweep signals (Appendix H). Crosscorrelation collapses reflections, refractions and other earth responses into wavelets that look much like the wavelets that are recorded directly in response to a dynamite explosion or a weight drop (Appendix H). All of this seems terribly complicated but the Vibroseis technique has gained much acceptance because of unique advantages. Some of these advantages are:

1. *Usually we may vibrate on paved roads and in densely populated areas.* With an impulsive source such as an explosion, which is over in a few milliseconds, the peak energy must be quite high to provide enough energy to record good reflections from deep interfaces. Also much of the energy from impulsive sources is wasted in frequency components that are outside the band of frequencies contained in the received wavelets reflected from depth. With vibrators, the required energy is injected into the earth over a period of seconds; therefore, the peak energy level can be relatively low. As a result the Vibroseis technique is much less likely to damage houses, pipelines, wells, etc. With vibrators it is frequently practical to record data through densely populated areas with source points near houses and other buildings. It is often possible to receive permission to vibrate on highways, roads and streets which are paved or unpaved. This makes it possible to explore areas where other techniques would not be allowed and to get more continuous coverage in many other areas.
2. *With vibrators there is less risk of damaging property.* With vibrators we can control which frequencies we inject into the earth and which frequencies we purposely do not inject. Because very low frequencies are more likely to cause damage, we can avoid damage by not injecting frequencies below our band of interest. In recent years the availability of good, nonlinear sweep generators has increased this control by allowing us to weight the sweep so as to alter the relative energy input as a function of frequency. With nonlinear sweeps we can further reduce the risk of damage to buildings, wells, etc. by reducing the amount of energy injected at the lower frequencies that are within our band of interest.
3. *We can select the frequency band that is most favorable for a high S/N ratio.* The control of frequencies injected can be used to avoid frequencies which create more undesired noise than they create reflected signals. Alternatively, or in addition, we can use our control on the relative amount of energy injected as a function of frequency to inject more energy at the higher frequencies and less energy at the lower frequencies, within our band of interest. It is advantageous to inject progressively more energy at progressively higher frequencies because attenuation within the earth is exponential and this exponent is proportional to frequency. Also we need to emphasize higher frequencies because the vibrators are less effective at the higher frequencies. For moderately low frequencies we need less energy in the earth because attenuation is less, but for the very lowest frequencies the vibrators may be less effective.

   We may or may not choose to take this equipment factor into account in designing sweeps that include very low frequencies, which may give us more noise and are likely to be more destructive to property. The design and use of nonlinear sweeps is discussed in Chapter 6. Nonlinear sweeps offer so many advantages that an appropriate nonlinear sweep should probably always be used instead of a linear sweep to achieve our goal of better seismic data.

4. *We can measure the effective upper frequency limit in the field.* With Vibroseis we can readily determine in the field the upper effective frequency limit for recording reflections from depths of interest. We can do this by recording first with a normal, wide-band sweep and carefully noting the time and moveout of reflections from depths of interest. We then record separate sets of data using several different sweeps that have very narrow bandwidths. Each of these narrow band sweeps should be at a high frequency that we wish to consider for the upper frequency limit. The correlated records from the narrow-band sweeps must be displayed at a high enough gain to reveal the very ringy results. By examining these ringy records at the times of the reflections of interest we can see if there are coherent events at these times that have the same moveout as the reflections on the records from the normal sweep. If so, then the frequency of that narrow band sweep is within the range of frequencies that can be recorded for those reflections.
5. *We can reverse the polarity of the signal injected into the earth.* Only with vibrators is it feasible to reverse polarity of the source, which allows us effectively to both push and pull vertically on the surface of the earth. Actually we apply a static load and then push more and push less, but the dynamic effect is the same as a push and pull. This unique capability provides several opportunities for solving important problems. It is a key to practical techniques for simultaneously recording multiple sets of data from multiple sets of vibrators and allowing subsequent separation of the data sets (Chapters 10 and 14). This saves expensive field recording time. For vertical seismic profiling with multiple source locations on the surface, the savings in time are increased in value because of the costs of drilling rig time and well service time. (Vertical seismic profiling is the name given to seismic measurements between surface locations and positions at closely-spaced depths within a borehole.) Another use is to reduce costs for mapping shallow objectives whether these objectives are related to petroleum exploration, to mining exploration, or to engineering needs. This application is discussed in detail in Chapter 14.
6. *We can discriminate effectively against many background noises.* The Vibroseis technique provides opportunities to discriminate against many noises that are not source-generated. The correlation itself does much in this regard. Also, elaborate field editing equipment and techniques provide very effective means for literally cutting strong noise bursts out of the data with minimal harm to the reflection data. (Indeed the results can be so dramatic that field crews are prone to become too lax and too sloppy in their operations and to then have little regard for the noise they create during recording).
7. *Vibrators are versatile sources.* With vibrators it is practical to use multiple units, to vibrate each unit at a number of earth locations, and to sum up the recorded results for a composited record. This is also practical with some other sources, but this capability makes feasible the use of rather elaborate

source patterns which contributes to the wide success of the Vibroseis technique and makes it more versatile. Because of this capability plus advantage (5), the Vibroseis technique lends itself to a method for canceling interference from electric power frequencies so that wide bandwidth data can be recorded quite successfully without notch filters (Chapter 16).

5.2 VIBROSEIS DISADVANTAGES

There is beauty and value in simplicity. The Vibroseis technique is sophisticated, and the equipment required is complex and complicated. These are significant disadvantages. It is amazing that anything so complex survived its infancy and grew and developed through adolescence to such vigorous adulthood. This achievement is a tribute to the vision, dedication, perseverance, technical competence, and ingenuity of many people and organizations and most particularly to the Continental Oil Company. With regret, the author can claim no part in this development, but congratulates those who are responsible, and hopes that he can contribute to the effective utilization of the technique.

The technique loses feasibility where terrain is too rough for vibrators to negotiate and operate in at least a semi-continuous manner. Vibrators cannot be used in dense woods without cutting a wide swath of trees. They cannot be used in marshes or swamps and efforts to develop and utilize the technique for water surveys have not met with much acceptance to date. Perhaps a highly successful marine Vibroseis will be available in the future. One problem is the need to use a minimum-phase deconvolution to de-reverberate marine data. One could use a minimum-phase correlation and minimum-phase processing, but there is resistance to this solution.

5.3 VIBROSEIS THEORETICAL CONCEPTS AND PRACTICAL CONCLUSIONS

Harmonic ghosts

With the Vibroseis technique the sweep can start at the lowest frequency and increase up to the highest frequency—an upsweep, or it can start at the highest frequency and decrease down to the lowest—a downsweep. Previously, downsweeps were used extensively because vibrator performance was better, but downsweeps can result in serious noise problems in the correlated records due to what are called 'harmonic ghosts'. Vibrators are imperfect and vibrator coupling to the earth is also imperfect. When waves with a frequency of f are injected into the earth, distortion creates harmonic frequencies which are injected simultaneously with the fundamental. The most serious harmonic is the second harmonic $2f$, which will not correlate with the primary frequency f in the sweep signal, but it will correlate with the frequency $2f$ generated earlier in the sweep. As a result, the correlated record will have second-harmonic events trailing behind every real event. These delayed, correlated events of the second harmonics are called harmonic ghosts (Appendix H) and are an unwanted noise to be eliminated or suppressed. (For shear-wave vibrators the third harmonic may be a problem and the second harmonic may be less significant.) Signal levels decrease drastically with time on a seismic record. Ghosts from strong, shallow events when delayed in time, then compete with relatively weak primary events from greater depth, which is why harmonic ghosts can be a serious problem. With modern recording equipment we can use long sweep times which will usually delay the harmonic ghosts to times greater than the length of the correlated records and thus eliminate the problem. If we used nonlinear downsweeps the harmonic ghost delay time would be more complicated.

With modern vibrators we can inject energy into the ground effectively using upsweeps which causes any harmonic ghosts present in the correlated records to precede the corresponding primary events. Now the ghosts from weak deep events must compete with strong primary events from the shallow earth. Consequently such ghosts are not a significant problem. With long sweep times the harmonic ghosts may be shifted far enough to precede time zero on the correlated records and hence totally suppressed.

Today there is no need to have a problem with harmonic ghosts. The simplest solution is to use upsweeps. With good, modern equipment upsweep records are just as good as downsweep records and with upsweeps you do not need to be concerned about harmonic ghosts even if you should decide to use short sweeps and to record events from very deep in the earth.

Wavelet phase and degree

Throughout this book we will characterize the type of phase spectra of wavelets by the words 'zero', 'minimum' or 'mixed'. When we say a wavelet is zero-phase we mean only that the phase angles are the same for all frequency components and this phase angle can be any constant value which we will refer to as the 'degree'. Hence a zero-phase wavelet of 90 degrees is one where the phase spectra is a constant 90 degrees for all frequency components.

A minimum-phase wavelet is one for which the phase spectra is that which yields a causal wavelet with minimum time delay after onset (and hence the minimun phase angles for the frequency components that will yield a causal wavelet). A causal wavelet is one that has a zero value until the time of some excitation plus the propagation time from the point of excitation to the point of observation.

Mixed-phase wavelets are those with phase spectra that are neither constant (like zero-phase wavelets) nor minimum for causality (like minimum-phase wavelets). Readers who are uncertain about the distinction between wavelet phase and wavelet degree may want to read Appendices G and H before proceeding with this chapter.

Originally, correlated Vibroseis wavelets were zero-phase, but 90 degree, because the available equipment yielded 90 degree wavelets; this is still the SEG standard for Vibroseis wavelets. The center of the energy of this wavelet is a zero crossing between a trough and a crest which have equal amplitudes (Appendix H). To pick the reflection time for an objective interface we have to pick the zero crossing between a trough and a crest of equal amplitudes. This is not easy to mark. Furthermore the difference between a reflection from an increase in acoustic impedance (positive reflection coefficient) and one from a decrease in acoustic impedance (negative reflection coefficient) is whether the trough precedes the crest or vice versa.

The author strongly recommends the use of zero-phase wavelets of zero degree over the old, historic SEG standard of the 90 degree wavelets. Must we continue forever with what is now a bad choice? With the zero degree wavelet the center of the wavelet is a crest for a positive reflection coefficient and the amplitude of this crest is greater than the amplitudes of the preceding and succeeding troughs, which have equal amplitudes (Appendix H). With the zero degree wavelets the center of the wavelet for a negative reflection coefficient is a trough preceded and succeeded by smaller crests, which have equal amplitudes. It is much easier on seismic records to spot prominent crests and prominent troughs and then to visualize the geology at depth than it is to look for paired troughs and crests of equal amplitude and then analyze which precedes which to visualize the geology, not a trivial problem. With Pelton* sweep generators set the degree switch on 0° for zero degree wavelets and on 90° for ninety degree wavelets. But with the Texas Instruments† sweep generator the switch position marked 0° means that the phase adjustment in the circuitry is 0° and the wavelets we get will be the old ninety degree wavelets. Conversely the 90° switch position means that a phase adjustment of 90 degrees is made in the circuitry, and with that adjustment the wavelets will be zero degree. This was so confusing that many people using the Texas Instruments sweep generators misunderstood (this included even supervisors, party chiefs, and observers with GSI which was a subsidiary of Texas Instruments), consequently they got one result when what they wanted was the other. This illustrates how easy it is to fail to communicate. For all sweep

*The Pelton Company, Ponca City, Oklahoma.
†Texas Instruments, Dallas, Texas.

generators you need to discuss this matter with technically competent employees of the company that makes the equipment. Even then you should ask more than one person, at more than one time, and check the answers for consistency.

The actual Vibroseis wavelet that is injected into the earth and propagates through the earth is the long and slowly varying quasi-sinusoidal wave generated by the sweep; this actual wavelet has a finite beginning time which means that it is not a zero-phase wavelet. The zero-phase wavelets (of whatever degree) we want for our seismic sections do not exist in the earth but rather they are the result of crosscorrelation of the earth response with the sweep. But correlation is a linear, mathematical operation. A change in the order of a series of steps does not change the result if all the steps are linear. Therefore, for a linear earth we can visualize the Vibroseis technique as if we applied correlated wavelets to the earth and recorded correlated wavelets as the response of the earth. (A linear earth is one where if you double the amplitude of an input signal, the output signal will double but will not be changed otherwise, i.e. not changed in shape.)

Suppose that we want a zero-phase wavelet of zero degree and we have good equipment with all the proper settings. Suppose that we recorded the response of a perfect sensor imbedded in a homogeneous earth at a depth of 100 ft (30 m) below a vibrator. If we crosscorrelated the response of the perfect sensor against the sweep would we get the desired zero-phase wavelet of zero degree? If the first answer were yes, and we had a second perfect sensor at a depth of 50 000 ft (15 000 m), would it also yield the desired zero-phase wavelet of zero degree after correlation? Previously the answer to both questions was no. Because of what are called near field effects, the ground motion immediately under the vibrator differs in phase from the far field ground motion. So if the phase angle of the motion of the earth directly in contact with the vibrator baseplate is made to be correct, then the phase will not be correct for the far field wavelet.

For many years the sensors and controls on vibrators sought to make the phase angle correct for the earth right under the baseplate (by making the phase angle correct for the baseplate that was hopefully tightly coupled to the earth). Theoretically what is needed is to make the phase angle of the force applied to the ground correct and then the far field phase angle will be correct. Therefore ground force phase locking systems have been developed in recent years and they do a better but imperfect job of generating the desired phase angle for the far field wavelets, (Appendix I and Chapter 6 give more information).

We need to extract as much good information as possible from our seismic data. We no longer use seismic data just to map structure. Therefore, ground force phase lock should be used to give us better control over our input signal to allow us to measure more accurately the effect of the earth in response to our input signal.

With ground force phase lock we should get approximately the wavelet we want from the perfect sensor at a depth of 100 ft (30 m). What happens as this wavelet propagates 50 000 ft (15 000 m) through a homogeneous earth to the second perfect sensor? As seismic waves travel through the earth there is some loss of acoustic energy and this loss is greater for higher-frequency components and is less for lower-frequency components. The earth acts like a low-pass filter that passes the lowest frequencies with modest attenuation but severely attenuates much higher frequencies. Any physically realizable filter (like the earth) must cause a shift in phase. This phase shift is called minimum-phase because the phase shift is the minimum-phase shift that preserves causality. Causality here simply means that you cannot get an output signal from the filter before you apply an input signal.

As was explained previously, correlation is a linear process; therefore we conceptually could correlate the sweep with the sweep (an autocorrelation) before applying the wavelet to the earth rather than after. If we visualize a zero-phase wavelet as if it were traveling through the earth, it is filtered by the minimum-phase earth filter. The result is a gradual change from zero-phase to mixed-phase, which means that the wavelet is no longer zero-phase but neither is it minimum-phase. The longer the wavelet travels in the earth the more it is filtered by the minimum-phase earth filter and the closer its phase approaches that of a minimum-phase wavelet. Therefore the response we measure at the second perfect sensor will not yield a zero-phase wavelet but a mixed-phase wavelet instead. Whenever earth attenuation is slight, the wavelets should be nearly zero-phase.

When we strive to record relatively high frequencies despite high earth attenuation of the higher frequencies, we can anticipate that the wavelets for the deeper reflections will deviate substantially from zero-phase. Particularly at the higher frequencies which are severely attenuated by the earth filter, phase will be shifted since all physically realizable filters must shift phase. It is disadvantageous not to know the phase angles of the wavelets we apply to the earth. Ignoring the problem is unwise particularly where the effects of the earth filter are quite significant. We seek to interpret the earth from the effects of the earth on our input wavelets. How can we do a good job of interpreting a complex earth if we do not know what our wavelets should look like in a homogeneous earth?

An entirely different approach is available which allows for better knowledge of phase. This approach is to use what is called a minimum-phase crosscorrelation rather than the normally used zero-phase crosscorrelation. With this approach the correlated Vibroseis wavelet is minimum-phase, the geophones and physical filters are minimum-phase, and the earth filter is minimum-phase. The wavelets on the final section should be minimum-phase if all data processing steps are minimum-phase. Digital filters can be minimum-phase, but many digital filters used in seismic data processing are zero-phase. Successful use of the all minimum-phase approach would require care to see that the desired result is not destroyed in processing. Nonlinear processes like automatic gain control (AGC) result in unknown phase changes. Mild AGC may not cause much phase distortion, but strong AGC should certainly be avoided when knowledge of phase is important.

Seismic wavelets from impulsive sources often are minimum-phase. If minimum-phase correlation is used with the Vibroseis technique, then the data should be more consistent with seismic data acquired with impulsive sources in the same area. Also, line ties between Vibroseis and impulsive source data should be much better with minimum-phase correlation. (The composite wavelets from some impulsive sources such as air guns are not minimum-phase because the composite excitation is the sum of two or more events which are separated in time.)

We often used vibrators along roads and negotiable trails where there were sizable sections of the line where the vibrators could not work. We used impulsive sources (mini-hole patterns of charges) in these line sections to maintain continuity and high CMP fold. If you use a minimum-phase correlation with the data recorded with vibrators, the data from the two sources should be much more compatible particularly if the amplitude spectra are made to match in processing. Then traces from both data sets within the same CMP gathers should be compatible and the stacked trace should be good. We had good success and usually the final sections showed no significant change where the sources were changed. The source points where impulsive sources were used should be prominently marked on the seismic section.

The bulk of Vibroseis data has been correlated to yield zero-phase data (either zero or ninety degree). But there are advantages to the use of minimum-phase correlations that should not be ignored. Indeed, final processed sections of both minimum-phase and zero-phase may be worthwhile for some special studies. Chapter 22 shows that under some circumstances there may be significant discrepancies in the record time for reflections from the same subsurface interface for minimum-phase versus zero-phase data. Such discrepancies in reflection times may be clues to geologic conditions that might be important in exploration.

5.4 PRACTICAL VIBROSEIS FIELD OPERATIONS

Optimum source power

The optimum source power is that which provides a satisfactory signal to background noise ratio (S/N_b) at the lowest, overall cost. With more vibrators

Table 5.1 Options for increasing the S/N_b ratio by 40 per cent

Previous parameters: 4 vibrators; 10 sweeps of 10 s for each vibrator; 100 s total sweep time.

Option	*Number of vibrators*	*Total sweep time (s)*	*Relative S/N_b*
(1)	6	88(8 × 11)	$(6/4)(88/100)^{0.5} = 1.41$
(2)	5	32(11 × 12)	$(5/4)(121/100)^{0.5} = 1.43$
(3)	4	196(14 × 14)	$(4/4)(196/100)^{0.5} = 1.40$

or with more powerful vibrators one can record data in less time by recording either shorter sweeps or fewer sweeps. The cost of more vibrator power must be weighed against the savings in recording time and hence in recording costs. The source energy is proportional to the source power times the total sweep time (the number of sweeps times the sweep length). The S/N_b ratio is proportional to the square root of the total sweep time and it is proportional to the source power.

Usually there is some Vibroseis data that you can relate to your planned project area. Suppose that the S/N_b ratio was inadequate in the previous data and that you wanted to increase it by 40 per cent. If the previous data were recorded with four vibrators using ten sweeps each and sweeping for ten seconds you might consider the options given in Table 5.1. If we used option (3) we would have to double our sweep time and this would probably be more expensive than increasing our source power by adding vibrators or by using more powerful vibrators.

Equipment required

Most vibrators are designed to have enough hydraulic force so that it is possible for the baseplate to decouple from the surface of the earth. Since we do not want decoupling to occur, why are they designed this way? The reason is that for frequencies that are not close to a resonant frequency of the shallow earth, decoupling will not occur and the large hydraulic force can be used effectively to inject more energy into the earth. Decoupling is less likely to occur on an earth with a thick soil layer and a soft surface because the soil layer will have a low Q value (i.e. the soil layer will attenuate sonic energy and any resonances will be broad but weak). But decoupling is very likely to occur on a hard outcrop of anhydrite because it will likely have some very high Q resonances. As the sweep frequency approaches one of these high Q resonances, the vibration of the anhydrite will build up in amplitude and decoupling will occur with most vibrators if a high drive level is used.

So we need to adjust the drive level so that decoupling will not occur, but do this with the vibrators working on a surface that is typical of the surface they will be working on while recording data. If the surface changes along the line, we may well need to readjust the drive level. It makes no sense at all to find a good spot where we can set a rather high drive level and obtain a good similarity with no decoupling and then proceed to use this drive level with the vibrators working on a different surface. If serious decoupling occurs, the vibrator operators will know it, but they cannot reliably detect less serious decoupling. This less serious decoupling may not hurt your data very noticeably, but it should be avoided nonetheless.

Modern equipment for checking vibrator performance will detect even minor decoupling, but if we do not have this equipment handy there is a simple test that will detect even minor decoupling. We lay a long strip of surveyor flagging down on the ground and set a vibrator baseplate on top with the end of the flagging sticking out several feet. We tug gently on the flagging while the vibrator goes through a few sweeps. If the flagging can be pulled out at any time during the sweep, then decoupling occurred. Automatic control to prevent decoupling is discussed in Section 6.7 and in Appendix I.

Every Vibroseis crew must have precise equipment that will allow the vibrators to be checked to see that they meet all performance specifications; once-a-day similarity checks are not sufficient. It is worthwhile to have and to use equipment to allow the observer to check similarities throughout the day during routine recording. See Appendix H for illustrations of the wavelets that should be observed for either zero-degree or ninety-degree operation. Note that the zero-degree wavelets have a mirror type of symmetry about the center of the wavelets which is called cosine symmetry. Conversely the 90° wavelets have a symmetry about the center of the wavelet which is like that of sinewaves. The correlated wavelets from the similarity check should show a high degree of symmetry of the applicable type. Also the main lobes of the correlated similarity wavelets should be as prominent relative to the side lobes as they are for the theoretical wavelets for the given frequency band (Appendix H). But visual inspection of the correlated similarity wavelets is inadequate; instead, precise, quantitative measurements are essential.

Frequently a nonlinear sweep may be used that is designed to boost high frequencies to partially offset the effects of significant attenuation of high frequency components by the earth. In such cases the high-frequency emphasis will result in a very ringy correlated similarity wavelet. It will have little visual appeal, but if precise instruments show that the similarities are good, proceed with your survey. We have had excellent result in many cases (Chapter 6).

Geophones with higher resonant frequencies have some advantages in certain applications. In the past one use has been to provide some low-frequency filtering in the field so that strong noises in the lower part of the sweep band do not obscure reflection events that have more of their energy at higher frequencies. But the availability of nonlinear sweeps provides a better means for achieving the desired result. We use a sweep that spends less sweep time at the lower frequencies and thereby gain the added advantage of saving recording time.

Adequate but practical vibrator specifications

Vibrators are extremely complex and must meet very rigid specifications to yield satisfactory results. As we seek to use higher and higher frequencies, the required precision becomes even more difficult to meet and maintain. Specifications that are reasonable for frequencies of less than 60 Hz may become unreasonable from a practical standpoint if arbitrarily extended to frequencies of up to 150 Hz. For example, if we specify a maximum phase error of $\pm 10°$, the time error allowed is ± 0.46 ms at 60 Hz, but only ± 0.19 ms at 150 Hz. If it is practical to adjust and maintain the vibrators to meet that specification, then so be it, but if this specification is too difficult, the cost of trying to meet it may be excessive.

In Appendix H, Table H-3 shows the error times in ms for various errors in degrees at various frequencies, while Table H-4 shows the phase error in degrees for a timing error of 0.35 ms. How serious would it be to have a timing error of 0.35 ms at high seismic frequencies? One way to look at this is what would be the maximum loss of amplitude for high-frequency components due to timing differences between vibrators. The worst possible case would be if half the sweeps that are summed together in a VP from all of the vibrators have a positive maximum timing error and the other half all have a negative maximum timing error. In this worst case for a frequency of 150 Hz and a ± 0.35 ms timing error limit the amplitude loss would be only 5 per cent. In 1984 we adopted the ± 0.35 ms limit for the timing error specification. But for the low frequencies (below 20 Hz) we adopted a timing error limit of ± 1 ms. We found these limits to be satisfactory from an operational standpoint and adequate to insure good data quality.

Polarity

How do you know that a Vibroseis crew is recording the right polarity? This is far from a trivial question because there are so many ways by which the polarity can be reversed. One method to determine polarity is to use a rugged, oil-filled geophone with low sensitivity and a long coil. Such geophones have low sensitivity but they can operate at very high energy levels. We plant this geophone in the earth very near a vibrator and record the direct arrival. The geophone should be connected in the normal manner to the recording cable in lieu of the normally used geophones. A better way is to examine the correlated response to a simple refraction from a thick, homogeneous refractor that is overlain by a thick, homogeneous layer. The response wavelet should be the same as the wavelet from an interface with a positive reflection coefficient, i.e. a positive central peak for a zero degree, zero-phase wavelet. (The reflection coefficient $= (\rho_2 V_2 - \rho_1 V_1)/(\rho_2 V_2 + \rho_1 V_1)$, where ρ_1 and V_1 and ρ_2 and V_2 are the density and velocity of the first layer and second layer respectively.)

Of course, you could also use the response from a reflection from any interface bounded by thick, homogeneous beds if you are sure you know whether the change in acoustic impedance is from lower to higher or higher to lower. One case where you can be sure is where with increasing offset the reflection event changes into a refraction. Unfortunately, you may not have handy a known earth situation that meets these requirements. Sometimes a shallow earth refraction will do if you use essentially a point source, point receivers, and a higher-frequency sweep (to reduce the required bed thicknesses so that the subject response wavelets will not be complicated by other events that overlap in time).

Noise editing and correlation

To be effective, a Vibroseis crew needs some special, sophisticated instruments. We must have an effective means for editing the uncorrelated data. Normal signal levels without interference from strong noises are measured. This means measuring and storing in separate memory locations normal signal levels over short time increments as a function of time throughout the sweep time plus listening time. This is done for each offset separately. During each time increment of each sweep if the energy level is significantly greater than the expected level for a given trace, then the editing system assumes that strong noise is present. The recorded signal for that time increment, and for that trace can then be either set to zero or it can be clipped at some prescribed level. When recorded signals from several sweeps are summed together, edited sections from one sweep will hopefully be filled in with good data from other sweeps.

When the data are correlated, the results will often be a record that is nearly free of noise from people, from cars, and even from tractors pulling plows in nearby fields! Even so, efforts should be made to minimize noise from crew operations and from all noise sources. For example if the line goes by a quarry or a municipal garbage land fill, perhaps it would be feasible to schedule the recording of that portion of the line on a day or at a time when noisy operations are not taking place (or are minimal). When we cannot avoid a strong noise source, it may be helpful to increase the number of sweeps for a section of the line. But the most critical locations for increasing the number of sweeps are not when the vibrators are closest to the noise source, but rather it is when active geophones near the noise source are at long offsets from the vibrators. Any change in operations should be noted on the observer's sheets. Also the presence and location of strong noise sources should be noted.

A Vibroseis crew without a high-quality correlator in the recording truck is severely handicapped. A blind man may be able to get along surprisingly well, but if he could buy good vision he would not hesitate. Similarly a good field correlator is worth the price. It needs to have enough dynamic range to avoid clipping strong events and to avoid loss of data for weak events. Also, the dynamic range should be sufficient so that the observer does not have to adjust the gain frequently to get acceptable correlations. The correlator must provide good correlations when the sweep is changed dramatically, particularly during initial tests to determine an effective upper-frequency limit using very narrow band sweeps (Advantage 4 in section 5.1). It would save much field time during tests if some manufacturer would allow us to sweep and correlate with a broad band sweep and then re-correlate a series of records with partial sweeps. From one field recording we could then generate a series of narrow band, correlated records with which we could determine the effective upper-frequency limit imposed by the earth for our objective. This would save the field recording time required for the narrow band sweeps and the field time required to check similarities for each of them. The Texas Instruments correlator allows several choices in defining the sweep pilot, but despite his best efforts the author found no way that it would let him change the pilot and re-correlate. It should be relatively easy for manufacturers to provide this option.

A very high quality field correlator will allow us to correlate records in the field and to use these correlated data for input at the processing center thereby saving a step at the processing center. For back-up, we would still want to record on tape in the field both the uncorrelated and the correlated data.

A correlated Vibroseis wavelet is band-limited in frequency and will ring at the cut-off frequencies if the cut-off is abrupt. In the past an amplitude taper has usually been applied to the sweep of the vibrator and to the sweep signal (pilot) used for correlation. Correlation then accentuates the taper with a consequent loss of energy near the upper and lower frequencies. This loss of energy may well be excessive with linear sweeps and can be quite severe for the lower frequencies with nonlinear sweeps used to boost higher frequencies. Use a minimum taper. If more taper should be needed it can be accomplished in processing. A taper of the vibrator sweep signal makes it easier for the vibrator electronics to control the vibrator sweep properly. With the better vibrator control systems available today, I am not sure that any taper should be applied to the vibrator sweep signal. The pilot should be tapered.

Field processing

Particularly with Vibroseis crews, field processing may be quite practical. Sometimes the necessary additional equipment is little more than what we already have in order to be able to edit and correlate effectively. Ideally we would have a full-time processor working continuously (at night if he used equipment in the recording truck) with the routine data. He could generate both brute stacks and final field stacks that might utilize very sophisticated programs and approach the quality produced by good processing centers. Sections and other displays of the data could routinely be sent by mail or by fast courier services to the office. A good processor should provide brute stacks that lag acquisition by a couple of days and final field stacks within a week. He could send you critical displays by facsimile via telephone lines. Advantages that should accrue from field processing are:

1. Initial tests can be better analyzed and displayed for better immediate decisions as to technique and parameters to use for the survey. This is particularly valuable for a Vibroseis crew because of the versatility of the method and the many options that are open to improve results. If you cannot be in the field at this critical time, facsimile reproductions of critical displays can be received and then discussed with the processor. Quick, effective decisions may be reached that would not be possible without field processing.
2. Any loss of quality of the processed data can be noted quickly and efforts can be made to find the problem and solve it to regain and maintain superior data. A good processor can learn to spot problems and inform you immediately by telephone. He can describe what he sees or he can send you a copy of critical displays via a telephone line.
3. The field program can be modified quickly while the crew is in the area to follow up any interesting features revealed by the stacks. You may want to extend a line, add a line, or move a line. There were occasions when a structure turned over sooner than expected and we shortened a line. At other times we extended lines that failed to show turnover where it had been anticipated.
4. We may be able from field stacks to justify land deals and get a jump on competitors. It has happened.
5. Final processing of data at the processing center can benefit from the field stacks, displays, velocity analyses, etc. This can help speed up final processing. The processors at the processing center are challenged to do a

first-class job or their sections may not be any better than the field final sections. The author has seen a few instances where the final section from the field was better than the final section from the processing center.

Field processing is not inexpensive, but it has the potential of adding considerably to the effectiveness of the crew and to the effectiveness of your exploration effort.

Vibrator operation

Sometimes vibrators will have to be operated at reduced drive levels to reduce the risk of damaging nearby buildings or other property or facilities. This should be noted on the observer's sheets. With reduced drive levels we will probably want to increase the total number of sweeps to keep from losing too much source power.

At times we will have one or more vibrators that need repair or adjustment and are not available for use. If we have no spare vibrators, we would still probably allow operations to continue with one less vibrator, and we should provide for such cases by specifying in writing and in advance exactly what adjustments should be made. This can best be done by filling out a Recording Parameter form that covers all of the parameters to be used in recording the line (Appendix A). We may or may not wish to change the spacing between the vibrators and we may or may not want to change the vibrator move-up between sweeps for each VP (Chapter 3). We should specify how many more sweeps should be summed for each VP whenever the number of vibrators used is less than standard. When we add sweeps we are also adding recording time which means that we are adding background noise as well as signal. If we want to maintain the same signal level relative to background noise, then the ratio of the number of sweeps needs to be equal to the inverse of the square of the ratio of the number of vibrators. For example, if we normally sweep eight times with four vibrators, then with just three vibrators we would need to sweep $8(4/3)^2 = 14.2$ sweeps to achieve the same signal to background noise level. We could decide to specify fourteen sweeps or decide to allow some decline in relative signal level to save recording time and might then specify twelve sweeps.

We may be tempted to use big, powerful vibrators. But smaller vibrators can cross bridges that are too weak for big vibrators. Permits for both on-road and off-road work may be easier to obtain for small vibrators. Six small vibrators are less likely to cause damage to nearby buildings than three or four larger vibrators (even if the total energy is the same). Six small vibrators can be used to create source patterns that would be more time consuming to do with three or four larger vibrators. Big, powerful vibrators should be used in some areas where much energy is required.

When operating vibrators on a hot, asphalt road, we risk damaging the road. The vibrator pads may indent the surface or the tacky surface may stick to the pads and be partially pulled up when the pads are raised to move up. A technique that works rather well is to wet-down the asphalt pavement ahead of the vibrators. A resourceful crew can improvise a tank on one of the crew vehicles to do the job. If the need is of long duration, a water truck may be justified.

As with any type of seismic crew, geophones need to be planted well, and arrays need to be laid out properly. Geophone arrays must be connected to the cable and must later be disconnected and picked up at the right time. Good coordination between operations is essential and requires good communication. Enough appropriate radios must be available to provide the opportunity for good communications, and they must be used effectively.

5.5 PRACTICAL VIBROSEIS FIELD TECHNIQUES

Where it is possible to meet exploration objectives with lines that are located along roads which can be permitted for vibrators, the Vibroseis recording technique is ideally suited. Under these circumstances data can be acquired rapidly and costs per mile should be relatively low even for high CMP fold data. But even where available roads exist along most section lines, we may need to acquire detailed subsurface information at off-road locations or along off-road lines. Often it is quite practical to vibrate along the roads but to lay out the cables and geophones off-road. One method is to use normal roll-along CMP recording except with a substantial lateral offset (Chapter 10).

Another need for off-road data may be to locate a well site. We may need an area of subsurface coverage to both sides of a road and centered at a location determined from a previous line along the road. In this case we could layout a fixed cable (no roll-along) perpendicular to the road and centered on the road and vibrate at source points at regular intervals along the road. The result would be single fold coverage over the area of a rectangle with a perpendicular dimension of half the total cable length and a longitudinal dimension of half the distance from the first source point to the final source point. If the single CMP fold results are not satisfactory, we may choose to sum-up groups of one trace with its nearest eight other traces (nine spots) to achieve something comparable to nine-fold data (Chapter 10).

We much prefer for our seismic lines to be straight, but where only crooked roads and trails are available, surprisingly good seismic data can often be acquired by operating vibrators along a road and running the cables along the road, but cutting across the sharper bends with the cable where it is feasible to do so (Chapter 17). In hilly and mountainous terrain the author's crews recorded hundreds of miles of data along very crooked roads. Most of the sections turned out quite good except in areas where data quality was also not good when recorded along very straight lines (with a portable crew using explosives).

Processing crooked line data requires capability for properly handling the geometry. Scatter displays should be made of the mid-point locations for each source position and trace position pair. Near tight bends in the roads we may show mid-point coverage that flares out to a considerable width (in a direction perpendicular to the general azimuth of the seismic line). If there is significant cross dip for some reflector or reflectors, then the CMP stack will be degraded for those reflectors with cross dip. As is often the case, this problem also presents an opportunity. We can try assuming different cross dips, correct for the assumed dips before stacking, and then see what cross dip assumption results in the best stack for a given event. Where data quality is relatively good, this can yield a valuable measure of cross dip. It is conceivable that data of this type might be so critical that we would purposely look for crooks in roads and trails where we could layout a program designed specifically to measure cross dips.

6

High resolution including practical use of nonlinear sweeps

In this chapter we discuss the need for high resolution seismic data and ways to improve resolution with either impulsive sources or vibrator sources with nonlinear sweeps.

6.1 NEED FOR HIGH RESOLUTION DATA

Ideally we would like for seismic data to provide us with the equivalent of continuous sonic logs or more accurately continuous acoustic logs sensitive to changes in sonic velocity and changes in density. Of course, it is not possible to record the range of frequencies required either at the low frequency end or at the high frequency end. So we have to settle for a much more limited band of frequencies in our seismic data and this limited band of frequencies limits the attainable vertical resolution. But if we can somehow record higher frequencies in our reflection data, we can improve the resolution of our data. Also, if we can improve the resolution we can map many more interfaces and we can detect upper and lower boundaries of thinner beds. This also gives us the potential for measuring interval velocities for thinner beds which increases our capability for estimating lithology and porosity of earth layers.

Often we cannot map objective layers directly or map objective intervals within broader layers and we have to settle for mapping overlying and underlying interfaces. With better resolution we may be able to map our objectives directly.

6.2 HIGH RESOLUTION SEISMIC DATA USING EXPLOSIVE SOURCES

For many years the best energy source for recording high resolution seismic data was a short (and not too heavy) explosive charge in a deep borehole. The explosion and nonlinear deformation around a short, light charge takes place in a fraction of a millisecond with the result that the source approximates a unit impulse for seismic frequencies. This source is very effective in generating high-frequency components for the input seismic wavelet. Small charges in deep holes are effective sources for high resolution seismic data, but this method is no longer unchallenged.

With an appropriate explosive source we can inject a seismic wavelet into the earth that has good energy at the higher seismic frequencies. If the earth filter is not too severe, and if the background noise at the higher frequencies is not too great, then we have the potential for recording high-resolution data. But we need to use arrays that are not too long, and the data processor will have to be able to correct accurately for static variations and for normal moveout. He may need our help so that he can extract the needed information from the data we record.

If our field records are dominated by strong noise at low seismic frequencies, then we may be able to help him by using geophones with resonant frequencies that are higher than the main frequencies of this strong noise. No doubt he could do the equivalent frequency filtering in processing, but the battle may already have been lost if the high-frequency signal level is too small compared to strong low-frequency noise. It is true that modern seismic recording systems have huge dynamic range capabilities, but it does not mean that we never have to worry about exceeding the dynamic range. By using geophones with higher resonant frequencies we can reduce the required dynamic range of our instruments in cases like the one described. But there is a penalty. We can reasonably compensate in processing for the amplitude suppression of the lower frequencies by geophones and you can approximtely correct for the effect of these geophones on the phase. Phase shift near resonance is very nonlinear so major phase compensation is necessary. But geophones vary and the loading and effective damping of the geophones varies, which means that the phase compensation will be only approximately correct. If we have enough dynamic range with geophones with 'standard' resonances of 8 to 12 Hz, then we should probably stick with these geophones. But if geophones with higher resonant frequencies can help us to achieve needed dynamic range, then we should use them.

We should ask our processor if he wants us to record some duplicate data with standard geophones and with higher frequency geophones. He can then determine from the data an amplitude function and a phase function to use for compensation. Or he can apply a theoretical correction and see how nearly he can make the two measurements match. Perhaps R & D or some other group in our company has already done this. But was the effective geophone damping the same as they will be in our data recording?

Field records may look better when recorded with geophones with higher resonant frequencies. But it may be possible to achieve the same effect by recording with standard geophones and then using a low-cut filter for playback records in the field, but not for recording. We may need to get low-cut filters with the cut-off frequencies we need. Preferably low-cut filters with higher cut-off frequencies should be used for playbacks only and not for recording data. But it may not be operationally feasible to record with one filter and playback with another.

In some cases we may choose to use more than one loaded hole at each shot point to provide a source pattern to suppress strong noises. With deep holes, this is expensive. Obviously source patterns must not be too long.

With charges in deep holes we may encounter a problem with surface ghosts. These ghosts result from energy traveling up toward the surface and then being reflected downward either at the surface or from some interface below the surface but above the shot. Frequently this reflecting interface is at the base of the weathered layer. From the shot you get a downward traveling wave directly from the source plus a second downward traveling wave that is delayed by the travel time from the charge depth up to the reflecting surface and back down to the charge depth. Usually ghosts are reversed in polarity because the ghost interface has a negative reflection coefficient. Ghosts set up interference patterns with the primaries which create notches and peaks in the amplitude spectra of the reflected wavelets. Often ghosts differ from one shot point to the next in terms of their amplitude and delay time. In this case the CMP stack of multi-fold data may adequately suppress the effects of ghosts. Chapter 9 provides more information about ghosts and about ways to suppress them.

If the surface or near surface is reasonably competent, we may sometimes be able to generate seismic wavelets with the needed high-frequency components using small charges in shallow holes or even with small charges on or slightly above the surface. In these cases a source pattern is more practical and may well be required. With charges at or above the surface the air wave will surely get your attention in more ways than one.

The basic procedure for initiating and conducting a high-resolution survey using explosive sources is the same as for any other survey. For initial tests we will want to have part or all of the holes drilled ahead of time so that recording can begin quickly when the crew arrives. We may want to test more than one source pattern and to test different charge sizes at different depths. We will want to record tests at offsets which will allow us to determine the most effective range of offsets for our production recording. Obviously geophone arrays and source patterns must not be so long as to seriously limit the high-frequency response due to differential NMO across the arrays and patterns.

We may have one or more drills working during the recording of the tests and these can be available to redrill some holes, if necessary, or to drill some special pattern that we decide to try based upon test results. Of course, the drills must be in radio communication with the recording truck so that they can be told when to shut down their operations so that data can be recorded with no drilling noise. In open country the wind may generate a lot of noise particularly at the higher frequencies. Burying the geophones can reduce high-frequency wind noise dramatically.

For high-resolution work the determination of the effective upper frequency limit for our objective is important. Yet this may be difficult to do unless we have good data processing capability in the field. In the absence of good data processing capability in the field, we should try to arrange for a crew break after the tests are completed to give ourselves time to get the data to a processing center for analysis, processing and display. Alternatively the crew may be able to record some other project if one is available that is not too far away. After we have made our decisions regarding the production recording of the seismic program, we will want the drills to begin drilling a few days before the recording crew returns.

Figures 19.1 and 19.2 show examples of the improvement in resolution that can be obtained in some cases using spectral whitening in processing.

6.3 HIGH RESOLUTION WITH THE VIBROSEIS TECHNIQUE

For a number of years the Vibroseis technique was not the best technique to use for high resolution work. There were problems in timing and problems with low energy output from the vibrators at the higher frequencies. High energy at high frequencies requires high acceleration of the reaction mass. High acceleration requires high force relative to the reaction mass. High force requires a high product of hydraulic pressure times the area of the piston (or pistons). If we limit pump pressure to 3000 lbs/in^2 (211 kg/cm^2 or 20.68 MPa) then we need a vibrator with a large piston area and a reaction mass that is not too large. But vibrators are made to operate over a rather wide range of frequencies and the requirements at low frequencies are different so compromises are required. At low frequencies you need a large reaction mass to try to push the huge earth mass with which we are interacting. (At a high frequency the wavelength in the earth is shorter and the effective volume of earth we have to move is much smaller since volume is proportional to the radius cubed.) At low frequencies the reaction mass must move farther back and forth to produce much energy. The required long displacement requires a high volume of hydraulic fluid and this required volume of fluid is increased if the piston area is increased to improve the high-frequency output. So we need a huge pump which requires a huge vehicle with a huge engine and a huge fuel tank full of diesel fuel. Also we need a reaction mass that is at least moderate in weight. We can get vibrators like this, but they cannot get across weak bridges. (Some of these super vibrators have so much acceleration capability that at moderate frequencies they can readily jump up off the ground and become an uncontrolled, reverberating weight drop—a poor seismic source.)

Appendix I and the last section of this chapter show ways to control the force at moderate frequencies and yet have maximum force available for more effective generation of the higher frequencies. Modern vibrators with proper adjustments can meet the timing requirements necessary for high resolution seismology (Chapter 5 and Appendix H).

One thing more was needed to make vibrators the best and most practical source for high resolution seismology. The development of good nonlinear sweep generators was the key. A major advantage of the Vibroseis technique is the ability to choose the band of frequencies we wish to use. Nonlinear sweeps add markedly to that advantage by allowing us to also choose the relative amplitude we want as a function of frequency. With nonlinear sweep generators we can tailor the amplitude spectrum to suit our fancy. We can spend more sweep time at higher frequencies to build up energy there and less time at lower frequencies so as to partially flatten the spectrum of the reflected wavelets. By doing this, we reduce the need for large pump volumes for large displacements at the low frequencies because we spend little sweep time there and hydraulic fluid stored under pressure in accumulators can meet the momentary need for high volume. But the main feature is that we can increase the relative energy output at the higher frequencies where earth attenuation is severe. Normally we do not need huge vibrators, in fact, some of the smaller vibrators work quite well when equipped with the best modern controls. Spending less time at the lower frequencies is like using a low-cut filter or using geophones with a high resonant frequency, but it is better than these because it means saving recording time. We can use this saved time to build up more energy at the higher frequencies, or speed up recording and save money.

6.4 PRACTICAL APPROACH TO THE USE OF NONLINEAR SWEEPS

Providing the geophysical industry with a practical nonlinear sweep generator was somewhat like giving a woodcutter a chain saw to replace his axe. We have to learn how to use our sharp and powerful tool without maiming ourselves. What is it that we really need to do in the field with this tool? The answers remain what they have always been, which are:

1. We need to achieve an adequate signal-to-background-noise ratio over the range of frequencies we want in our reflected wavelets at the record times of our objectives, and we would like to accomplish this in the minimum amount of recording time;
2. We need to suppress source-generated noises in the field well enough so that the data processor can extract the information he needs to make static corrections and to correct for normal moveout;

3. We would like an adequate S/N ratio at non-objective times to help the processor solve the problem of static variations and provide more geologic information to the interpreter.

Do we want to achieve a precisely flat amplitude spectrum for our reflections from our objective? No, we do not! If we knew enough about earth absorption and about the effectiveness of our equipment as a function of frequency to allow us a ghost of a chance of doing this in the field we would certainly know enough to do it better and easier in processing. Even if we could precisely flatten the spectrum in the field we would not want to do so if it required more recording time. Furthermore we could flatten the spectrum for reflections from only one objective. Reflections from shallower depths would be over-compensated at the higher frequencies and those from deeper depths would be under-compensated. If we suspect that earth attenuation for our objective will be more severe than normal, we may want to increase the boost of the high frequencies. Conversely, where the earth is well consolidated and low attenuation is expected, we may choose to use less boost for reflections from a shallow objective. But let the processor flatten the spectra. We should concentrate on getting him data that will better meet his needs.

6.5 TYPE OF NONLINEAR SWEEP AND BOOST NEEDED

Appendix J includes a derivation of the type of nonlinear sweep that is required to offset the effects of earth attenuation. The answer is that we need a logarithmic sweep function. This would be the ideal type of sweep function to use if background noise level does not change with frequency and if vibrators and recording instruments were just as effective at all frequencies. We know vibrators will be less effective at rather high frequencies, but probably background noise level will often be lower at higher frequencies. Earth attenuation of high-frequency noises will limit background noise to relatively nearby sources. If it is windy you may need to bury the geophones to reduce the background noise at high frequencies. In open grass and bush land wind noise has strong components at high frequencies, but in forested land, trees swaying in the wind couple mainly low-frequency noise into the earth.

Our biggest problem is to decide how much boost we want our logarithmic sweep function to give to the highest sweep frequency as compared with the lowest sweep frequency. In the case of the Texas Instruments nonlinear sweep generator our choice is quite limited.* The amount of boost for high

*The Texas Instruments sweep generator also gives you the option of reducing high-frequency boost by creating a sweep that is between their logarithmic sweep and a linear sweep. We weight the linear sweep in 1% steps (with 100 being completely logarithmic and 0 being completely linear). Whereas we had this capability on a contract crew, we never found a situation where we chose to use it.

frequencies increases slowly with sweep length (Appendix J), so we can adjust our high-frequency boost somewhat by changing the sweep length. Also if we want more or less total input, we can change the number of sweeps. Fortunately this worked quite well for the author. The high-frequency boost as a function of sweep length with the Texas Instruments sweep generator is as follows:

Sweep length (seconds)	*High-frequency boost (TI Sweep Gen.) (ratio of time at highest frequency to time at lowest frequency)*
8	7.1
16	11.4
24	14.8

The Pelton sweeps generator provides a different sweep function from the logarithmic function derived in Appendix J to compensate for earth attenuation. The sweep function provided by the Pelton sweep generator is the function that we would derive if we were trying to compensate for loss of efficiency in generating high frequencies due to a limited force that limited the amount of acceleration of the reaction mass. If that were our only purpose you would set the Pelton sweep generator to provide a boost of 6 dB per octave. But elasticity, principally of the hydraulic fluid, may require more high-frequency boost. We need to be concerned about the total loss of energy at the higher frequencies from both equipment limitations and earth attenuation. The Pelton sweep generator allows us to set the boost in 1 dB/octave steps up to a maximum of 15 dB per octave, which gives us the capability of boosting the highest frequency as much as we would with the Texas Instruments sweep generator. But then intermediate frequencies are not boosted the same amount. However, both types of sweep generator have yielded excellent results. (Appendix J gives equations and more information: Appendix F, programs D and D′ can be used to calculate sweep functions.)

The author much prefers using sweep generators with dial adjustments for the frequency range and the sweep length over using 'erasable, programmable, read-only memories' which are called EPROMS for short. Some of the programs created to program EPROMS did some wild things in trying to fit a user-defined sweep function. The author had some problems with EPROMS and gladly gave all of them away (within the company) when the Texas Instruments and Pelton sweep generators with dial controls became available.

There is every reason to believe that nonlinear sweep generators produced by companies like SSC, Geosource, Sercel and others are quite satisfactory.

Boost high frequencies by a factor of about 10 and the results will probably be quite good in most cases. Do not use the ridiculous 'T^2' sweep function which spends an inordinate fraction of the sweep for the last one per cent of the band of frequencies swept.

The relative increase in higher-frequency components injected into the earth improves the signal-to-background noise level at the higher frequencies. When we correlate the earth response with the same sweep we are effectively filtering the data with a filter that again boosts the highest frequency by the same factor relative to the lowest frequency. Therefore the apparent boost is squared, which helps to bring out higher frequencies so that their presence can be seen in the field data. That is good. In all probability a ten-fold boost is far from adequate in many cases, to offset the total loss of higher frequencies from all causes, particularly for high-resolution surveys where we often try to record frequency components which are severely attenuated by the earth and try to record a rather wide band of frequencies. A ten-fold boost is generally sufficient to bring higher frequencies up above the background noise. Then the additional, ten-fold boost from the filter effect from correlation may be sufficient to make their presence visible.

For years the processors have been trying to flatten spectra that were drastically out of balance due to severe attenuation of the higher frequencies and for other reasons. So what we do with nonlinear sweeps is to make the processor's job easier, not harder. Deconvolution has long been used to balance (flatten) spectra, but even good data processors have often been too timid in the use of deconvolution to balance spectra. Probably every explorationist has been 'burned' by deconvolution at some time or another. But if we cut a finger with a sharp pocket knife do we then dull the blade of the knife? No, and neither should we dull our deconvolution tool. With properly used nonlinear sweeps we give the processors better data and he needs to adjust his thinking as to what can and cannot be done with these improved data.

A spectrum flattening program (called 'spectral whitening') developed by Western Geophysical Company does a superb job of taking data acquired with nonlinear sweeps and balancing spectra (Chapter 19). It is not a timid program and it really does the job. After a processor sees what can be done by this spectral balancing program, he will likely find he can do about as well or as well with an appropriate* deconvolution when he really lets it work. The author recommends the use of the spectral whitening program as an excellent partner with nonlinear sweeps to drastically improve the vertical resolution of

*The appropriate deconvolution depends upon what phase relationship we want for our reflection wavelets. If we use the standard zero-phase correlation and are trying to get zero-phase wavelets, then we should use a zero-phase deconvolution. But if we use a minimum-phase correlation and use minimum-phase processing to get minimum-phase wavelets, then we should use a 'spiking' deconvolution (which assumes minimum phase).

your seismic data. For final sections you may want to use deconvolution or deconvolution followed by spectral whitening. The reason is that the spectral whitening program uses AGC and this has an unknown effect on the phase of the data. If we use a slow AGC the phase change will be minimized. If we flatten the spectra first with deconvolution and then follow with spectral whitening just for final flattening, then the phase of our reflection wavelets should not be seriously affected by the spectral whitening program.

A nonlinear sweep amounts to a zero-phase filter. There is a disadvantage to the use of nonlinear sweeps in those cases where we want a minimum-phase reflection wavelet and the high-frequency boost is used to offset earth attenuation rather than just to offset loss of effectiveness by vibrators at the higher frequencies. In this case a phase correction should be applied in processing to offset part of the zero-phase effects of the nonlinear sweep.

6.6 PRACTICAL USE OF NONLINEAR SWEEPS

We do not necessarily have to increase sweep time with nonlinear sweeps. For example, suppose in an area we previously used a 14–56 Hz linear sweep over 12 s and swept twelve sweeps with four vibrators for each VP. The rate of change of frequency was a constant 3.5 Hz per second. Presumably we needed all this sweep time at 56 Hz but could have used less sweep time at lower frequencies. With a nonlinear sweep we need to provide about the same rate of change of frequency at 56 Hz.

Suppose we plan to use a 12–72 Hz sweep over 12 s with a Texas Instruments logarithmic sweep. We calculate the frequency and rate of change of frequency as a function of time using program D′ from Appendix F and list the results in Table 6.1. At a sweep time of 5.5 s the frequency is 55.79 Hz and the rate of change of frequency is 3.6 Hz/s which is very nearly the same as with the prior linear sweep. We would need to test to see if twelve sweeps of 12 s each with four vibrators provides sufficient energy over the 12–72 Hz band. Based upon the author's experience, he anticipates that it would be satisfactory. If more energy were needed, we might add a vibrator or increase the number of sweeps. But we must not assume that we will need more source energy.

In some areas raw field data acquired with linear sweeps look poor because of strong, low-frequency, ground roll (or other source-generated noise). In such cases there is a tendency to increase source power to be sure we have enough energy. With nonlinear sweeps raw field data frequently look much better and we may find that we can actually decrease sweep time! On the other hand, if you are trying hard to get as much resolution as possible, and trying to extend the high frequencies by an octave or more; you may need to increase the source energy by using more sweeps, more vibrators, or both.

Can you get hurt with nonlinear sweeps? Yes, but if you can see reflections

Table 6.1 Frequency and rate of change of frequency as a function of time for a Texas Instruments logarithmic sweep

	Sweep band	*12–72 Hz*
	Sweep length	*12 s*
Time	*Frequency*	*Rate of change of frequency*
0	12.00	23.39
1	28.20	11.70
2	37.70	7.80
3	44.43	5.85
4	49.65	4.68
5	53.91	3.90
5.5	**55.79**	**3.60**
6	57.52	3.34
7	60.64	2.92
8	63.40	2.60
9	65.86	2.34
10	68.09	2.13
11	70.13	1.95
12	72.00	1.80

on the field records near objective time, you can proceed with confidence. You must determine the effective upper frequency limit that the earth allows for your objective. And you can do this accurately at your test site (or on your line) using narrow band sweeps at selected high frequencies. (The earth filter is not extremely sharp, but we reach a point of diminishing returns for the effort required which establishes an effective upper-frequency limit, chapter 22.) Responses will be very ringy, but if there is coherence near or after objective time that is consistent with primary reflections on wider band data, then reflections with that frequency component can be successfully recorded from depths as great as your objective. Where high vertical resolution is critically important, you may well choose to try to record frequencies slightly higher than those that pass the narrow band frequency test. The rationale is that with stacking and with spectral whitening you may be able to push the upper-frequency limit slightly higher. But if you go too far you will spend much of your field recording time generating frequencies that you cannot get back in reflections from objective depth. (With nonlinear sweeps that boost the higher frequencies if the upper-frequency limit is 10% above the effective upper frequency allowed by the earth for our objective, then the wasted field time will be much more than 10%.) This is a risk that should be carefully evaluated before making your decision.

With nonlinear sweeps to boost higher frequencies, you should use a sweep band of more than the two octaves that have been used traditionally. There are two reasons for increasing the sweep bandwidth. First, any taper on the sweep at the lower-frequency limit will affect more of the lower frequencies because we will sweep through these much quicker. Hence the effective lower-frequency limit may be somewhat higher than we think it will be. Second, the effective upper-frequency limit that the earth allows may change along the seismic line. If this effective upper limit should decrease to a significantly lower limit it will reduce the bandwidth. For these reasons we should widen the bandwidth to two and a half or three octaves. At times we have used a sweep band of three and a half octaves and recorded good data. The greater the effective bandwidth the greater the amount of information we can extract from the data about the earth section, but as we extend the bandwidth to lower frequencies we may add more noise than signal.

The mistaken concept that we should use nonlinear sweeps to try to flatten the spectrum precisely in the field leads to other fallacies. One such other fallacy is that we must not, or should not, use nonlinear sweeps on a regional line because things are bound to change along the line and such changes would be sure to upset the precise balance that is 'needed'. One of the author's early lines recorded with a nonlinear sweep was a long, regional line. Along the line the surface varied from Permian to Devonian and included shales, sandstones, limestones and dolomites, and the geology was far from 'layer cake'. The same logarithmic, nonlinear sweep of 10–100 Hz was used for the whole line. The resulting section was superb from shallow to deep and throughout the length of the line. The signal-to-noise ratio was excellent and the resolution was far better than we had achieved previously in the area (Fig. 6.3).

In summary you do not need to have a PhD to use nonlinear sweeps quite successfully, you do not need to know the Q of the earth and you do not even want to use the nonlinear sweep to offset precisely the effects of earth attenuation or the effects of equipment limitations. What we want to do is simply to determine the effective upper-frequency limit, select a reasonable bandwidth (at least 2.5 octaves) and then achieve a satisfactory signal-to-background-noise ratio at objective time over the selected sweep band. Also we want to record data in the minimum time required to achieve the adequate S/N ratio. With the recommended techniques and approach we can do the job quite well. In fact, the vertical resolution of our data may be so much better that the interpreting explorationist will be lost in terms of tying in the new data with his old, low-resolution data. It has happened. Figures 6.1, 6.2 and 6.3 give comparisons between earlier Vibroseis data and data acquired with nonlinear sweeps and processed with the spectral whitening program. But you can get the processor to filter your data back to the limited band of frequencies of the old data so your explorationist can make his correlations. He will like his new data—after he gets over being lost.

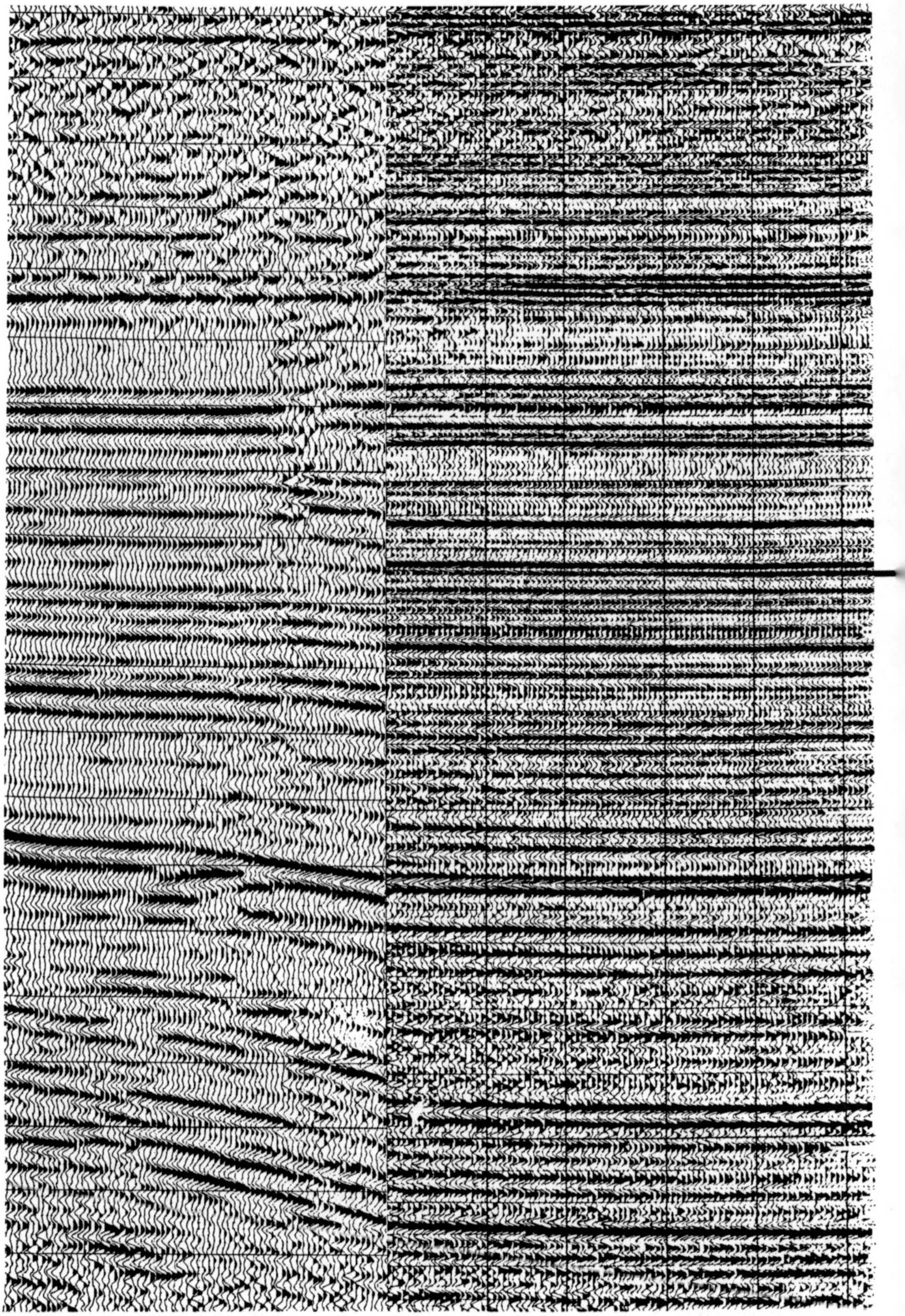

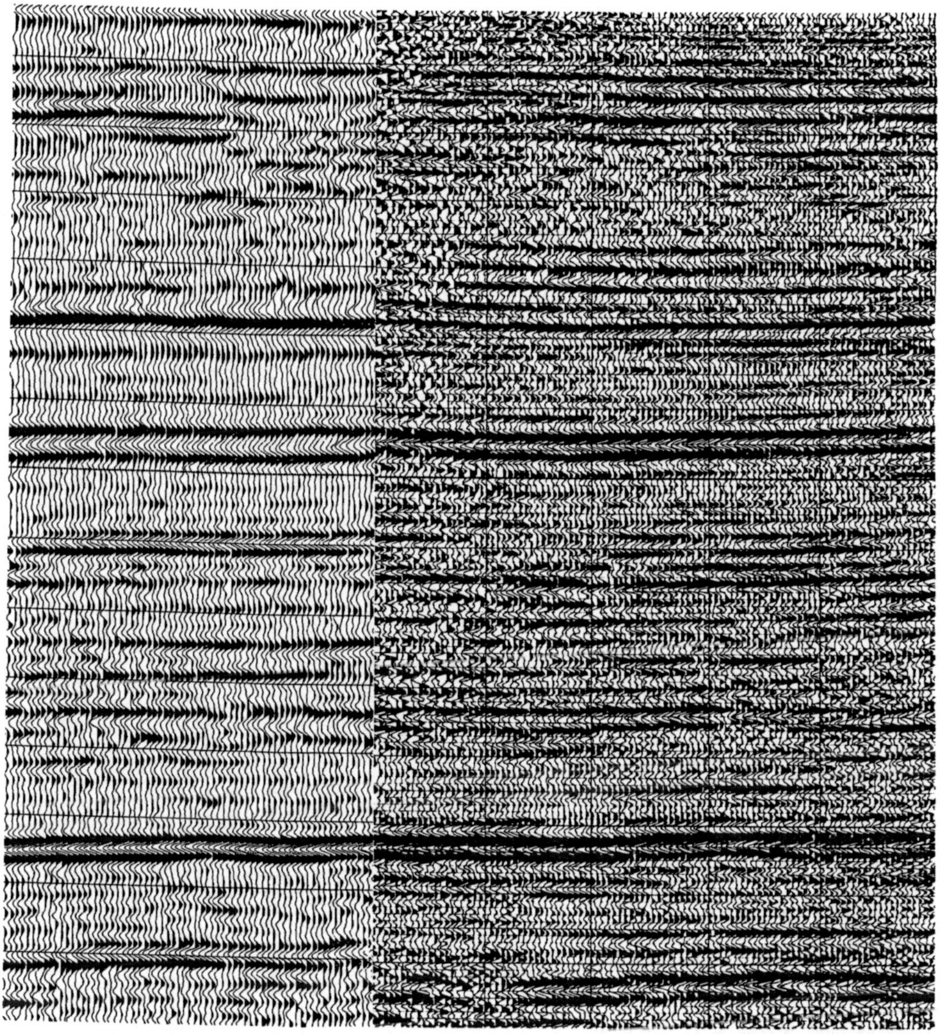

Fig. 6.2 Comparison of high-resolution data acquired with a nonlinear sweep with previous linear sweep data—Example B. These two lines nearly cross. The high-resolution line is on the right.

Fig. 6.1 Comparison of high resolution data acquired with a nonlinear sweep with previous linear sweep data—Example A. These two lines do not meet, but are from the same general area and the samples are representative of many lines in the area. The high resolution line is on the right.

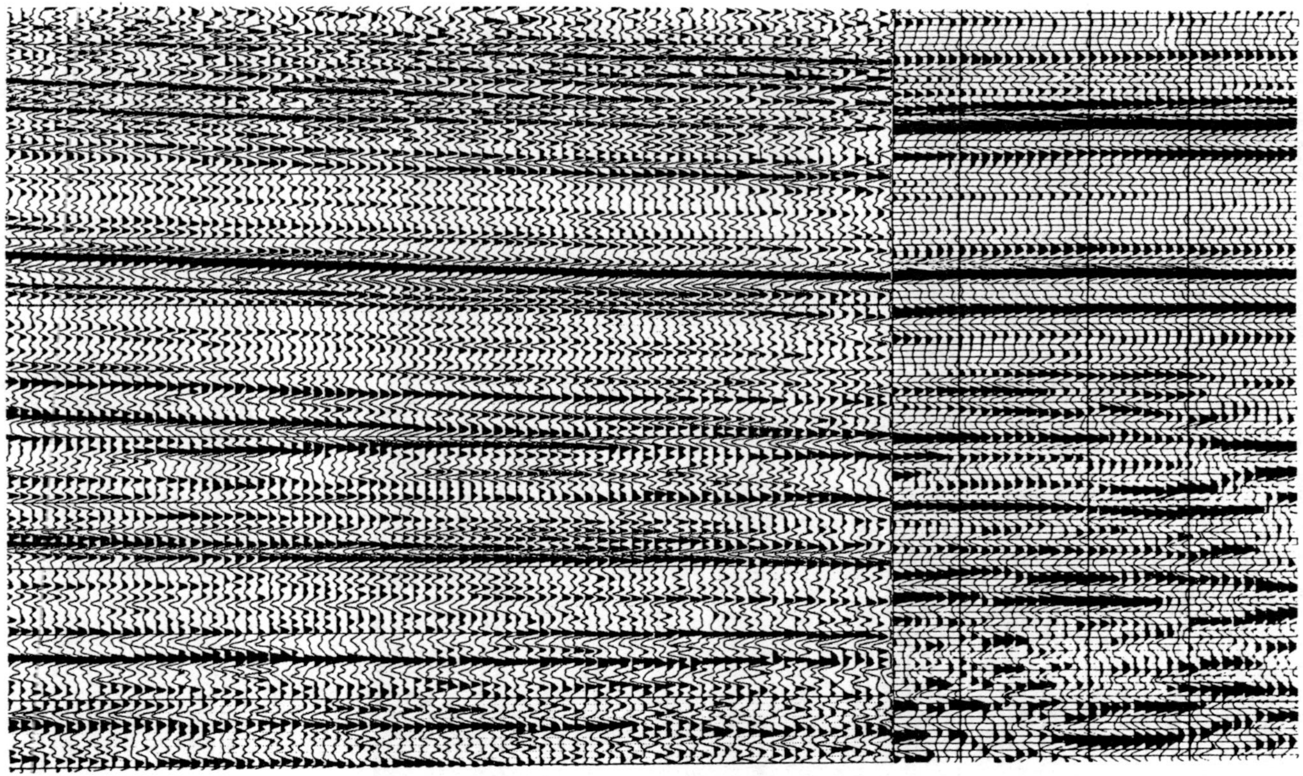

Fig. 6.3 Comparison of high-resolution data acquired with a nonlinear sweep with previous linear sweep data—Example C. These two lines are nearly parallel at the shown location. The high-resolution line on the left was a long, regional line with many geologic changes and structures. Outcrops varied widely in age and in lithology yet data quality, resolution and continuity were excellent throughout the length of the high-resolution line for reflectors ranging from shallow depths to basement.

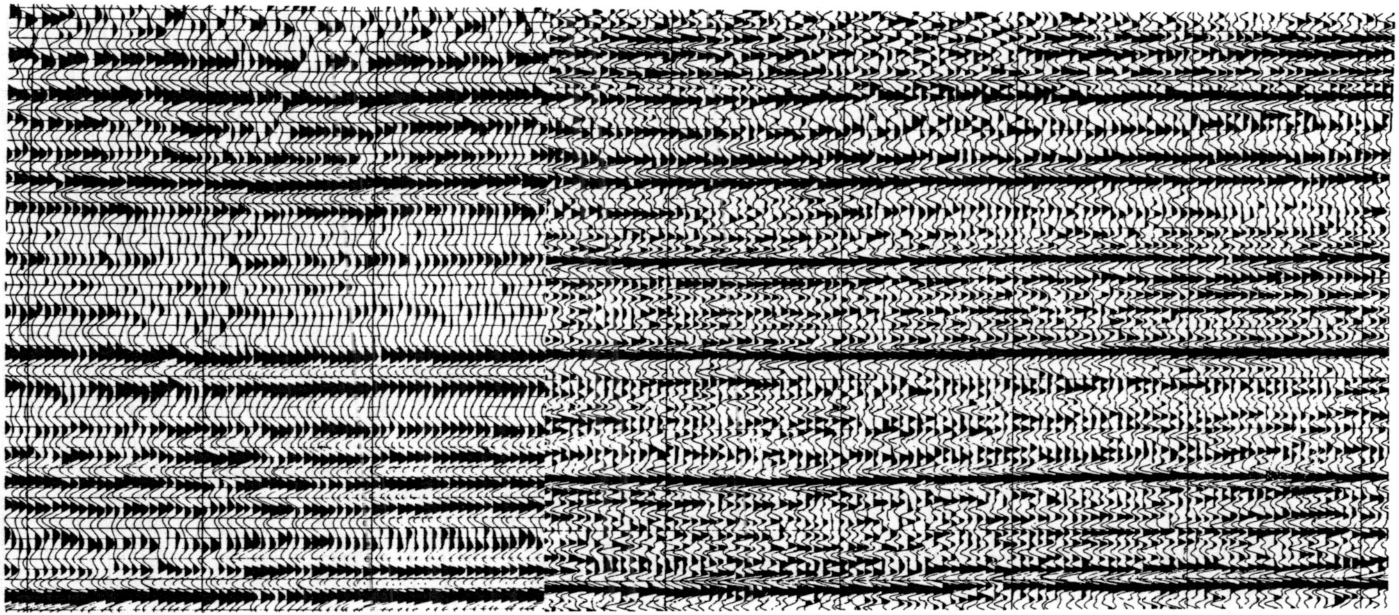

Fig. 6.4 Comparison of high-resolution data acquired with a nonlinear sweep with previous mini-hole dynamite data—Example D. The high-resolution Vibroseis data on the right were recorded along precisely the same line as the previously acquired dynamite data. We recorded hundreds of miles of equally good, high resolution data in this area.

Example A in Fig. 6.1 compares high-resolution data (right side) acquired with nonlinear sweeps and processed with spectral whitening, with nearby data acquired previously with linear sweeps and processed with deconvolution. Both data sets are typical of many miles of data acquired in the area with the two techniques.

Example C in Fig. 6.3 is in yet another basin and in another state. The high-resolution data (right side) were acquired with a nonlinear sweep and processed with spectral whitening whereas the previous data used a linear sweep and were deconvolved in processing. In this area the high-resolution data were not as consistent as in the first area, but they were significantly better than nearby previous data.

Example C in Fig. 6.3 is in yet another basin and in another state. The high-resolution data (left side) were acquired on a long regional line that crossed many different outcrops. The data quality was excellent along the whole line with greatly improved resolution and excellent continuity. The low-resolution data from the same area are typical of the data previously acquired.

In the preceding three examples all of the recording parameters were carefully selected based upon field tests. The improved resolution in each case is attributable directly to the use of nonlinear sweeps.

Examples D and E in Figs 6.4 and 6.5 compare high-resolution Vibroseis data acquired with nonlinear sweeps and processed with spectral whitening, with data acquired from patterns of explosive charges in shallow holes. In these two cases precisely the same lines were repeated. In Fig. 6.4 the high-resolution data on the right is spliced on to the low-resolution data. No wonder the interpreters got lost! But they were reassured when a high-cut filter was used to degrade the high resolution data to match the old dynamite data.

The lateral changes in the high-resolution Vibroseis data of Fig. 6.5 disclose known earth changes. These earth changes are hidden in the low-resolution data that were shot previously along the same line.

In the last two examples the mini-hole dynamite data are typical of data recorded by the industry in the area at the time the author's crew recorded the high-resolution data.

6.7 AMPLITUDE CONTROL

As a part of the nonlinear sweep development, some manufacturers of equipment also allow us to alter the drive level during the sweep. But there is a fallacy here if the purpose is to reduce the amount of low-frequency output to partially balance the spectra of reflected wavelets. The problem is that a reduction in the drive level will have an unknown and variable effect and may or may not significantly affect the energy output of the vibrators. Even if this would work, it would be better to reduce the time spent sweeping these lower

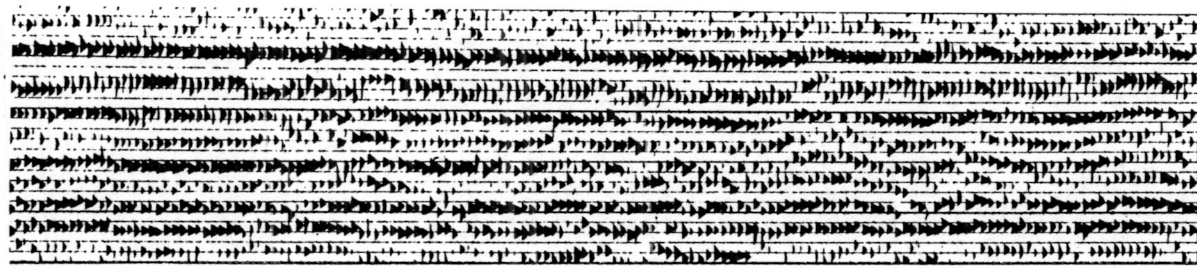

(a) Conventional data acquired with explosive charges in mini-hole patterns

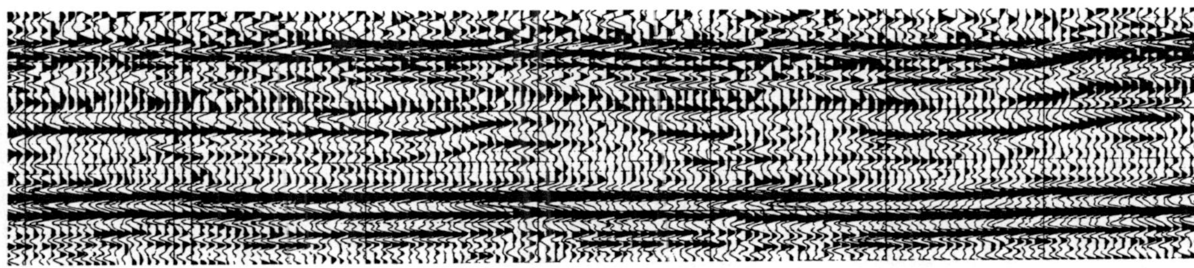

(b) Vibroseis data acquired along the same line with nonlinear sweeps

Fig. 6.5 Comparison of high-resolution data acquired with a nonlinear sweep with previous mini-hole dynamite data—Example E. The high-resolution Vibroseis data on the bottom were recorded along precisely the same line as the previously acquired dynamite data.

frequencies in order to save recording time (or use it advantageously to increase signal output at higher frequencies). The author does not recommend the use of programmed amplitude control during sweeps.

Amplitude control may be used advantageously in conjunction with what is called ground force phase lock. Ground force phase lock was discussed briefly in Chapter 5 and more information is included in Appendix I. With ground force phase lock we use the ground force measurement to sense when the vibrator is approaching a condition when it would decouple from the earth. With the ground force phase lock system we can also get an automatic control for the drive level. In this automatic system the drive is set at a very high level. But when the ground force signal senses that decoupling is imminent, the drive level is reduced until the ground force signal no longer senses that decoupling is about to occur. Then later in the sweep at the higher frequencies, the full force is available to generate maximum high-frequency energy. With this automatic system the vibrator is pushed to yield maximum output at all frequencies without decoupling. This can be a significant advantage particularly for high-performance vibrators with high force capability. The need for ground force amplitude control is greatest where the earth surface has a high Q resonance within the frequency band of the sweep. Under these conditions, as the sweep frequency approaches the resonant frequency of the surface, the amplitude of the vibration of the earth surface increases markedly and decoupling will occur unless the drive level is quite low. But if we have to set the drive level quite low on a high-performance vibrator it loses much of the high performance capability that you are paying for.

Depending upon circumstances ground force amplitude control may be worthwhile and in some situations it is quite important. But programmed changes of the drive level during the sweep are not recommended.

7

Practical field tests

The purpose of this chapter is to help you design and carry out practical field tests to determine the optimum recording parameters to meet your exploration objectives. Instead of talking about field tests and how to design them, we go through example cases for (1) a Vibroseis crew; (2) a shot hole crew and (3) a portable crew using mini-hole shot patterns. The two goals are to show you some techniques that should be useful and to help you to develop an overall approach and an overall philosophy.

7.1 VIBROSEIS CREW FIELD TEST

Figure 7.1 is a prospect form for a hypothetical seismic program in an area where our company has not worked before. Three lines with a total length of 30 miles (48 km) have been assigned along roads and trails which are suitable for vibrators. Permits for working on roads should not be difficult to obtain. More program may or may not be assigned in the area at a later date. The decision has been made to use a vibrator crew that is already under contract to your company. The crew is equipped as follows:

Number of traces (for data)	96
Correlator is high quality (32-bit, floating point)	
Sweep generators:	
Linear and logarithmic (nonlinear) with dial settings for sweep parameters.	
Filters:	
Low-cut frequencies	9, 12, or 18 Hz
Alias filters	64, 90, or 128 Hz
Power line notch	60 Hz
Geophones:	
Number of geophones per string	6
Number of strings available	1000
Resonant frequency	12 Hz
Damping	70 per cent

Cable:	
Number of signal pairs	72
Segment length	250 ft (76 m)
Signal takeouts every	120 ft (37 m)
Number of segments available	150
Vibrators:	
Moderate size mounted on trucks	
Number	5

How should we go about planning the initial tests? First, we need to go over the prospect form and discuss the project with the geologist and the interpreting geophysicist to see what the exploration objectives are and to see what is known about the area.

The estimated depth of our primary objective is 11 000 ft (3353 m) with a secondary objective at 8000 ft (2438 m). This secondary objective requires frequency components at least as high as 75 Hz in the reflected wavelets in the final, stacked section. Some commercial, 24-fold data in the area are not good enough to meet our exploration objectives either in terms of quality or in terms of required vertical resolution for the secondary objective. Therefore, we will want to use all 96 available channels to increase the maximum fold to 48. We will likely choose a group interval of not more than 165 ft (50 m) otherwise too many traces might be at offsets that are secondary too long for our objective (tables of cables in Chapter 4).

We sent the crew surveyor ahead to check out the planned line routes, look for alternate routes, and report back any problems. Since he reported power lines along the roads, we anticipate that 60Hz pick-up will be a problem and that we will need to use notch filters, since we do not have any field means for canceling this pick-up. We do not want to have to change the same rate or the alias filter during the test; therefore we plan to use a 2 ms sample rate and the 128 Hz alias filters. For production we will change to the 4 ms sample rate and alias filters of 90 Hz unless we choose a sweep that includes frequencies higher than 90 Hz, which is certainly a possibility with our nonlinear sweep capability.

From available commercial data we can gain limited information. There is some coherent noise near objective times but mostly at frequencies of 20–25 Hz. No tapes are available to make displays of shot and CMP gathers nor for filter tests. Reflections from the objectives can be followed across the sections, but continuity is poor in some sections. There may be some problem from multiples particularly for the secondary objective at 8000 ft (2438 m). The stacking velocities used were 12 000 to 13 000 ft/s (3658 to 3962 m/s) for reflections from 8000 ft (2438 m) and 11 000 ft (3353 m) respectively. Steep dips are not expected. The commercial data were acquired with 10 pound (4.5 kg) explosive charges in one-hole patterns at depths of 70–90 ft (21–27 m).

Onshore Seismic Prospect Form

Prospect Gopher Line Prefix G Auth. No JO-131-3
Miles of Program 30* Date 2 – 25 – 1988 By Wm. Brown
Country USA State of Prov Wyoming Counties Carbon
Quads

General Objectives 1) Map Tensleep structure.
2) Map updip wedgeout of the Sundance. Map deeper horizons if feasible.

Topography rolling plain Vegetation grass Culture pasture
Seismic datum 4200 ft Correctional Velocity 6300 ft/s

Specific Objectives

Priority/ Horizon	Record Time(s)	Depth (ft)	Stacking Velocity (ft/s)	Lithology	Dip	Resolution Vertical horiz.
1 Tensleep	1.7	11 000	13 000	sandstone	low	not critical
2 Sundance	1.3	8000	12 000	sandstone	low	need 75 Hz min.
3						
Basement	2.5?	18 000?	16 000?	lm & sh/grn		

Geology

Surface clay and sandy soil

Near Surface shales and sandstones

At Depth shales, sandstones, limestones and evaporites

Key Wells Ajax #1 Smith is about 14 miles NW of tie center of the prospect

Previous Seismic Data

Line	Prospect Name	SP Range	Tie Points	Quality	Problems	Improvement Required
CD-3	Gopher	137–254	192	poor	S/N & res.	better S/N & res.
CD-5	Gopher	120–386	—	poor	" "	" "

Additional Comments or Suggestions *Possibly more program after this work is interpreted and evaluated. Multiples may be a problem.
Power lines along roads may be a problem.

Preferred order for recording lines? Yes/No. If 'yes', give details including reasons, on a separate sheet.
Please fill out this form as completely as is feasible. Attach additional sheets if necessary.
Distribution: Prov. Mgr. Interpreter Processor WPC KMK JPB SRS

Fig. 7.1 Prospect sheet for a hypothetical seismic program.

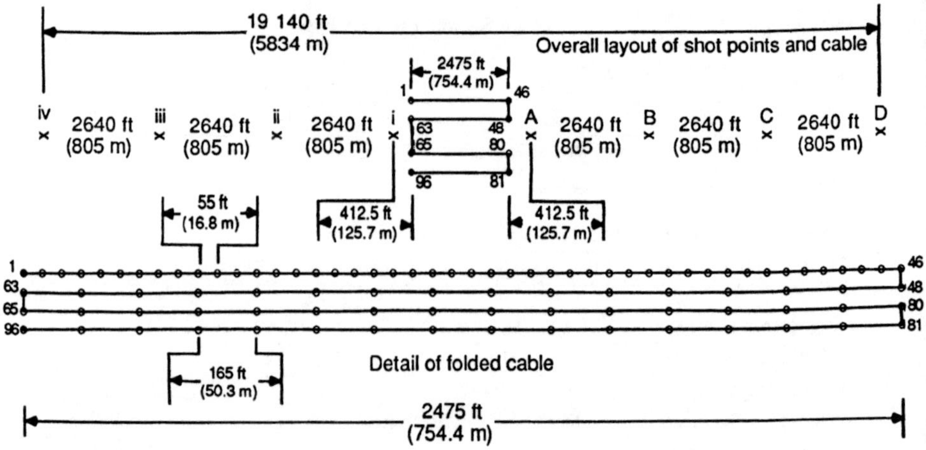

Traces	Group interval	Geophone array	Geophone spacing	nd	Purpose
1-46	55 ft (16.8 m)	6 podded	0+	0+	For the noise test
48-63	165 ft (50.3 m)	24/158 ft (48 m)	6.9 ft (2.1 m)	165 ft (50.3 m)	For the highest frequencies from the 8000 ft (2438 m) objective with a 165 ft (50.3 m) group interval array
65-80	165 ft	36/257 ft (109 m)	7.3 ft (2.2 m)	264 ft (80.5 m)	For fair high frequencies for a 264 ft (80.5 m) double group interval array
81-96	165 ft	36/321 ft (97.8)	9.2 ft (2.8 m)	330 ft (100.6 m)	For the primary objective with good noise suppression

Test plan

Sweep Frequency	Sweep Type	Source Pattern (5 vibrators in line) Spacing	Move-up	VPs Definite		VPs Probable	VPs Possible	
14-56 Hz	Linear	30 ft*	0	i, ii, iii, iv, A, B, C				
14-56 Hz	Linear	50 ft (15 m)	25 ft (7.6 m)	ii,	B			iii, C
12-80 Hz	Log	30 ft*	14 ft (4.3 m)	i, ii, iii, iv,	B		i,	A, C, D
20-80 Hz	Linear	30 ft*	14 ft	ii, iii,	B	iv	i,	A, C, D
14-100 Hz?	Log	?	?	?		?		?

*in line, bumper to bumper

Fig. 7.2 Test layout with folded cable—Vibrator crew

Since our assigned program is quite limited and future program in the area is uncertain, we do not want to spend much time on initial tests. One of the primary needs from our initial tests is to determine whether or not it is feasible to try to attain the secondary objective, with its frequency requirements, without sacrificing data quality for the primary objective or greatly increasing the cost of the program. In addition, we need to:

1. Ascertain the effective upper frequency limit particularly for reflections from depths near the secondary objective at 8000 ft (2438 m);
2. Determine the maximum effective offset for our primary and secondary objectives and we will likely need to use offsets up to near the maximum effective offset to suppress the possible multiple problem;
3. Define the major noises and select geophone arrays and source patterns to suppress coherent noise;
4. Determine the best type of sweep and define the sweep parameters;
5. Determine the amount of sweep time required to build up an adequate signal-to-background-noise (not source-generated noise) ratio for reflections from the objectives.

First, can we effectively use the 96 channels available to define the source-generated noises and with the same layout evaluate some geophone array and source pattern combinations that may be effective? Figure 7.2 shows a test layout with the cable folded back and forth to provide the opportunity for recording a noise spread and three sets of data with three different geophone arrays. Traces 1–46 are spaced 55 ft (16.8 m) apart. These are for recording noises with only one string of geophones planted only inches (10 cm) apart in a 'pod'. Traces 48 to 63, 65 to 80, and 81 to 96 are for recording reflection data with carefully selected geophone arrays.

We already concluded that we might use a group interval of 165 ft (50 m) in our final recording plan and that is why we considered it for the group interval for the arrays in the test. But before finalizing this selection of the 165 ft (50 m) group interval we check the high-frequency response of a group length array and of a double group length array for the primary and the secondary objectives. We also plan a possible complementary source pattern to go with these arrays specifically for the primary objective and also for trying to meet both objectives satisfactorily.

For the double-group length array we calculate the moveout for the primary objective for an offset of approximately 8500 ft (2591 m) and then calculate the frequencies for relative responses of 64, 30, and 0%.

$D = 11\,000$ ft (3353 m)
$V = 13\,000$ ft/s (3962 m/s)
$t_0 = 2D/V = 1.692$ s

$$\Delta L = 330\,\text{ft}\ (100.6\,\text{m})$$
$$L = 8500\,\text{ft}\ (2591\,\text{m})$$

You calculate Δt from the exact and approximate equations (Chapter 4) as follows:

$$\text{Exact:}\quad \Delta t = 0.00915\,\text{s}$$
$$\text{Approximate:}\quad \Delta t = 0.00981$$

$$f_{0.64} = 1/(2\Delta t) = 55\,\text{Hz}$$
$$f_{0.3} = 3/(4\Delta t) = 82\,\text{Hz}$$
$$f_0 = 1/\Delta t = 109\,\text{Hz}$$

The double-group length array [$nd = 330$ ft (100.5 m)] appears to be quite suitable for recording the primary objective, particularly if noises with wavelengths in the range of 300 ft (90 m) are a significant problem. (Example in Chapter 4.) We then calculate the notches and peaks for this array and design a complementary source pattern as shown in Table 7.1. We may or may not actually plot these peaks and notches now and sketch out a composite response.

If only shorter wavelength noises are a problem, then a shorter, group length array 165 ft (50 m) long might work well. Before designing a complementary source pattern for the group length array we check to see how we could try to meet both objectives. First we check to see how the group length array and the double group length array would work for the secondary objective. For the group length array:

$$D = 8000\,\text{ft}\ (2438\,\text{m})$$
$$V = 12\,000\,\text{ft/s}\ (3658\,\text{m/s})$$
$$t_0 = 2D/V = 1.333\,\text{s}$$
$$\Delta L = 165\,\text{ft}\ (50\,\text{m})$$

$L = 8000$ ft (2438 m). (Longer offsets are not likely to be effective for this objective.)

We calculate Δt:

$$\text{Exact:}\quad \Delta t = 0.0062\,\text{s}$$
$$\text{Approximate:}\quad \Delta t = 0.0069\,\text{s}$$

$$f_{0.64} = 1/(2\Delta t) = 81\,\text{Hz}$$
$$f_{0.3} = 3/(4\Delta t) = 121\,\text{Hz}$$
$$f_0 = 1/\Delta t = 161\,\text{Hz}$$

Certainly the single group length array would allow the recording of the higher-frequency components that we seek from the secondary objective. If noise is not

Table 7.1 Double group interval array and complementary source pattern

Geophone array nd = 330 ft		*5 vibrators spaced 50 ft nd = 250 ft*		*8 positions 7 move-ups of 25 ft nd = 200 ft*	
Notch (ft)	*Peak (ft)*	*Notch (ft)*	*Peak (ft)*	*Notch (ft)*	*Peak (ft)*
330					
		250			
	220				
				200	
165			167		
	132				133
		125			
110					
			100	100	
	94				
83		83			80
Geophone array nd = 101 m		*5 vibrators spaced 15 m nd = 76 m*		*8 positions 7 move-ups of 7.6 m nd = 61 m*	
Notch (m)	*Peak (m)*	*Notch (m)*	*Peak (m)*	*Notch (m)*	*Peak (m)*
101					
		76			
	67				
				61	
50			51		
	40				40.6
33.5		38			
			30.5	30.5	
	29			11	
25		25			24

Notches at: 330, 250, 200, 165, 125, 110, 100, 83,... ft
(101, 76, 61, 50, 38, 33.5, 30.5, 25,... m)

a serious problem at wavelengths in the range 200–300 ft (61–91 m), then this array could be used to try to meet both objectives. We calculate the notches and peaks for the group length array and then design a complementary source pattern for this array for the secondary objective in the hope that it would be satisfactory also for the primary objective. Table 7.2 gives the results.

Next, we check the double-group length array for the secondary objective to see if it is possibly usable. You find that the differential arrival time across the

Table 7.2 Group interval array and complementary source pattern

Geophone array		*5 vibrators*		*8 positions*	
nd = 165 ft		*spaced 30* ft*		*7 move-ups of 14 ft*	
		nd = 150 ft		*nd = 112 ft*	
Notch (ft)	*Peak (ft)*	*Notch (ft)*	*Peak (ft)*	*Notch (ft)*	*Peak (ft)*
165					
		150			
	110			112	
			100		
83					
		75			75
Geophone array		*5 vibrators*		*8 positions*	
nd = 50 m		*spaced 9* m*		*7 move-ups of 4 m*	
		nd = 46 m		*nd = 34 m*	
Notch (m)	*Peak (m)*	*Notch (m)*	*Peak (m)*	*Notch (m)*	*Peak (m)*
50					
		46			
	33.5			34	
			30.5		
25					
		23			23

*Note: The 'bumper-to-bumper' spacing of about 30 ft (9 m) is a minimum for work along roads where lateral separation is impractical.
Notches at: 165, 150, 112, 83, 75,... ft
(50, 46, 34, 25, 23,... m)

array for an offset of 8000 ft (2438 m) is:

$$\Delta t = 0.0125\,\text{s (exact—approximate equation yields 0.0137 s)}$$
$$f_{0.64} = 40\,\text{Hz}$$
$$f_{0.3} = 60\,\text{Hz}$$
$$f_0 = 80\,\text{Hz}$$

This means that offsets as long as 8000 ft (2438 m) will not provide the needed resolution for the secondary objective if arrays with effective lengths of 330 ft (101 m) are used.

We then check for an offset of 6000 ft (1829) with this array. We calculate:

$$\Delta t = 0.00989\,\text{s (exact—approximate equation yields 0.0103 s)}$$
$$f_{0.64} = 56\,\text{Hz}$$
$$f_{0.3} = 76\,\text{Hz}$$
$$f_0 = 101\,\text{Hz}$$

The effective CMP fold would be reduced from 48 to 33 if we discounted offsets of more than 6000 ft (1829 m). By comparing with the example in Chapter 4 we recognize that the 75 Hz relative response of the CMP stack for offsets no longer than 6000 ft (1829 m) would be about 70% which should be acceptable. But we doubt that a fold of 33 would yield satisfactory results. This doubt is increased by the possible multiple problem and the consequent need for good data from long offsets (but not longer than the effective maximum offset imposed by the earth). But if noise in the wavelength range of 300 ft (90 m) is so strong it jeopardizes our primary objective, we might decide to use double group length arrays of 330 ft (101 m). If so, it is possible that the secondary objective could also be achieved. We better take our data processor out to lunch tomorrow!

We can try yet one more array in our test. It is possible that we could meet our second objective with an array somewhat longer than 165 ft (50 m) and that this longer array may be needed to suppress some strong noises with wavelengths in the range 175–275 ft (53–84 m). Alternatively, we could stay with the 165 ft (50 m), group length array and lengthen our source pattern to put some notches in the possible noise range. But we might also want to consider decreasing the group interval to 132 ft (40 m) and using a double group length array with an effective length of 264 ft (80 m). This would have the advantage of reducing the maximum offset to about 7000 ft (2134 m) (tables in Chapter 4) which would allow full CMP fold for the secondary objective without exceeding the probable maximum effective offset. We would have to judge from the test results whether or not this would still allow us to meet the primary objective fully (the principal concern here may be with multiples and an inadequate offset range). Use of a shorter group interval would increase the number of VPs from 32 to 40 per mile (20 to 25 per km) which would add to the cost of the survey. But this could be the key to meeting both objectives satisfactorily.

First, you check to see if a 264 ft (80 m) array would too seriously filter the higher-frequency components required for the secondary objective:

$$D = 8000\ \text{ft}\ (2438\ \text{m})$$
$$V = 12\,000\ \text{ft/s}\ (3658\ \text{m/s})$$
$$t_0 = 1.333\ \text{s}$$
$$\Delta L = 264\ \text{ft}\ (80\ \text{m})$$
$$L = 6800\ \text{ft}\ (2073\ \text{m})$$

We calculate the exact and approximate equations:

$$\text{Exact:}\quad \Delta t = 0.00875\ \text{s}$$
$$\text{Approximate:}\quad \Delta t = 0.00935\ \text{s}$$

$$f_{0.64} = 57\ \text{Hz}$$
$$f_{0.3} = 85\ \text{Hz}$$
$$f_0 = 114\ \text{Hz}$$

By comparing this with the example case in Chapter 4 we see that for the stack of the CMP gathers we would lose some energy for higher-frequency components, but the loss should not be serious and the approach is sound.

So now we have the last geophone array for the test. We check to see how well it would work with the same source pattern we were planning to use for the single group length array with an effective length of 165 ft (50 m). Table 7.3 gives the results. This looks pretty good for a test. We may want to adjust the source pattern if, from the test results, we decide to use the 132 ft (40 m) group interval and the double group length array. For example, we may want to increase the vibrator spacing to 35 ft (10.7 m) to put a notch at a wavelength of 175 ft (53 m).

Table 7.3 Double group interval array with shorter group interval

Geophone array nd = 264 ft		*5 vibrators spaced 30 ft nd = 150 ft*		*8 positions 7 move-ups of 14 ft nd = 112 ft*	
Notch (ft)	*Peak (ft)*	*Notch (ft)*	*Peak (ft)*	*Notch (ft)*	*Peak (ft)*
264					
	176				
		150			
132					
				112	
	106				
			100		
88					
		75			75

Geophone array nd = 80 m		*5 vibrators spaced 9 m nd = 46 m*		*8 positions 7 move-ups of 4 m nd = 34 m*	
Notch (m)	*Peak (m)*	*Notch (m)*	*Peak (m)*	*Notch (m)*	*Peak (m)*
80					
	53.6				
		46			
40					
				34	
	32				
			30.5		
27					
		23			23

Notches at: 264, 150, 132, 112, 88, 75,... ft)
(80, 46, 40, 34, 27, 23,... m)

For the tests we have laid out eight vibrator points (VPs). We shall want to record noise measurements with a minimum source pattern (no move-up and probably fewer than five vibrators bumper-to-bumper.) For the rest of the tests we shall need to try the two different source patterns, and want to try probably two or more linear sweeps and two or more nonlinear sweeps. Actually, based upon his experience the author would not use any linear sweeps except for the noise test. But you may need to gain that confidence for yourself. So now we have eight VPs, three source patterns, and four or more different sweeps. We could easily try 100 combinations or more and that would take a lot of crew time. So we make out a plan with combinations we want to try, combinations we shall probably try, and combinations we probably won't try. We shall examine results as the data are recorded and make our decisions on which tests to do based upon prior results. We shall probably try some combination of source pattern and sweep that we had not originally considered. Our plan is shown in Fig. 7.2. We know we must see how many sweeps are needed to overcome background noise, but we hope that eight sweeps of 10 s each will suffice. We shall use a 14–56 Hz linear sweep as a basic, two-octave sweep that would meet our primary objective. This sweep would also do quite well to define the stronger, source-generated noises that we would need to attack with geophone arrays and source patterns. (Higher-frequency noises will have shorter wavelengths and will be well suppressed by the planned geophone arrays.) We shall try the 20–80 Hz linear sweep as a two-octave sweep that is capable of allowing us to meet our secondary objective. We shall use logarithmic sweeps with about a ten-to-fifteen-fold boost (Chapter 6 and Appendix J) in the sweep time at the highest frequency as compared with the sweep time at the lowest frequency. We shall do this to try to bring out the required high-frequency components in reflections from the secondary objective. We do not even try to estimate Q or to offset precisely earth attenuation effects for either objective. All we want is to achieve an adequate signal-to-background-noise ratio for the reflections over the sweep-frequency range. Hopefully this will make the presence of the needed high-frequency components visible without frequency filtering (other than the filtering effect of correlation).

These nonlinear sweeps will also suppress much of the source-generated, low-frequency noise. We shall first try a broad band 12–80 Hz logarithmic sweep and then select probably one other nonlinear sweep based upon the test results. Perhaps we shall try a 14–100 Hz logarithmic sweep. After we have pretty well decided on our final plan for the production work, we may want to confirm our decision by trying some different combination of sweep and source pattern not included in our plan. Certainly, we shall want to try one or more narrow-band sweeps at some VP with some source pattern to determine the effective upper-frequency limit for our two objectives (Chapter 6 and Appendix J). These choices of source pattern and VP can best be made after the other tests have been conducted and evaluated. We shall need to study each record as soon as it is completed and keep up so as to be ready with our decisions as soon as we can. Crew time is expensive, but not as expensive as a mistake. They can wait a little while, if it is necessary. We plan to get the crew to layout the tests ahead of time and we expect to complete the test in one recording day.

We must pick out a location for the tests. Usually we want an area that is typical of the terrain of the survey area as a whole. (Where the main survey area includes two, quite different terrains, we may consider testing in both terrains.) Preferably the test location should be near the middle of one of the three lines we are to record. We may prefer a location on a dip line so as to test over a wider range of earth conditions. It is desirable to use a test location where there is a relatively straight road over the entire range of offsets to be used, but particularly on dip lines we may have to use a crooked road. The folded cable should be laid out where the road is reasonably straight and where there is not an excessive amount of cultural noise. We must check to see where planned source location offsets would fall relative to the road or roads and we would very much like our source positions to lie on or near an extension of a straight line defined by the cable.

We must draw and write up the plan and then communicate it to our field crew, making very sure the party manager understands our instructions and that the instructions are not ambiguous. You will most likely be amazed at some point as to how apparently clear instructions can be totally misunderstood. Redundancy in the instructions can be helpful if carefully done. One of the key advantages of working with the same field crew over a substantial period of time is that we can develop much better communication and understanding. The crew must ask for clarification if there is any doubt as to what we want them to do. Pictures in the form of clear drawings and sketches can be worth a thousand words—and thousands of dollars too.

It helps to send copies of instructions to the crew, to the field support people with our company, to the supervisor of the crew, to our interpreting explorationists, and to the data processor. If there is a mistake or if something is not clear, then hopefully it will be brought to our attention in time to prevent any serious consequences.

But the plan may have hardly left the mail room before Helpful Harry is in our office telling us that we cannot record all the noises with a noise spread spacing of 55 ft (18 m) and that we forgot about spatial aliasing of our reflection data. But we calmly reply that our arrays will take care of short wavelength noises and we do not need to record them if they are not going to hurt. And we explain that expected dips are moderate. He insists that there might be steep dips or steep fault planes and that we would miss them. We reply that if we get good data and see any evidence of steep dips or faults that need better definition that we can try beam steering in processing (Chapters 11 and 19) or come back later. As he turns to go, he wheels back around and asks if the angle of arrival at a 10 000 ft (3048 m) offset would alias 100 Hz components of reflections. We reply that if we get any 100 Hz components from a depth of 11 000 ft (3353 m) that it is more likely to be on the shorter offsets which would arrive at small angles of incidence from low dip reflectors. And in any case when the CMP gathers are summed, the data from the shorter offsets can provide unaliased higher-frequency components. Still to get him off our backs we do a quick check with one of our handy tools as follows:

$$\lambda_a = \frac{V\sqrt{(2D)^2 + L^2}}{fL} = \frac{13\,000\sqrt{(22\,000^2 + 10\,000^2}}{f(10\,000)}$$

$$\lambda_a = 31416/f$$

We laugh about the ten thousand pies as we calculate wavelengths of 314 ft (96 m) for 100 Hz and 392 ft (119 m) for 80 Hz. And with a group interval of no more than 165 ft (50 m) he agrees that for the stacked data the alias problem would only arise with steep dips. You wonder out loud if we have stumbled onto a method of calculating π to 10 000 significant figures.

7.2 SHOT HOLE CREW TEST

Figure 7.3 is a hypothetical prospect form for a seismic project to be recorded by a shot hole crew. Eighty miles (129 km) of program has been assigned in an arid and unpopulated area with little surface relief. This is an area that has excellent potential for hydrocarbons, but exploration has been stifled by poor seismic data. If this program is successful in acquiring usable data and delineating expected structures, a much larger seismic program will be undertaken later. Your objective is at a depth of about 10 000 ft (3048 m) with an estimated moveout velocity of 16 000 ft/s. Drilling costs are high and the management does not think it could pursue a major exploration program that required more than about eighteen deep shot holes per mile (11.2 per km). It is probably not economic to drill deeper than about 125 ft (38 m). The surface is a calcareous sandy soil with thin hard streaks that become thicker and more numerous with depth. This soil is underlain by bedded limestone at a depth of a few hundred feet (about 100 m).

Crew vehicles can go pretty well wherever they need to go. The lines are straight and the area has been permitted. Previous data in the area were recorded by a shot hole crew with a multi-conductor cable. These data are almost useless with only hints of reflections mostly below one second. The previous recording parameters were:

Number of data channels	96
CMP fold	12
Shot pattern	Single hole

Shot depth	100–120 ft (30–37 m)
Charge size	20 lb (9 kg)
Geophone arrays:	
Type	In-line, equally spaced
Number of geophones	48
Length of arrays	480 ft (146 m)
Group interval	220 ft (67 m)
Shot point interval	880 ft (268 m)
CMP fold	12
Cable:	
Type	Symmetrical split
Offset range	11 440–1100–0–1100–11 440 ft 3487–335–0–335–3487 m
Filters:	
Low-cut	16 Hz
High-cut	48 Hz
Power line notch filter	60 Hz

Displays of shot gathers and CMP gathers show a wide range of coherent noises and background noise seems to be strong. The background noise tends to be worst through the middle of the recording day and may be related to a few notations on the observer's sheets about wind. Also, particularly through the middle of the day, there are short noise bursts that show up across whole shot gathers at the same time. These are called 'static slashes.'

The first breaks are very strong and reverberate down the record for about 0.5 s with an apparent velocity of about 15 000 ft/s (4572 m/s) with periods ranging from 0.055 to about 0.065 s. The strongest noises are given in Table 7.4.

There is little or no evidence of an air wave, but since they used such a long minimum offset and used a geophone spacing of only 10 ft there may be a rather strong airwave that they minimized with their technique.

A shot hole crew will be available to us in a few months, using a telemetry system from boxes on the line to the recording truck. Each box handles eight traces. The crew has 25 boxes and normally records 120 traces of data when operating in a 'roll-along' mode. But for tests where the layout will not be changed, it is feasible to record up to 192 channels of data simultaneously. The crew has 12 000 geophones that are strung twelve to a string with a spacing of 30 ft (9 m) between geophones. Strings can be tied together in long lengths or connected together at a point to form a star array. Each string also has a center takeout. Low-cut filters have cut-off frequencies of 12, 16 or 24 Hz. Alias filters have cut-off frequencies of 90 or 180 Hz. In telemetry systems these filters are in the boxes out on the line and to change filters we would have to go to each box. This would mean that we could not record data with one filter setting and then

Onshore Seismic Prospect Form

Prospect Rattlesnake Line Prefix RA Auth. No JO-109-7
Miles of Program 80* Date 1–17–1988 By James Rhodes
Country USA State or Prov New Mexico Counties Lea
Quads
General Objectives 1) Map any reflectors near objective depth with reasonable continuity. Map deeper horizons if feasible.

Topography arid valley Vegetation desert Culture none
Seismic datum 2800 feet Correctional Velocity 7000 ft/s

Specific Objectives

Priority/ Horizon	Record Time (s)	Depth (ft)	Stacking Velocity (ft/s)	Lithology	Dip	Resolution Vertical Horiz.
1 Montoya	1.3	10 000	16 000	limestone	mod.?	not critical
2						
3						
Basement	1.9?	16 000?	18 000?	lm & sh/grn		

Geology

Surface calcareous sandy soil with hard streaks (Tertiary)

Near Surface bedded limestone (Cretaceous) at a depth of a few hundred feet

At Depth large structures with major faults

Key Wells none

Previous Seismic Data

Line	Prospect Name	SP Range	Tie Points	Quality	Problems	Improvement Required
KR-1	Knight	100–595	145 280	NR to very	high noise	less noise & usable
	Ranch		415	poor	(coherent & backgnd	reflections

Additional Comments or Suggestions *If we can map structure at or near objective depth, then a much larger program is anticipated. But we must keep costs reasonably low.

Preferred order for recording lines? Yes/No. If 'yes', give details including reasons, on a separate sheet.
Please fill out this form as completely as is feasible. Attach additional sheets if necessary.
Distribution: Prov. Mgr. Interpreter Processor WPC KMK JPB SRS

Fig. 7.3 Prospect sheet for a second hypothetical seismic program.

Table 7.4 Strong noises in shot crew program

Velocity (ft/s)	Velocity (m/s)	Period (s)	Frequency (Hz)	Wavelength (ft)	Wavelength (m)	Remarks
15000	4572	0.055	18	825	251	Strong first breaks
15000		0.065	15.4	975	297	Strong first breaks
11000	3353	0.06	16.7	660	201	Strong
10500	3200	0.06	16.7	630	192	Strong
9200	2804	0.065	15.4	598	182	Strong
9200		0.04	25	370	113	Strong
9200		0.03	33	276	84	Moderate
8500	2591	0.065	15.4	553	169	Moderate
8500		0.04	25	340	104	Strong
8500		0.035	29	298	91	Moderate
5500	1676	0.06	16.7	330	101	Strong
5500		0.035	29	193	59	Moderate

play back the same data through a different filter. This is unfortunate and we shall inquire today if there is any feasible way to provide means for filtering playbacks with various low-cut filters. We plan to use a 4 ms sample rate and the 90 Hz alias filters.

First we plot record time versus offset and draw a line representing the air wave arrival time with offset distance. We also draw a line for the first breaks arrivals at a slope corresponding to the velocity of 15 000 ft/s (4572 m/s), Fig. 7.4. Since the first breaks reverberate, we draw in another line that is delayed by 500 ms and is parallel to the first breaks arrival line. Next we calculate the time of the objective reflection for a few offset distances and draw in the reflection arrival time. As we have no shallow objectives, we conclude from our sketch that the noise from the reverberating first breaks is not a problem we have to deal with and the air wave will be no problem for offsets of more than about 1400 ft (427 m). We also note from displays of shot gathers that the noise with a velocity of about 5500 ft/s (1676 m/s) is not a problem at record times later than one second. But you must deal with the other coherent noises and the background noise in an effort to make the survey successful.

From past experience we know that the telemetry system will be much less susceptible to serious disturbance from static slashes. In arid areas static slashes are associated with windblown sand. But telemetry probably will not help in regard to wind noise in general, but burying the geophones would help. We decide to attack the wind-noise problem by having two shooters with helpers so that data can be recorded rapidly when conditions are favorable. We decide also to try to get more telemetry boxes so that the line crew can layout more stations ahead when recording is held up by strong wind. We also

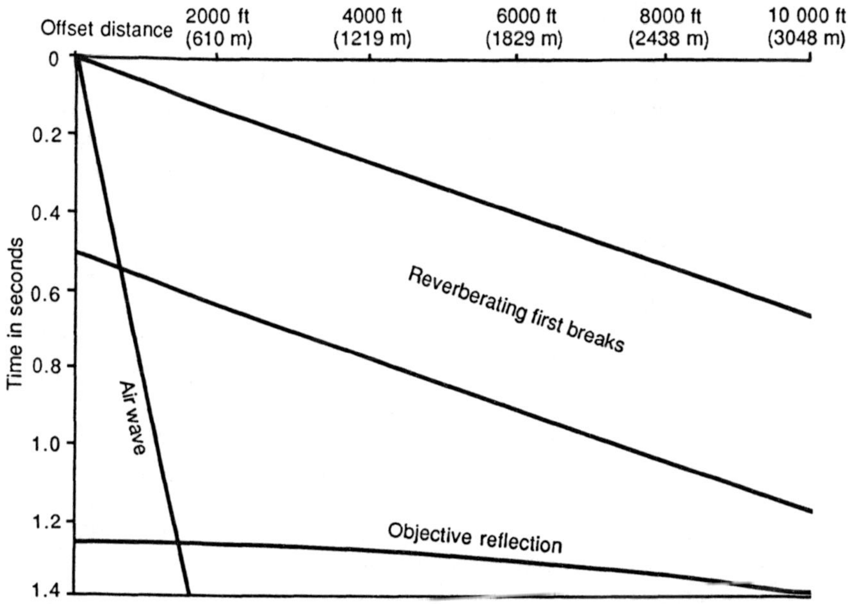

Fig. 7.4 Sketch of air wave, reverberating first breaks, and objective reflection. The reverberating first breaks are a serious problem for shallow reflections, but not for reflections from our deep objective.

plan to bury geophones to see if we can then work successfully even when the wind is strong. But we plan to leave the geophones in a few, selected arrays unburied so that we can ascertain the improvement and hence the justification for burying geophones.

How can we handle the coherent noise problem that remains? We would gladly settle for a low resolution, final section with a definite reflection from near objective time that had reasonable continuity. Hopefully the relative bandwidth would not be so limited that the reflection would be very ringy, but even that would be acceptable in this tough area. One approach would be to record with high CMP fold to suppress the noise. But if the noise is too strong on the raw data, we may very well not be able to extract the needed velocity function and not able to make necessary static corrections. As a result we may not achieve the noise suppression capability of a high fold CMP stack. Also we would have to drill more than 18 holes per mile (11.2 per km) to achieve a high CMP fold for your objective. For example, if we used 120 traces, a group interval of 160 ft (49 m), and a near offset of 560 ft (171 m); the far offset would be 10 000 ft (3048 m). We probably could not use longer offsets than that. Call this Geometry A. If we shot full 60 fold we would need 33 holes per mile (20.5 per km). If we shot half maximum fold (30) we would need 16.5 holes per mile (10.25 per km). Perhaps we would try this technique on a short line with 30 fold recording if nothing else appeared to offer much chance of success.

Whereas full fold CMP recording is not economic with deep hole shots, it might be economic if we could be successful with somewhat smaller charges in holes only 50 ft (15 m) deep or in patterns of holes only 10 ft (3 m) deep. To keep the number of shot points down, we could increase the group interval to 176 ft (53.6 m) and use only 96 traces with minimum and maximum offsets of about 616 ft and 8888 ft (188 m and 2709 m) respectively. Call this Geometry B. This would require 30 shot points per mile (18.6 per km) which might not be as costly to drill as 18 deep holes per mile (11.2 per km). This would yield 48 fold data. You could use a double group length array, 36 over 342 ft (104 m) and spaced 10 ft (3 m) apart, or a triple group length array, 48 over 517 ft (157.6 m) and spaced 11 ft (3.35 m) apart. To better suppress long wavelength noise in the data before stacking, we might even consider a quadruple group length array of 48 geophones over 689 ft (210 m) spaced 15 ft (4.6 m) apart. But all of this is contingent upon the success of either the 50 ft (15 m) holes or the shallow patterns. The source must provide sufficient seismic energy without swamping us with much more noise.

Another approach is to use geophone arrays that are long enough to attack at least some of the problem noise better than what was done before. Note first that the previously used geophone array would have notches at wavelengths of 480, 240, 160, 120 ft, etc. (146, 73, 49, 36.6 m, etc.). There may be strong noises near these wavelengths that will hurt us if our array does not have similar notches. We shall have to see in a test what works and what does not work. So we mentally try a few geophone arrays and come up with an array of 48 geophones in two parallel lines of 24 geophones each. The spacing between the geophones in each line is 25 ft (7.6 m), so each line of 24 geophones is then 575 ft (175 m) long. But the two lines are slipped 175 ft (53.3 m) relative to each other so that the entire array is 750 ft (229 m) long. We plan to use a lateral separation of 100 ft (30.5 m) between the two lines to help suppress broadside noise from wind or other causes. Figure D-8 gives a drawing of this parallelogram array and of an equivalent trapezoidal array. For noise travelling along the seismic line the geophone array would have notches at wavelengths of 600, 350, 300, 200, 150, 120, 100 ft, etc. (183, 107, 91, 61, 46, 37, 30.5 m, etc.). This array is analyzed in a later paragraph and in Appendix D. We realize that the 25 ft (7.6 m) spacing between geophones would leave us with no attenuation of a 44 Hz component of the air wave, but this would not be a problem for the objective for offsets greater than 1400 ft (427 m). This array response and the noises from Table 7.4 that we must deal with are shown in Fig. 7.5 along with the response of the array used to record the previous data.

We have one final hope. We shall try some short array and then filter the data with a low-cut filter that cuts moderately at a frequency of 35 Hz and begins to cut strongly at 25 Hz. We then hope to find some high-frequency

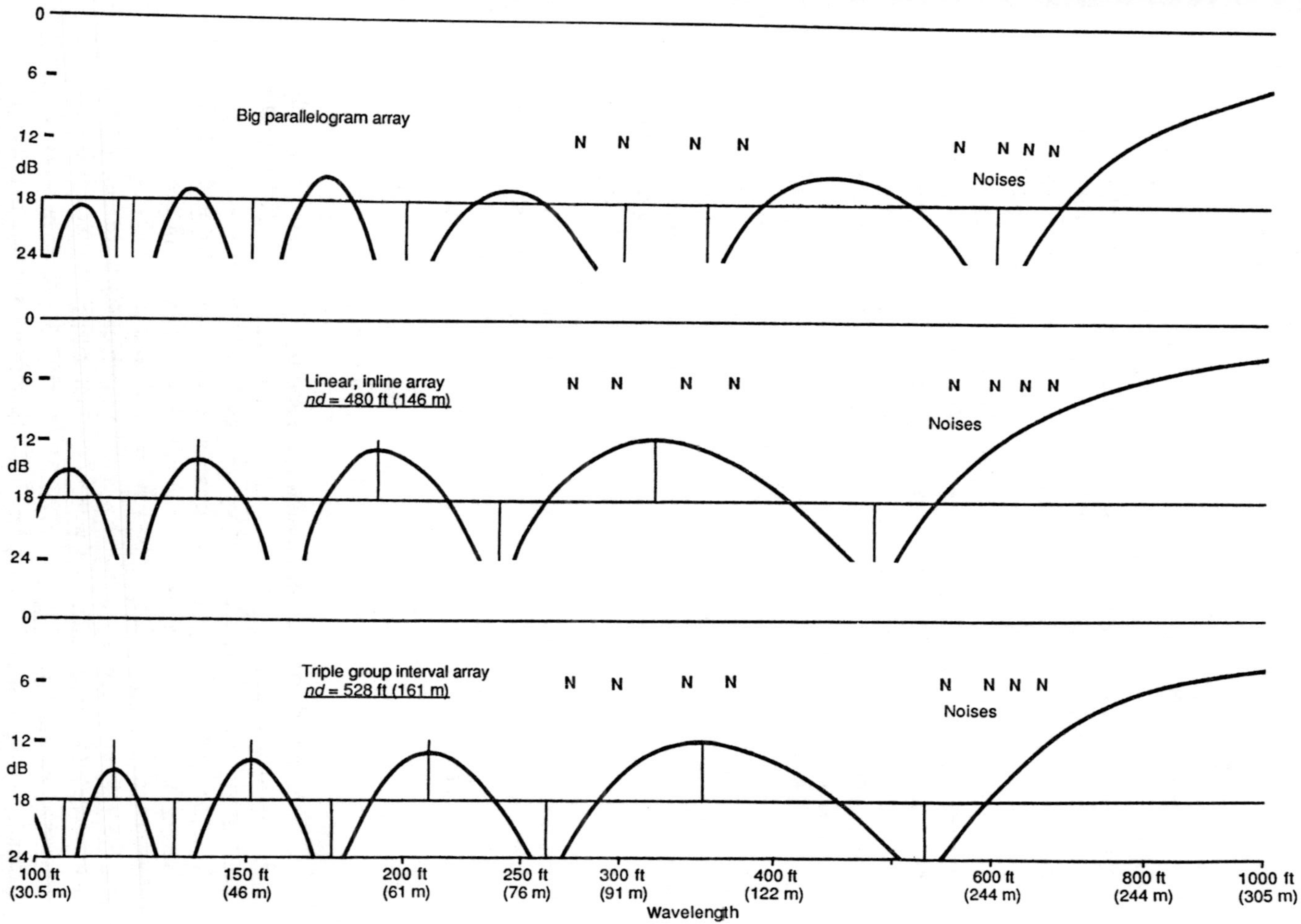

Fig. 7.5 Responses of a parallelogram array and other arrays.

reflections that will be visible above all the noise; this probably will not happen, but we are going to look for it anyway. If we can filter playbacks in the field, we shall try to get 40 Hz low-cut filters to use but if this is not feasible, then the filtering will have to be done later at a processing center. It certainly would be useful to have processing capability in the field, especially for tests.

We plan to conduct some tests at the same locations with a deep hole shot, a shot in a hole only 50 ft (15 m) deep, and a pattern of charges in holes only 10 ft (3 m) deep. Since we can layout 192 channels for a test, we plan to try more than one array and include a noise test. We want to see what the most effective offset range is. For the near offset we can judge by the offset where the noise diminishes with a given array, but for the far offset it will be difficult to guess if we cannot see in the field data any definite reflections from our objective (or from near our objective). If we cannot see any definite reflections how can we judge the effective upper-frequency limit? Perhaps we shall be lucky enough to find some reflections, for some array, at some offset. We had better make some overlays for the field displays for reflections near expected reflection time with moveout based upon velocities near the expected moveout velocity. These may help us to find a reflection, or to confirm an event that we find without the overlays.

We plan to get the crew to do the tests and then have the recording crew take a one-week break. But we need to decide on the next step quickly to give the drill crews time to drill a reasonable number of shot points before the recording crew returns to work. So we shall have to make some decision in the field at the time the tests are run, but we shall have all night to study the results and make our decisions so we can get the drill crews started the next day. If we are stumped, we plan to drill and record part of one of the lines with full CMP fold using deep hole shots and send the data to the processor. He can stack the data for that portion with full fold. If we get satisfactory data, he can stack it using fewer shot points and see if he can get satisfactory data for a shot point spacing that would be economic for a big exploration program. We would have to select some array, group interval, near and far offsets, and number of channels to use. These decisions should be based upon our best guess after reviewing all of the test data. But we hope to see some reflections and be able to come up with a recording plan that we are reasonably confident will succeed.

This parallelogram responds just like a trapezoidally weighted, in-line array to noise that travels along the seismic line. You will have all the notches previously noted for nd values of 600 and 350 ft (183 and 107 m). [For $n = 2$ and $d = 175$ ft (53 m), you will get a notch at a wavelength of 350 ft (107 m), but not one at 175 ft (53 m) (because that will be the first zero peak). You will have notches at wavelengths of 350, 117, 70, 50 ft etc. (107, 36, 21, 15 m etc.)]. You calculate the response of this array for a reflection from your objective at some offset and for some modest frequencies. (The method is given in Appendix D and the example given there is precisely the same array.) You soon realize that you will have to limit your offsets substantially to use this geophone array. You could shorten it by changing all the spacings to 20 ft (6.1 m) instead of 25 ft (7.6 m), but then your longest notch would be at a wavelength of 480 ft (146 m) which was what was used unsuccessfully before. So instead, you start cutting back on the maximum offset.

Even though you are only planning a test at this time you are looking ahead to see how you could use whatever you test in a viable, production mode. You realize that to use the original parallelogram you would also need to cut back on the number of data channels to keep from getting the group interval too small which would shorten the shot point interval and add more holes to be drilled. If you drop back to 96 traces, and shoot half maximum fold; you would still get 24 fold or twice what was used before. You finally settle on a possible group interval of 160 ft (49 m) and possible near and far offsets of 640 and 8160 ft (195 and 2487 m). Call this Geometry A′. You decide to use an odd trace gap since you are contemplating half fold shooting (Chapter 9). Now you would be down to 16.5 shot points per mile (10.25 per km). For this cable you calculate the high-frequency response for a double group length array, $nd = 320$ ft (98 m), a triple group length array ($nd = 480$ ft (146 m), and a

Table 7.5 Differential time across arrays and high-frequency response to reflections from the objective for Geometry A′

Note:
$L = 8160 - nd/2$ ft or $2487 - nd/2$ m
$\Delta L + L = 8160 + nd/2$ ft or $2487 + nd/2$ m
$D = 10\,000$ ft or 3048 m
$V = 16\,000$ ft/s or 4877 m/s
$t_0 = 2(10\,000)/(16\,000) = 1.25$ s

Item	*nd = 320* (*97.5 m*)	*nd = 480* (*146 m*)	*nd = 640* (*195 m*)	*nd = 775** (*236 m*)	*nd = 425** (*130 m*)
Δ(s)	0.007555	0.011332	0.015109	0.018296	0.010034
f_0 (Hz)	132	88	66		
$f_{0.3}$ (Hz)	99	66	50		
$f_{0.64}$ (Hz)	66	44	33		
$(\Delta t)(50)(\pi)$ (θ in radians)	1.1868	1.7801	2.3734	2.8739	1.5761
$(\sin\theta)/\theta$ (for $f = 50$ Hz)	0.781	0.550	0.293	0.09205	0.63445
$(\Delta t)(40)(\pi)$ (θ in radians)	0.9494	1.4241	1.8987	2.2991	1.2609
$(\sin\theta)/\theta$ (for $f = 40$ Hz)	0.856	0.695	0.499	0.3246	0.7553

*Needed to calculate responses of a trapezoidal array that is equivalent to the big, parallelogram array.

quadruple group length array, $nd = 640$ ft (195 m). We also calculate the high-frequency response to reflections for the big parallelogram. We do this by calculating the differential moveout across the two component lines of the trapezoidally weighted, inline array that is equivalent to the parallelogram. In this case the nd values for the two lines of the trapezoidal equivalent are 775 ft and 425 ft (236 m and 130 m) with 31 geophones and 17 geophones respectively. Table 7.5 summarizes the results for Geometry A′.

The $(\sin\theta)/\theta$ values in Table 7.5 are the responses of the arrays for the frequencies given. But to calculate the response of the trapezoidal equivalent of the parallelogram we must weight the two responses by the number of geophones in each, sum, and then divide by the total number of geophones. For this example we get:

$$\text{Response at 50 Hz} = \frac{(31)(0.09205) + (17)(0.63445)}{48} = 0.284$$

$$\text{Response at 40 Hz} = \frac{(31)(0.3246) + (17)(0.7553)}{48} = 0.477$$

We are a bit disappointed by the loss of higher frequencies with the big parallelogram, but we compare this case with the example in the Chapter 4 and conclude that the relative response at 50 Hz for stacked traces should be about 70 per cent which is not really all that bad. Actually the earth filter may not let us get 50 Hz energy back from our objective anyhow. Not all the ideas we have considered can be included but, depending on the test results; they could be used later. We deveop the test plan shown in Fig. 7.6. We shall have the holes drilled prior to the test. The holes that are 50 ft (15 m) deep or deeper will be loaded when they are drilled. For the patterns of holes only 10 ft (3 m) deep we shall consider further whether or not they will be loaded when drilled. We do not want anyone to recover explosives from shallow holes and most certainly not for criminal purposes. If we pre-load the shallow holes we shall have someone stay overnight to see that the charges are not disturbed. The guard should have a radio with which he can contact the sheriff. We may choose to use an off-duty deputy sheriff for the guard. We plan to have one drill crew standing by during the tests so that we could drill additional holes to add shots to follow up any interesting results from the planned shots.

The six and twelve-hole patterns were designed to provide a notch at a wavelength of 240 ft (73 m) since the double group length array (Geometry B) has a peak at a wavelength of 235 ft (71.6 m). Figure 7.7 is a sketch of the source pattern response and the double group length array response for Geometry B.

This is a rather elaborate test, yet only eighteen records are planned. We plan to layout the cable and arrays, and have everything ready the evening before and to get an early start the next day. We plan to bury the geophones (for all but a few arrays) so that we may be able to record satisfactory data even

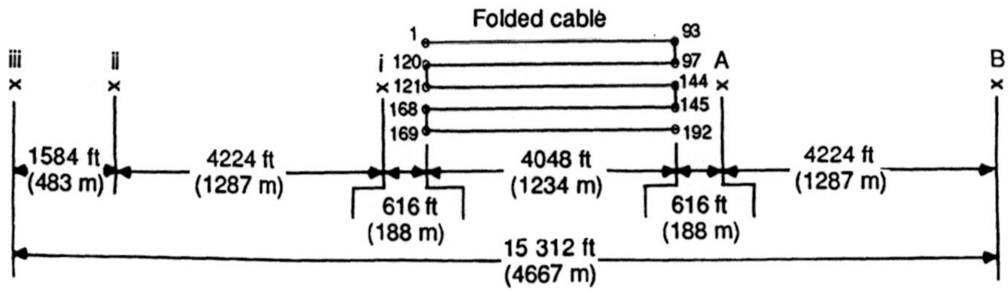

Traces	Group interval	Geophone array	Spacing	nd	Purpose
1-93	44 ft (13.4 m)	12 in pod	0+	0+	Noise test
97-120	176 ft (53.6 m)	24/169 ft (51.4 m)	7.3 ft (2.24 m)	176 ft (53.6 m)	Good high frequency response
121-144	176 ft (53.6 m)	36/342 ft (104 m)	9.8 ft (3 m)	352 ft (107 m)	Good to fair suppression of coherent noises
145-168	176 ft (53.6 m)	48/517 ft (158 m)	11 ft (3.4 m)	528 ft (161 m)	Good suppression of all coherent noises
169-192	176 ft (53.6 m)	Big parallelogram 48 in 2 lines			Excellent suppression of all coherent noises

Test plan

	Location				
Source description	iii	ii	i	A	B
20 # at 100 ft (30.5 m)		yes	yes	yes	yes
10 # at 100 ft (30.5 m)		yes		yes	yes
20 # at 125 ft (38 m)	yes	yes			
10 # at 50 ft (15.24 m)		yes			yes
5 # at 50 ft (15.24 m)		yes			
6 holes, inline, spaced 40 ft (12 m), 1# ea.		yes	yes	yes	yes
6 holes, inline, spaced 40 ft (12 m), 1/2 # ea.					yes
12 holes, inline, spaced 20 ft (6 m), 1/2 # ea.		yes		yes	

Have one drill crew on standby during tests to drill and load any other test source that may be desirable.

Fig. 7.6 Test layout with folded cable—Shot hole crew

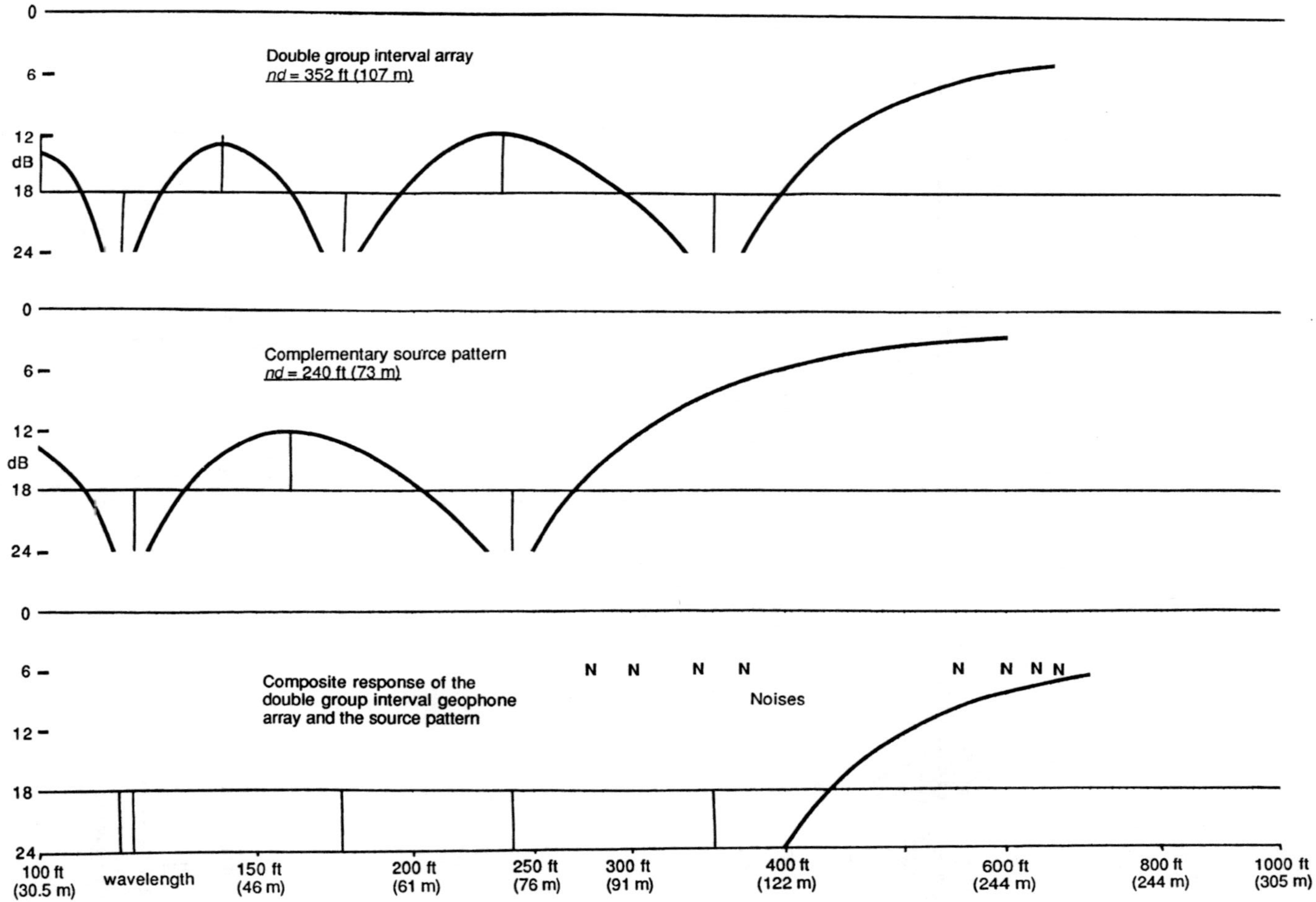

0
Double group interval array
nd = 352 ft (107 m)
6
12
dB
18
24
0
Complementary source pattern
nd = 240 ft (73 m)
6
12
dB
18
24
0
N N N N N N N N
Composite response of the
double group interval geophone
array and the source pattern
Noises
6
12
18
24
100 ft
(30.5 m)
wavelength
150 ft
(46 m)
200 ft
(61 m)
250 ft
(76 m)
300 ft
(91 m)
400 ft
(122 m)
600 ft
(244 m)
800 ft
(244 m)
1000 ft
(305 m)

when the wind picks up. We plan to have two shooting crews and with luck we should complete the tests before the wind gets up. If noise becomes excessive, we shall suspend recording until the wind subsides in the late afternoon. We may have to complete the recording after dark and should arrange for some good, battery-operated lights mounted on two drill truck masts to illuminate the shot points.

We hope to finish early so as to have the rest of the day to study the results and make decisions. If we have to suspend recording during the middle of the day, we can study all the data we have recorded up to that time. We plan to record all the data with a low-cut filter of 12 Hz and if we are successful in providing a means for filtering playback records in the field, we plan to filter and playback appropriate data sets with low-cut filters of 16 and 24 Hz. If we are also able to get a low-cut filter near 40 Hz, we also plan to use it for some playbacks. Any significant delays during the day should be used to make playbacks. If we do not finish the tests early, we may be up late that night. Better plan on taking a small electric immersion water heater, a ceramic mug, and some instant coffee.

7.3 PORTABLE FIELD CREW TEST

Figure 7.8 is a prospect form for a third hypothetical seismic program for a pair of dip lines in rugged country with very limited access for vehicles. Previous data were acquired in this area with vibrators working along very crooked roads. The new lines are to firm up a prospect that has been outlined from the vibrator data. These new lines are between old lines and are near possible crests of the structure along strike. A strike line may be recorded at a later date through the structural crests of the dip lines that will be recorded in this phase of the program. The two lines are each about six miles (9.7 km) in length.

The primary objective is at a depth ranging from 8200 ft (2500 m) up to about 7200 ft (2195 m). The stacking velocity is about 14 000 ft/s (4267 m/s). Dips of up to about 15 degrees are expected. Part of each line will be on a dolomitized limestone. The previous Vibroseis data were poor on this outcrop, but were satisfactory elsewhere. (Chapter 12 deals in depth with problems that cause poor data quality under high velocity outcrops. In this chapter we will confine discussion to a rather narrow approach to the problem.) We shall have

---

Fig. 7.7 Responses of a double group interval array and a complementary source pattern (Geometry B). The source pattern is designed to provide a notch to effectively suppress noise at a wavelength of 240 ft (73 m) where the geophone array is less effective. The responses of the geophone array and source pattern are independent and the composite response in dB is the sum of the dB responses of the array and pattern.

Onshore Seismic Prospect Form

Prospect Shady Lane Line Prefix SLP Auth. No JO-427-8
Miles of Program 12 Date 4–17–1988 By Jane Gonzales
Country USA State of Prov Pennsylvania Counties Bedford
Quads
General Objectives 1) Map structure and define crest of the objective with satisfactory data beneath Knox dolomite outcrop. Migrated section needed.

Topography steep hills Vegetation trees Culture none
Seismic datum 1100 ft Correctional Velocity 6000 ft/s

Specific Objectives

Priority/ Horizon	Record Time (s)	Depth (ft)	Stacking Velocity (ft/s)	Lithology	Dip	Resolution Vertical Horiz.
1 Oriskany	1–1.2	7200–8200	14 000	sandstone	1.5°	not critical
2 Treton	1.05–1.25	7600–8600	14 500	limestone	15°	not critical
3						
Basement						

Geology

Surface 1/2 soil & 1/2 Knox Dolomite (karst)

Near Surface limestones, hard sandstones & shales

At Depth thrusts and compressional structures

Key Wells Brown #1 Staples is 10 miles north of the prospect

Previous Seismic Data

Line	Prospect Name	SP Range	Tie Points	Quality	Problems	Improvement Required
SLV-2	Shady Lane	210–305	245	good to poor	poor data under Knox outcrop	better data under Knox outcrop
SLV-5	Shady Lane	100–275	—	" "	" "	" "
SLV-5	Shady Lane	145–417	375	" "	" "	" "

Additional Comments or Suggestions May add strike line later along structural crests.

Preferred order for recording lines? Yes/No. If 'yes', give details including reasons, on a separate sheet.
Please fill out this form as completely as is feasible. Attach additional sheets if necessary.
Distribution: Prov. Mgr. Interpreter Processor WPC KMK JPB SRS

Fig. 7.8 Prospect sheet for a third hypothetical seismic program.

to use a portable crew to record these two dip lines. A small, four-wheel drive, recording truck will be able to get to each line at enough points to record the data. Most of the area is heavily wooded which will limit helicopter support somewhat.

The Vibroseis data were recorded a few years ago with 48 traces and 24 CMP fold. They used a 14–56 Hz linear sweep of 12 s duration. There were four vibrators spaced 32 ft (9.8 m) apart (pad to pad). They summed data from ten sweeps per VP with 5 ft (1.5 m) of move-up between sweeps. The cable was a symmetrical split of 5610–550–0–550–5610 ft (1710–168–0–168–5610 m). The group interval and VP interval were 220 ft (67 m). The geophone array was 36 geophones over 440 ft (134 m).

In the previous data the air wave was weak but visible despite the 5 ft move-ups. With the geophone arrays and source patterns that were used, coherent noise was apparently not a severe problem on the unstacked data even in the bad record area on the dolomitized limestone. But there was some coherent noise in the bad record area at wavelengths of about 160, 270, and 550 ft (49, 82, and 168 m).

A portable crew has been selected that has drills for drilling mini-holes which are only 5 ft deep. This crew has had reasonable success in somewhat similar areas using 0.5 lb (0.25 kg) charges with patterns of these shallow holes. The number of holes per pattern has ranged from six to fifteen. The crew has 300 strings of geophones with twelve geophones per string 25 ft (7.6 m) apart with center takeouts. They use a multi-conductor cable. The available low-cut filters have cut-off frequencies of 12 or 18 Hz. Alias filters have cut-off frequencies of 90 or 180 Hz. The crew does have 60 Hz notch filters.

We look at the tables of cables in our handy tool box and tentatively decide on a group interval of 132 ft (40 m), 96 traces and offsets of 7128–924–0–924–7128 ft (2173–282–0–282–2173 m). We then consider how a double group length array, $nd = 264$ ft (80.5 m) and how a triple group length array, $nd = 396$ ft (121 m) might work in suppressing the noise observed in the Vibroseis data. We like the way the double group length array handles the observed noise, but not the triple because of its poor suppression of the noise with a wavelength of 270 ft (82 m). So we try a quadruple group length array, $nd = 528$ ft (161 m) and tabulate our results in Table 7.6.

The double and quadruple group length arrays have a notch near the noise wavelength of 270 ft (82 m), but the triple has a peak near that wavelength. None of the geophone arrays has a notch near the noise wavelength of 160 ft (49 m), therefore a source pattern was chosen that provides a notch there. The quadruple group length array is the only one long enough to attack the noise with wavelengths near 550 ft (168 m). But is this array too long in terms of suppressing the higher-frequency components of reflections from our objective? First we check for reflections from a horizontal reflector at a depth of 7200 ft (2195 m) received at an offset of 7182 ft (2173 m) and calculate the

Table 7.6 Notches and peaks for three geophone arrays and a source pattern

nd = 264 *ft* notch (*ft*)	peak (*ft*)	*nd* = 396 *ft* notch (*ft*)	peal (*ft*)	*nd* = 528 *ft* notch (*ft*)	peak (*ft*)	Source pattern *nd* = 160 *ft* notch (*ft*)	peak (*ft*)
				528			
		396					
					352		
264			264	264			
					211		
		198					
	176			176			
			159		150	160	
132		132		132			
			113		117		
	106			106			106
nd = 80.5 *m* notch (*m*)	peak (*m*)	*nd* = 121 *m* notch (*m*)	peal (*m*)	*nd* = 161 *m* notch (*m*)	peak (*m*)	*nd* = 48.8 *m* notch (*m*)	peak (*m*)
				161			
		121					
					107		
80.5			80.5	80.5			
					64		
		60					
	54			54			
			48.5		45.7	49	
40		40		40			
			34.4		35.7		
	32			32			32

Table 7.7 Double and quadruple group interval geophone arrays and a complementary source pattern

Quantity	*Geophone array* *nd = 264 ft (80 m)*	*Geophone array* *nd = 528 ft (161 m)*	*Source pattern* *nd = 160 ft (48.8 m)*
Δt (s)	0.007921	0.01589	0.0048
f_0 (Hz)	126	63	208
$f_{0.3}$ (Hz)	95	47	156
$f_{0.64}$ (Hz)	63	32	104

zero offset reflection time:

$$t_0 = \frac{2(7200)}{14\,000} = 1.029\ \text{s}$$

$$\left[= \frac{2(2195)}{4267} = 1.029\ \text{s} \right]$$

We then calculate and tabulate the results in Table 7.7.

From Table 7.7 we conclude that the quadruple group length arrays suppress higher frequencies more than we like, but they are usable if that is what is necessary to suppress the longer noise wavelengths to get good data. For dipping reflectors, the updip arrays would have less moveout across the arrays. But the down-dip arrays, at much offset, would be ineffective. If the longer-wavelength noise is not too severe, we would like to use the double group length array with its much better high-frequency response. For source patterns down-dip from the receivers the effective wavelength would be decreased. But the source pattern with a number-spacing product (*nd*) of 160 ft (48.8 m) looks pretty good. But, you decide also to try a shorter source patterns with *nd* product of about 105 ft (32 m) to provide a notch where there is a peak for the double group length array.

We want to test both on and off the outcrop where the data may be poor, but we do not want to set up two separate tests. Therefore, we plan to lay out the cable with the contact between the outcrops bisecting the cable. The location will be along one of the two lines to be recorded. We shall have shot points located off both ends of the cable and the longitudinal distance between shot points will be short enough so that we shall have continuous offset coverage for the arrays using just half of the cable. This will allow us to make continuous offset displays with both shot points and receivers in the good area and with both in the bad area. Also it will allow us to see results with shot points in the bad and receivers in the good areas and vice versa. On single-shot records we hope to be able to follow reflections in the good area right up to the contact and see where the reflections should continue into the bad area. For the noise test data we can make similar displays. Hopefully, all of these data together will provide clues to the reason for poor data on the dolomitized limestone outcrop and, hopefully, we shall find a set of recording parameters which will allow recording better seismic data in this area.

Our test plan is illustrated in Fig. 7.9. You plan to use traces 1–47 with a

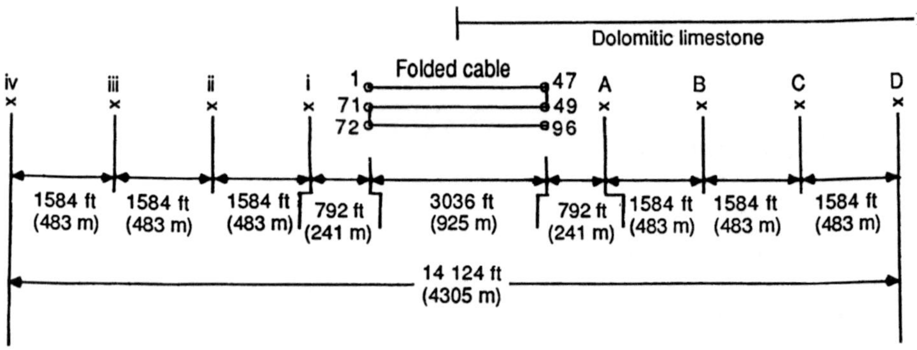

Traces	Group interval	Geophone array	Spacing	nd	Purpose
1-47	66 ft (20 m)	12 in pod	0+	0+	Noise test
49-71	132 ft (40m)	24/253 ft (77 m)	11 ft (3.35 m)	264 ft (80.5 m)	Double group interval array [group interval of 132 ft (40 m)]
72-96	132 ft (40 m)	24/506 ft (154 m)	22 ft (6.7 m)	528 ft (161 m)	Quadruple group interval array to suppress long wavelengths

Test plan

Number of holes	Spacing of holes	nd	Locations of patterns	Purpose
6	5 ft (1.5 m)	30 ft (9 m)	iv, iii, ii, i, A, B, C, D	Noise test
6	25 ft (7.6 m)	150 ft (46 m)	iii, ii, B, C, D	Minimum source with nd = 150 ft (46 m)
7	15 ft (4.6 m)	105 ft (32 m)	iv, ii, i, A, B, C, D	Possible source with nd = 105 ft (32 m)
8	20 ft (6.1 m)	160 ft (49 m)	iv, iii, ii, i, A, B, C, D	Probable source with nd = 160 ft (49 m)
15	7 ft (2.1 m)	105 ft (32 m)	iii, B, C, D	Maximum source with nd = 105 ft (32 m)
16	10 ft (3 m)	160 ft (49 m)	iii, B, C, D	Maximum source with nd = 160 ft (49 m)

Have one drill crew on standby during tests to drill and load any other test source that may be desirable.

Fig. 7.9 Test layout for a mini-hole seismic program with a portable crew.

group interval of 66 ft (20 m) for a noise spread for noises with moderate to long wavelengths. Traces 49–72 will be used for the double group length array with a group interval of 132 ft (40 m). Traces 73–96 will be used for a quadruple group length array. We plan to have the holes for the tests drilled ahead of time. We want to avoid any standby time for the expensive recording crew, therefore we shall commit ahead of time to a group interval of 132 ft (40 m) and we shall have the surveyor lay out the first few miles of the line with that group interval. We are also going to have the drill crews start drilling on the line with an eight hole pattern with a spacing of 20 ft (6 m) between holes. After the tests are completed the recording crew will take a break of a few days to let the drill crews get far enough ahead so that the recording crew will not be waiting on the drills.

If the tests show that the eight hole shot patterns are not satisfactory, then some adjustment will have to be made. Of course, we could always add more holes (possibly with some lateral displacement). Our initial plan is to record half fold by putting a shot point at alternate positions, 264 ft (80 m) apart. Depending upon the test results, we may have the drills go back and drill patterns at the skipped positions for shot points on and near the dolomitized limestone outcrop so as to increase CMP fold in the bad record area.

We decide to use helicopter support when weather permits but only for hauling equipment. It is one thing to try to snake equipment in and out of the trees on long lines and quite another to take personnel in and out. We hope the helicopter pilot is extremely careful—better still that he is lucky. Come to think about it, everyone in exploration needs to be lucky. But it helps to augment some good luck with good geophysics, good geology, good planning, and hard work.

8
Marine seismology

The subject of this chapter is seismic surveys in water-covered areas. Marine seismology is extremely important and will be increasingly important in the future as exploration opportunities on land are depleted. Seismic operations at sea are very different from those on land, but the geophysical problems are essentially the same. However the severity of certain problems is quite different with some problems being more difficult for sea but others more difficult for land work. The limitations of wide gun arrays for suppressing off line scattering is illustrated. Beam steering with a gun array is also illustrated as a means for reducing the loss of higher-frequency components of reflected events.

Marine seismic surveys are expensive on a per-day basis but usually are quite inexpensive on a per-kilometer (or per-mile) basis, because of very high data-acquisition rates.

8.1 GHOSTS

When we shoot below the surface on land, we sometimes have a significant ghost problem from seismic waves from the shot traveling upward and being reflected back downward either from the base of the weathering or from the surface (Chapter 9). In marine work we must always contend with a severe ghost problem. The ghost problem occurs not only at the source from guns below the surface but also at the receivers with hydrophones (or other detectors) also located below the surface. The ghost problem is most severe when the water surface is smooth. Because of the large difference in density (and in velocity of sound) between water and air, the reflection coefficient for water ghosts is -0.9994 or essentially -1.0. When the water surface is rough, there is some scattering, but ghosts remain strong particularly at the lower seismic frequencies where wavelengths are long.

For some frequency, the travel time from the gun or hydrophone to the surface will be one-half of a period (and the depth will be one-half of a wavelength) so that the total travel time up and back down to the gun or hydrophone depth is a full period (and the total added path is a full wavelength). Since the reflection coefficient is negative, the ghost will be out of time phase with the direct wave and the two will nearly cancel. This will result in a near-zero notch in the spectrum for every reflection. At all integral multiples of this frequency there will also be notches in the spectrum of every reflection since the total added time will be an integral number of periods; and the added path length for the ghosts will be an integral number of wavelengths. (Appendix B supplies equations and more discussion.) For very long offsets the ray paths in the water for reflections from depth will be slightly oblique and the center frequency of the notches in the spectra will decrease slightly. Rough seas will result in some variation in the effective depth and therefore will tend to broaden the spectral notches and to fill them in somewhat.

What is sometimes overlooked is the fact that there is also a notch at zero frequency since the reflection coefficient is negative. As a result the effective bandwidth for marine seismic surveys is usually limited to frequencies between the notch at zero frequency and the first notch above zero frequency. Figure 8.1 illustrates this effect. For this case we assumed a water velocity of 5000 ft/s (1524 m/s), and a ghost reflection coefficient of minus one. We also assumed that the guns and hydrophones were at the same depths. From this figure we see that if we want high frequencies in our reflected wavelets we should use a shallow depth for our guns and our hydrophones. (This is equally true with other sources and with other receivers.) But if we use a shallow depth for our guns and hydrophones we suppress low frequencies that we may also want in our reflected wavelets. What if we put our guns at one depth and our cable at another? Figures 8.2, 8.3, and 8.4 show contours of the total response with either the guns or the cable at depths of 20, 30, or 40 ft (6, 9, or 12 m) and

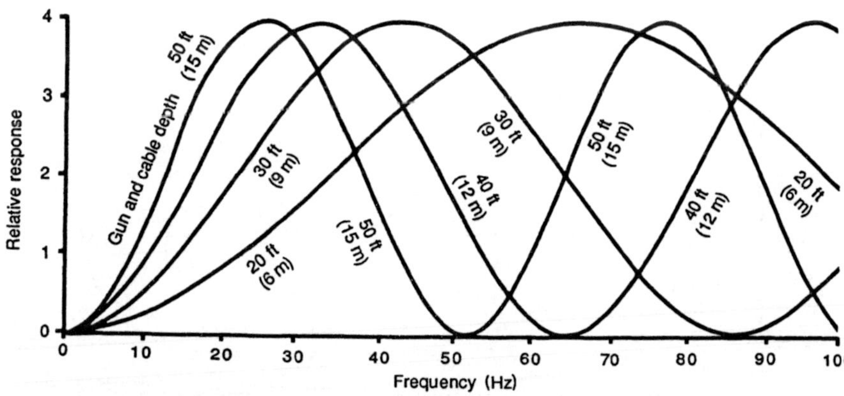

Fig. 8.1 Filtering effect of ghosts with both guns and cable at the same depth. If we operate the guns and cable at relatively shallow depth, we improve the response at higher frequencies but decrease the response at low to moderate frequencies.

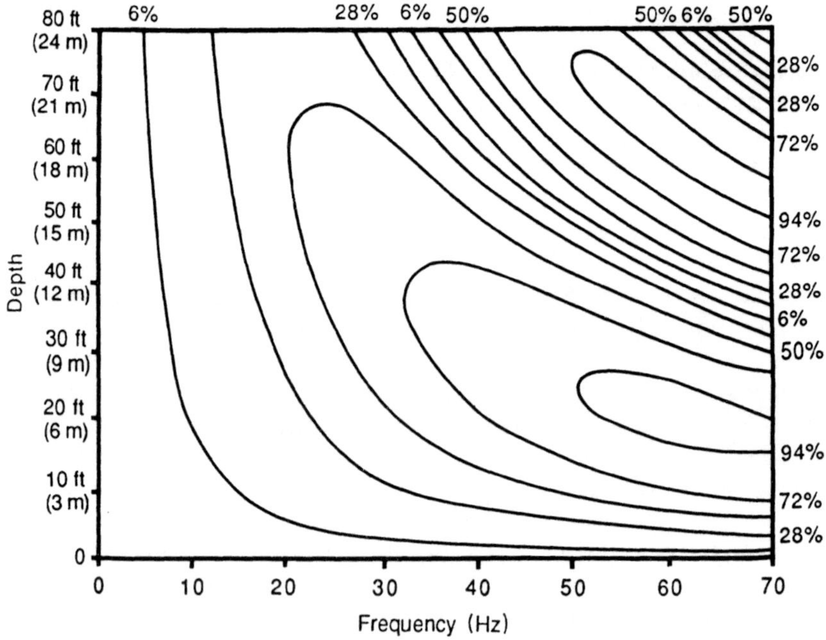

Fig. 8.2 Filtering effect of ghosts with guns or cable at 20 ft (6 m). The contours show the composite ghost effect with the cable or guns at 20 ft (6 m) and the other element at any depth from 0–80 ft (24 m).

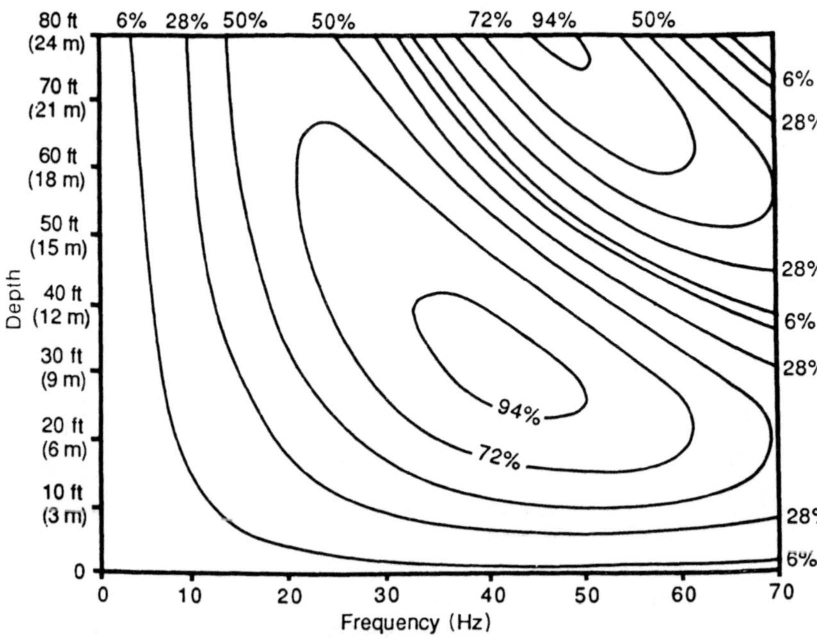

Fig. 8.3 Filtering effect of ghosts with guns or cable at 30 ft (9 m). The contours show the composite ghost effect with the cable or guns at 30 ft (9 m) and the other element at any depth from 0–80 ft (24 m).

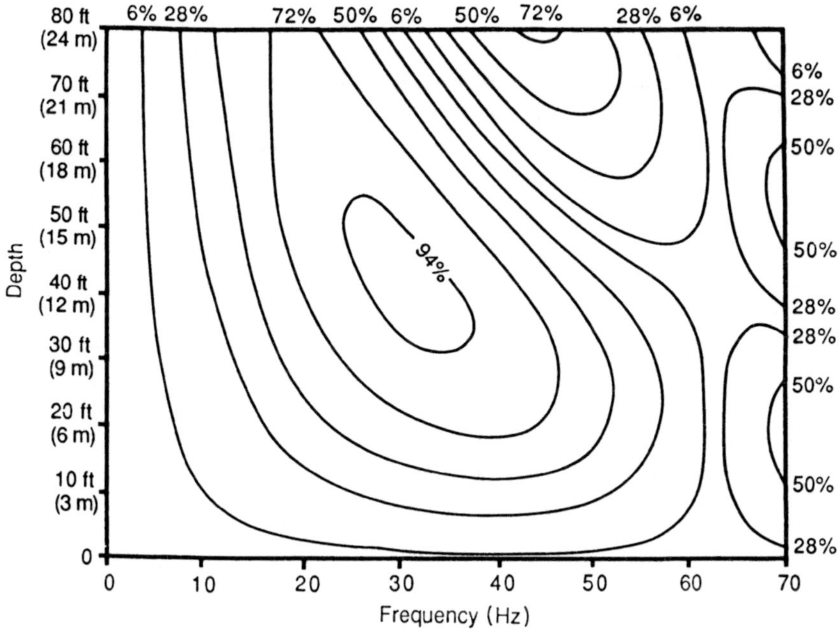

Fig. 8.4 Filtering effect of ghosts with guns or cable at 40 ft (12 m). The contours show the composite ghost effect with the cable or guns at 40 ft (12 m) and the other element at any depth from 0–80 ft (24 m).

the other element at other depths. Whatever we do, we shall have some spectral notches in marine seismic surveys which will limit bandwidth.

We could operate sets of guns at different depths and could tilt the cable so that one end is deeper than the other to flatten out the spectra and to partially fill in the deep notches, but this is not generally done. (Fairfield Industries, Inc. of Houston, Texas has developed and used a system for recording using a tilted cable.) Operating guns at different depths should not be difficult and could provide an additional advantage if the guns groups nearest the boat are deeper than the gun groups farthest from the source. This would allow favorable beam steering during data acquisition (Figure 8.6, Section 8.9 and Chapter 11).

How serious is the ghost problem in terms of bandwidth limitation? Suppose that we have both guns and cable at a depth of 30 ft (9 m). Figure 8.1 shows that the relative response is 50 percent or greater for frequencies between 21.5 Hz and 65 Hz. We shall probably have plenty of signal below 21 Hz down to at least 10 Hz since earth attenuation is low for these low frequencies. So perhaps the problem need not be too serious if we use a gun and cable depth that is appropriate for the frequency range needed. But suppose one of our explorationists decides that he needs some very high-frequency marine data and he wants us to put our guns and cable at very shallow depths. Most sources will not work well at shallow depths. Shallow depths would cause more suppression of the lower frequencies and at some point this would become a problem. The ghost problem in marine seismology can be handled satisfactorily, but it should not be overlooked. Usually the solution is to select a reasonable frequency band and an appropriate depth for the guns and cable for the selected frequency band.

8.2 STATICS

Static variations on land are often a very serious problem due to severe variations in the shallow earth section near our sources and receivers. At sea the water is homogeneous and static problems are usually minimal. However, in shallow water areas such as deltas, we may encounter a problem due to pockets of gas trapped in very shallow sediments which can cause a severe statics problem. In addition, we can experience serious statics problems at sea due to a very irregular sea floor. It is highly advantageous to record lines that are nearly perpendicular to channels in the sea floor. A sparker survey may be helpful in modeling sea floor geometry and in finding a data processing solution to the problems caused by an irregular sea floor. (A sparker survey is a seismic survey using very high-frequency seismic waves generated by an electrical spark between electrodes immersed in the water. These high-frequency compressional waves do not penetrate far beneath the sea floor. Sparker sections provide a detailed map of the sea floor and of the very shallow earth beneath the sea floor.)

8.3 WATER WAVES AND WATER BOTTOM MULTIPLES

Seismic waves travel through water at a rather low velocity but with little attenuation and almost no scattering. Therefore the direct wave through the water arrives at the detectors with high amplitude. The direct arrival does not usually create a serious problem. Sometimes there is a transition at the sea floor from water to low density mud that gradually increases in density with depth. In this case there is no sharp change in velocity or acoustic impedance at the sea floor. Furthermore the low density mud attenuates seismic waves. Reflections from the sea floor are weak. This situation is described as a 'soft bottom' sea floor.

In other cases there is an abrupt increase in velocity and in acoustic impedance at the sea floor. The reflection coefficient can be quite high. Even an unconsolidated sand bottom can result in a reflection coefficient of 0.5 or more, and the angle of refraction would then be 35° or less. In this case all of the energy emitted at angles greater than 35° from the vertical would be reflected and more than 25 per cent of the energy emitted at angles of less than 35° would also be reflected. The reflected waves would then travel upward where they would be totally reflected at the sea surface. Therefore most of the energy emitted by the source would be trapped in the water layer where it would be noise that would be picked up by the receivers in the cable. Where a sharp transition from water to a firm sea bottom occurs, the situation is described as a 'hard bottom' sea floor. If the sea bottom were a consolidated rock, the reflection coefficient would be even higher, the angle of refraction would be even less, and even a greater fraction of the shot energy would be trapped in the water layer to generate noise which is called reverberations. De-reverberation of the data in processing is essential and this is best accomplished with a minimum-phase deconvolution (Section 19.2).

8.4 COHERENT NOISE

Larner *et al.* (1983) in an excellent paper dealt with three categories of noise in marine seismology. The first of the three categories included:

(a) Direct arrivals of energy from the shot to the receivers through the water;
(b) Reflections and multiple reflections from the sea floor;
(c) Refractions and multiple refractions from the sea floor.

Larner concluded that in stacked sections of field data these first category noises resulted in strong, horizontal events at early record times (from unmuted refractions) and in 'odd, even' noise patterns at rather early record times. But in the stacked data these noises were not a major problem later in the record.

The second noise category included noises resulting from the mechanical motion of the cable (in response to variable tugging force). The noises in this category were dominantly very low in frequency and these noises were not a serious problem in stacked sections.

The third category of noise included noise scattered from irregularities in the sea floor or irregularities in the very shallow earth beneath the sea floor. They concluded that this category of noise was the source of a major problem in stacked data at late record times in two field cases with hard sea bottoms. A velocity analysis of raw CMP gathers for one of the field cases was dominated by peaks near the water velocity of 1500 m/s at record times extending from 0.2 s to 3.5 (the longest time displayed). After dip filtering, these peaks were relatively minor and strong peaks emerged for primary reflections with reasonable velocities ranging from 3000 m/s to more than 4000 m/s.

Where there is a sharp transition from water to a firm or hard sea floor, there will be a lot of energy from the source trapped in the water layer; if the sea floor is irregular, scattering of this trapped energy gives rise to strong events which exhibit moveout at the velocity of sound in water. This moveout will vary from nearly linear to nearly hyperbolic. Scatterers that are not laterally displaced and are either forward of the boat or beyond the tail buoy (beyond the far end of the cable) will result in noise with linear moveout. Scatterers that are between the source and the receivers and which are displaced a considerable distance, either laterally or vertically, will display a more nearly hyperbolic moveout. (Obviously most of the scatterers will not be directly below the seismic line.)

Larner (1983) also set up a computer model with a velocity of 1500 m/s with scatterers randomly placed over the surface. This model represented the effect of scatterers on a shallow sea floor interacting with seismic waves that travel only in the water layer. The scatterers produced events on synthetic shot and CMP gathers that looked very similar to reflection events. Velocity analyses on these synthetic data showed peaks for velocities near the water velocity of the model but also showed peaks at higher velocities ranging up to 4000 m/s and more. Yet the only velocity used in the model was 1500 m/s. When these synthetic data were stacked using a stacking velocity from one of the two field cases, the result was a dominant criss-cross pattern of linear events. The slopes of these events were equal to the 1500 m/s velocity of the model (where the slope was defined as half of the group interval divided by the trace-to-trace time difference for the events).

Field examples exhibited a very similar criss-cross pattern on the stacked data, but the slopes indicated somewhat higher velocities. They concluded that this criss-cross noise in the stacked field data was from scattering at very shallow depths below the sea bottom rather than scattering strictly within the water by irregularities of the sea bottom itself. (This conclusion was based upon the higher than water velocity of the criss-cross pattern.) The waves traveling in the shallow earth layer might travel horizontally (refractions) or nearly horizontally (reflections that enter the water layer at angles only slightly less than the angle of refraction).

Irregularities on the sea bottom such as channels, faults, etc. would extract energy from the seismic waves trapped in the water layer. (This is because trapped waves traveling largely horizontally encounter some irregularities at small angles to the normal to the surface such that they would not be totally reflected as they would be by a smooth sea floor.) Waves would then be set up in the earth layer below the sea bottom. Such waves would travel at a faster velocity in a hard sea bottom. Energy could then be transmitted back into the water particularly at other surface irregularities. This mechanism was not modeled by Larner *et al.* but it might play a part in producing the higher-velocity criss-cross pattern observed on stacked field data.

Larner *et al.* were quite successful in suppressing the criss-cross noise by using a dip filter (passing dips between −5 and +5 ms/trace) on the shot gathers and then using predictive deconvolution after stack. (Predictive deconvolution involves the use of the noise in the early part of a seismic trace to predict and deconvolve the later part of the same trace.)

8.5 MULTIPLES

Multiples other than the water bottom multiples can be a serious problem in marine seismology (just as they can be on land). If the multiple velocity is not much different from the primary velocity at a given record time, we need to provide the maximum possible difference in the two moveouts. This means using a cable as long as the maximum effective offset for the primary reflection from the objective. Chapter 19 gives processing methods which can be used to suppress multiples. Sometimes a long gun array is used to suppress strong multiples (Section 8.10).

8.6 ADVANTAGES FOR SEISMOLOGY AT SEA

With all the preceding discussion of problems it may seem that marine seismic data are seldom good. Quite the contrary, they are often very good. The relative homogeneity of the water as compared to the shallow earth on land is a big factor. This makes for a very uniform coupling and a uniform waveform from the seismic source and a very uniform coupling and uniform response at the detectors. In fact the coupling is nearly perfect. The sea floor is often smooth and relatively homogeneous. Obstacles are relatively few and far between so we can readily run lines where we want them and in the most favorable directions (dip or strike and perpendicular to faults to be mapped). Frequently it is practical to record many traces and acquire very high CMP fold data. Permitting is not always easy, but when a permit is obtained it is usually for a very large and contiguous area. The biggest advantage of marine seismology is economic—high production and low unit cost.

8.7 RECORDING TECHNIQUES AND EQUIPMENT

As on land, we start with a prospect form and discussions with the geologist and the interpreting geophysicist to define the exploration objective or objectives of the survey. We need to study any nearby data that are available. If time permits and it is feasible, we should display samples of previous data in shot point and CMP gathers with and without AGC and with and without normal moveout corrections. We should mark coherent noises and measure the periods and the apparent velocities and calculate the frequencies and apparent wavelengths. We must pay particular attention to noises present at or near objective times.

We need an estimate of the moveout velocity of the objective reflections and we may want to make some overlays for the field shot gathers that show the expected arrival times of reflections from the objective as a function of offset. If very little is known about the area, we can still make a set of overlays with simulated reflections every 500 ms with the moveout determined by various guesses of reasonable moveout velocities.

Our choices for the hydrophone cable and the hydrophone arrays are limited and sometimes we really have no choice. The boat has one or possibly two cables available and that is what we use. We may be able to shorten the cable but this may be achieved by taking off some data channels.

Originally the marine cables were analog but digital cables have been used quite successfully for several years. Advantages attainable with digital cables include less loss of high-frequency components of seismic reflections, and less cross-talk between data channels. But digital cables must have many, in-cable, amplifier sections dispersed at relatively short intervals along the cable. These amplifier sections are likely to be larger in diameter than comparable analog cables and may result in more towing resistance and in more cable noise due to towing. Even so, digital cables (including fiber optics cables) appear to offer enough advantages so that they are likely to displace analog cables for the majority of marine seismic operations.

In marine work we would prefer to use Anstey's stack-array approach (Anstey, 1986), where the detectors in the cable are uniformly spaced along the entire cable and the effective array length is equal to the group interval. Most marine cables provide group interval arrays with uniform spacing of the hydrophones. With a towed cable we have an end-on spread and we would prefer to record with a shot interval equal to half of the detector group interval. This provides the continuous and uniform sampling in the CMP gathers that is advantageous and it provides for the maximum possible CMP fold. But some cables have so many traces that the group interval is short and it is impractical to shoot twice every group interval. (We must wait some time between shots to record data and allow the waves to decay and we do not want to slow the boat speed and lose production.) In these cases we would probably shoot every group interval. But we may very well add adjacent traces in processing anyway and end up with the stack-array. If this is what is ultimately done, we could have used a cable with only half as many channels in the first place. If we have a choice, we select a cable with a group interval that is long enough to allow two shot points per group interval at normal boat speed; by doing so we would save tape and some processing costs.

Theoretically, with more channels we have the option of going back and fully processing the data using each channel separately. Or in processing we can form more different detector arrays and possibly use receiver beam steering to advantage (Chapters 11 and 19). But how often are the unsummed data used and how worthwhile is this when they are used? For routine production recording you probably do not need a group interval of less than 50 m. Nor do you need more channels than are required at a group interval of 50 m to get an offset that is equal to the maximum effective offset or equal to the maximum operational length from a practical standpoint.

Two choices that we need to make involve the seismic source: how much power, and which gun array? On the question of power, we shall probably use all that is practical on a continuous basis. We shall be limited by the compressor capacity and by the recharge time between shots, but if we are not overdriving the dynamic range of the system, what harm does it do to use maximum power? If we encounter a problem with background noise, the extra source power will help, but you need to have some excess power capacity in reserve and you certainly need spare guns in reserve.

Water guns are an effective source for marine seismology, but tuned air-gun arrays are the dominant choice. After each air-gun impulse the air bubbles from the guns alternately expand and contract producing a source waveform of long duration. In tuned air-gun arrays the bubble noises from several guns largely cancel one another yielding a composite source wavelet of short duration and high power level. The author likes, the approach of having short, identical sets of guns where each set is individually tuned, and these sets of guns can be put together to form various composite arrays. If we have problems with one set of guns we can substitute a spare set relatively quickly and proceed with the survey with minimal delay.

The research-oriented people nearly always want a *point source* array and many processors agree. But if there is a problem from long-wavelength noises or multiples it may be necessary to use a long gun array. If we do use a long gun array, we shall certainly limit the high-frequency response for many of the offsets in the CMP gathers. This loss of higher frequencies can be reduced by beam steering (Section 8.10). Earth attenuation may limit the higher frequencies anyway. It would make it easier to choose a point source gun array in tough areas if we could shoot some data with a point source and quickly see a satisfactory result after the data were processed and stacked, but unless we have onboard processing it will be weeks before we shall know if we acquired satisfactory data. By then we may well have recorded several hundreds of miles of data and be at another location. If data have previously been acquired in the area with a point source array, we should check to see if these data were satisfactory when processed. If so, look at the field data in shot gathers and be aware of how bad data can look and still turn out all right. In many cases we probably should use a point source array and depend upon competent processors with good processing tools to extract quality data from noisy data

(Chapter 19). But this can be a hard, risky decision to make when all we see before processing is strong noise and strong multiples.

Wide arrays have been used in efforts to acquire better data where the water bottom is incised by canyons or other irregularities. In the author's limited experience wide arrays were not effective for suppressing noise caused by water bottom irregularities. To be effective the lateral separation must be sufficient to achieve a significant difference in the path lengths for the noise.

The practical separation of guns that can be achieved for wide arrays depends upon the size of the paravanes that pull the guns apart and the drag of the gun arrays. The bigger the gun array (on each line) the more the drag and the less the separation that is feasible. Paravanes add to the drag on the boat which increases noise from the boat engines, boat propellers and the paravanes themselves. It is feasible to achieve a maximum, total, lateral separation of

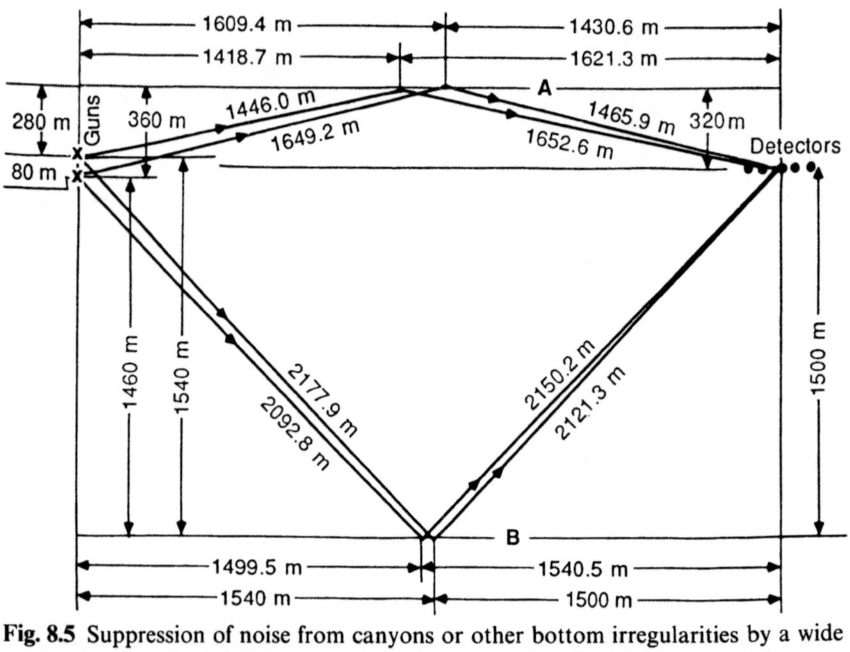

Fig. 8.5 Suppression of noise from canyons or other bottom irregularities by a wide gun array (plan view). This figure illustrates the use of a wide gun array with just two gun groups to suppress noise reflected or scattered from laterally displaced, submarine canyons or bottom irregularities. Noise suppression by a wide gun array depends upon having enough difference in the path lengths from the two sources. The difference in path lengths is greater for reflectors or scatterers with lateral offsets that are at least comparable to the longitudinal distance to the source. The noise reflected from canyon A has a path length difference of only 16.47 m which gives a time difference of only 11 ms. The noise reflected from canyon B has a path length difference of 56.19 m which gives a time difference of 37.5 ms. and much better suppression of low-frequency components of the noise.

about 80 m. For energy that travels only horizontally and only perpendicular to the seismic line this translates to a time difference of 80/1500 = 0.053 seconds. That is a substantial time difference, but most of the noise will not travel substantially horizontal and substantially perpendicular to the cable to arrive at most hydrophone positions.

Figure 8.5 illustrates noise that reflects from two channels designated as 'A' and 'B' that are parallel to the seismic line at lateral offset distances of 320 m and 1500 m respectively. In this example the longitudinal distance from the sources to the receiver position is 3040 m and the lateral separation between the sources is 80 m. If we neglect water depth, then for noise reflected from the channel at an offset of 320 m, the difference in path length from the two sources to the receiver is 16.47 m which would yield a time difference of 10.98 ms (using a water velocity of 1500 m/s). This time difference is half a period for a frequency of 45.5 Hz which is the lowest notch frequency for the suppression of the noise. A wide array with a maximum separation of 80 m would not be very effective in suppressing noise reflected from channel 'A' for the case shown.

For noise reflected from a channel at a lateral offset of 1500 m, the difference in path lengths from the two sources is 56.19 m resulting in a time difference of 37.46 ms. This time difference is half a period for a frequency of 13.35 Hz. This is much more promising for allowing us to suppress noise reflected from channel 'B' with a wide array with a maximum lateral separation of 80 m between the source elements.

We can better relate to our previous development of arrays as wavelength filters by considering the apparent velocity and apparent wavelengths in a lateral direction at the sources. The mean, azimuthal angle of emission from the sources for a noise reflection to the receiver from channel 'B' is the arctangent of (1500/1519.73) which is 44.626 degrees. The apparent lateral velocity is the water velocity divided by the sine of this angle which is 1500/0.70247 = 2135.3 m/s. The apparent lateral wavelength is then $2135.3/f$ meters. Solving for frequency we get:

$$f = 2135.3/\lambda$$

With this relationship we can design lateral source arrays in the usual manner and convert the wavelengths to frequencies of the noise reflected from channel 'B' (Tables 8.1 and 8.2).

These tables illustrate that we can use wide arrays of a few gun groups to suppress noises that arise from an irregular bottom if the azimuthal angle of emission from the source that produces the noise is large. If a maximum separation of 120 m were feasible then the wavelengths shown would all be increased 50 per cent and the frequencies would be 2/3 the values shown.

It is true that with incised canyons and other water bottom irregularities there may be times when energy from a source would have difficulty in

Table 8.1 Wavelengths of notches and peaks in the suppression of noise reflected from channel 'B' for a wide gun array

			Note: $(n-1)d = 80$ m	
Number of gun groups	*d (m)*	*nd (m)*	*Notch wavelengths (m)*	*Peak wavelengths (m)*
2	80	160	160, 53.3, 32,...	80*, 40*, 26.7*,...
3	40	120	120, 60, 30,...	80, 40*, 26.7,...
4	26.7	107	106.7, 53.3, 35.6,...	71.1, 42.7, 26.7*,...
5	20	100	100, 50, 33.3,...	66.7, 40, 28.6,...
6	16	96	96, 48, 32,...	64, 38.4, 27.4,...

*Peaks with no suppression of the noise

Table 8.2 Frequencies of notches and peaks in the suppression of noise reflected from channel 'B' for a wide gun array

			Note: $(n-1)d = 80$ m	
Number of gun groups	*d (m)*	*nd (m)*	*Notch frequencies (Hz)*	*Peak frequencies (Hz)*
2	80	160	13.3, 40.1, 66.7,...	26.7*, 53.4*, 80*,...
3	40	120	17.8, 35.6, 71.2,...	26.7, 53.4*, 80,...
4	26.7	107	20, 40, 60,...	30, 50, 80*,...
5	20	100	21.4, 42.7, 64.1,...	32, 53.4, 74.7,...
6	16	96	22.2, 44.5, 66.7,...	33.4, 55.6, 77.9,...

*Peaks with no suppression of the noise

penetrating the water bottom at angles that would result in reflections from deep interfaces back to the cable. With laterally displaced gun groups it is more likely that energy could penetrate at a suitable angle for reflection back to the cable from at least one source. However, in high CMP fold recording we will have many different gun locations within each CMP gather and the advantage under this consideration for a wide array does not appear to be very significant. But some improvement should result.

It can be advantageous in three-dimensional surveys to use two laterally displaced sets of guns which are fired alternately to produce two closely spaced lines.

8.8 TESTS AND PARAMETER SELECTION

Particularly in new survey areas and in tough areas, some tests should be conducted at the start of the survey. It will probably not be feasible to make changes in the cable during the tests (except to change the depth and this may not be necessary). The main parameter we can change is the source. We should probably try a point source gun array and probably shall want to try a moderately long gun array and possibly a long gun array. We may want to try source beam steering which is discussed in Section 8.9. If the water bottom is very irregular, we may want to try a wide array, but may find that it is not very advantageous.

Before choosing the gun arrays to be used in the tests, we consider the noise suppression of the hydrophone arrays in the cable. For example, we can assume that we have a group interval of 50 m and the arrays consist of 24 hydrophones evenly spaced with an *nd* value of 50 m. The noise suppression by the hydrophone arrays would have notches and peaks as follows:

Notches	50 m, 25 m, 16.7 m, 12.5 m, 10 m, etc.
Peaks	33.3 m, 20 m, 14.3 m, 11.1 m, etc.

We should favor gun arrays that would provide noise suppression notches at wavelengths of previously identified noise and at wavelengths where the hydrophone arrays are less effective. The first test should employ a point-source gun array and we should mark coherent noises on the records and note particularly the wavelengths of noises near objective times. If we evaluate which noises are likely to present the most problem at and near objective times, we can then design gun arrays to attack these noises.

We shall want to ascertain the maximum effective offsets for our objectives. For the tests we should use a cable that is longer than the expected maximum effective offset if a sufficiently long cable is available and if it is feasible to use the cable.

If we do not have data processing capability on board ship, then we may want to try cutting back on the bandwidth with filters particularly if the records are dominated by either low-frequency or high-frequency noise. Even if filters improve the data quality, we should probably use a wide bandwidth for the production data since the data can be filtered as needed in processing.

If we do have data processing capability on board, then there is much that can be learned and applied immediately to improve data acquisition for the current project. If not, then we shall have to work with the field monitors and do the best we can. Overlays with curved lines for various moveout velocities can help in the search for primary reflections in records that may be dominated by multiples and noise.

We should make records with various fixed gains with no shots but with the compressors operating and the boat running at the normal speed of about five knots (9.26 km/h), at a somewhat lower speed, and at a higher speed. This will help us to decide on a suitable boat speed for recording. If background noise is a serious problem, we can try recording with a slower boat speed.

8.9 BEAM STEERING

If we choose a long gun array, then we would prefer to shoot dip lines downdip so that the long gun array is updip from our short receiver arrays (Chapter 11). Unfortunately, this is not operationally attractive since we would spend too much time traveling back updip without recording data. One solution would be to beam steer by using an appropriate difference in the firing time between guns groups (Chapter 11). Equipment to provide suitable time differences is available, but the author does not know of anyone who has used the technique except for limited tests. The gun group farthest from the boat would be delayed the most, with progressively less delay for gun groups progressively closer to the boat. Ideally there would be no time delay for the center gun group and a progressive advance in firing time for gun groups closer to the boat. The firing time differences when shooting updip should be greater than the firing time differences shooting downdip. We probably may want just enough firing time difference to beam steer toward the middle of the cable for reflections from the main objective horizon, with due consideration of any dip of the objective.

As was suggested earlier in the ghost discussion, it is also possible to beam steer the source during acquisition by setting gun groups at different depths. The source beam can be steered at angles of emission that will favor reflections from low-dip interfaces by putting the gun group nearest the boat deeper than the gun group farther from the boat, with intermediate gun groups at

Fig. 8.6 Beam steering by use of multiple gun groups at different depths. By running gun groups at different depths it is possible to steer the source beam in a direction to favor reflections from an objective. The preferred beam angle will depend upon the depth and dip of the objective. For low-dip objectives the deepest gun group should be nearest the boat as shown and the gun groups should be progressively shallower with greater distance from the boat. Another advantage is that the notches in reflection spectra due to source ghosts will be suppressed by having the gun groups at different depths.

intermediate depths (Fig. 8.6). If we were shooting alternate lines updip and then downdip we may choose to adjust the beam angle between lines to maintain an optimum beam angle for an objective. Adjustment would involve changing the relative depths of gun groups or changing the horizontal spacing between gun groups. Changing the horizontal spacing would be quicker and easier, but this would change the source pattern and the suppression of noise. Again, the desired beam angle would be such as to favor reflection from an objective interface to the middle of the cable.

The depth range required to achieve a desired beam angle may be more than is practical to use. The water may be too shallow. Also the guns are attached to floats and the drag on the attaching cables becomes excessive if the guns are too deep. In some cases it may be desirable to use some time difference in the firing times in addition to some difference in gun depth to achieve a desired beam angle.

8.10 AN EXAMPLE DESIGN FOR BEAM STEERING A LONG GUN ARRAY

An example design for beam steering is based upon an actual problem that was encountered, but some changes have been made in the equipment specifications to better illustrate some points. We did not use beam steering in the actual case—perhaps we should have. Our primary objective was at a record time of 2.1 s and the moveout velocity was about 4300 m/s. A very troublesome multiple at this record time had a velocity of about 2200 m/s.

Consider a gun array with an effective length of 300 m, consisting of 5 gun groups spaced 60 m apart. Assume that the cable has a group interval of 50 m with 72 traces and a near offset of 150 m. The far offset is then 3700 m. Suppose that we wanted to record frequencies from the objective of at least 55 Hz. Table 8.3 was calculated using program A′ of Appendix F. This table shows the differential time of arrival for the reflection from a gun array with an effective length of 300 m. It shows also the frequencies where the response drops to 64, 30 and 0 per cent due to this differential time. Instead of calculating for all 72 different offsets we use only 24. The offset of 200 m represents offsets of 150, 200 and 250 m and the offset of 350 m represents offsets of 300, 350 and 400 m, etc. The sampling is sufficiently close to provide good answers. The average differential time for all of the offsets is 14.287 ms. Suppose that we made a 14 ms correction by changing the firing times between gun groups, by changing the depths of the gun groups, or some combination of both. Table 8.3 also shows the new differential times and the new cut-off frequencies where the relative response drops to 64, 30 and 0 per cent. It is evident from this table that the 14 ms shift would increase the high-frequency response for the middle of the

Table 8.3 Differential time across a gun array and the high-frequency cut-off for a primary reflection

Gun array	*Five groups spaced* 60 *m apart* (*nd* = 300 *m*)
Reflection time	2.1 *s*
Primary velocity	4300 *m/s*

	Without beam steering				*With beam steering*			
Offset (*m*)	Δt (*ms*)	$f_{0.64}$ (*Hz*)	$f_{0.3}$ (*Hz*)	f_0 (*Hz*)	$\Delta t-14$ (*ms*)	$f_{0.64}$ (*Hz*)	$f_{0.3}$ (*Hz*)	f_0 (*Hz*)
200	1.55	324	486	647	−12.46	40	60	80
350	2.70	185	278	370	−11.30	44	66	89
500	3.86	130	195	259	−10.14	49	74	99
650	5.01	100	150	200	−8.99	56	83	111
800	6.16	81	122	162	−7.84	64	96	128
950	7.30	69	103	137	−6.70	75	112	149
1100	8.44	59	89	119	−5.57	90	135	180
1250	9.57	52	78	105	−4.44	113	169	226
1400	10.69	47	70	94	−3.31	160	226	302
1550	11.80	42	64	85	−2.20	227	341	455
1700	12.91	39	58	78	−1.09	457	686	914
1850	14.00	36	54	71	0.00	∞	∞	∞
2000	15.09	33	50	66	1.09	461	691	922
2150	16.16	31	46	62	2.16	232	348	463
2300	17.22	29	44	58	3.22	155	233	311
2450	18.27	27	41	55	4.27	117	176	234
2600	19.30	26	39	52	5.30	94	142	189
2750	20.32	25	37	49	6.32	79	119	158
2900	21.33	23	35	47	7.33	68	102	136
3050	22.32	22	34	45	8.32	60	90	120
3200	23.30	22	32	43	9.30	54	81	108
3350	24.26	21	31	41	10.26	49	73	98
3500	25.21	20	30	40	11.21	45	67	89
3650	26.14	19	29	38	12.14	41	62	82
Total	342.88				6.88			
Average	14.29				0.29			

cable and for the cable as a whole. But this does not give us a quantitative answer as regards the response at 55 Hz. We can calculate the 55 Hz response using the $(\sin\theta)/\theta$ approximation where:

$$\theta = 55(\Delta t)\pi \text{ radians}$$

Table 8.4 shows the results both with and without the 14 ms time shift. The

Table 8.4 Relative response of the 55 Hz component of the primary reflection using a long gun array

	Gurn array		*Five groups spaced 60 m apart (nd = 300 m)*	
	Reflection time		2.1 *s*	
	Primary velocity		4300 *m/s*	
	Without beam steering		*With beam steering*	
Offset (m)	Δt *(ms)*	$(\sin\phi)/\phi$	$\Delta t-14$ *(ms)*	$(\sin\phi)/\phi$
200	1.55	0.988	−12.46	0.388
350	2.70	0.964	−11.30	0.475
500	3.86	0.928	−10.14	0.561
650	5.01	0.880	−8.99	0.644
800	6.16	0.822	−7.84	0.721
950	7.30	0.755	−6.70	0.791
1100	8.44	0.682	−5.57	0.853
1250	9.57	0.603	−4.44	0.905
1400	10.69	0.521	−3.31	0.946
1550	11.80	0.438	−2.20	0.976
1700	12.91	0.354	−1.09	0.994
1850	14.00	0.273	0.00	1.000
2000	15.09	0.196	1.09	0.994
2150	16.16	0.123	2.16	0.977
2300	17.22	0.056	3.22	0.949
2450	18.27	−0.005	4.27	0.912
2600	19.30	−0.058	5.30	0.866
2750	20.32	−0.103	6.32	0.813
2900	21.33	−0.140	7.33	0.753
3050	22.32	−0.170	8.32	0.695
3200	23.30	−0.192	9.30	0.622
3350	24.26	−0.207	10.26	0.553
3500	25.26	−0.215	11.21	0.482
3650	26.14	−0.217	12.14	0.412
Total		7.276		18.282
Average		0.303		0.762

average relative response is only 0.303 without beam steering but is 0.762 with beam steering. This is quite an improvement. Figure 8.7 shows the relative response for the 55 Hz component of the primary reflection as a function of offset both with and without beam steering. If we really need frequencies as high as 55 Hz, then we had better beam steer if we use the long gun array.

Should we nearly always beam steer when we are using a single-ended cable on land or at sea? As usual, there is some good news, but there is some bad

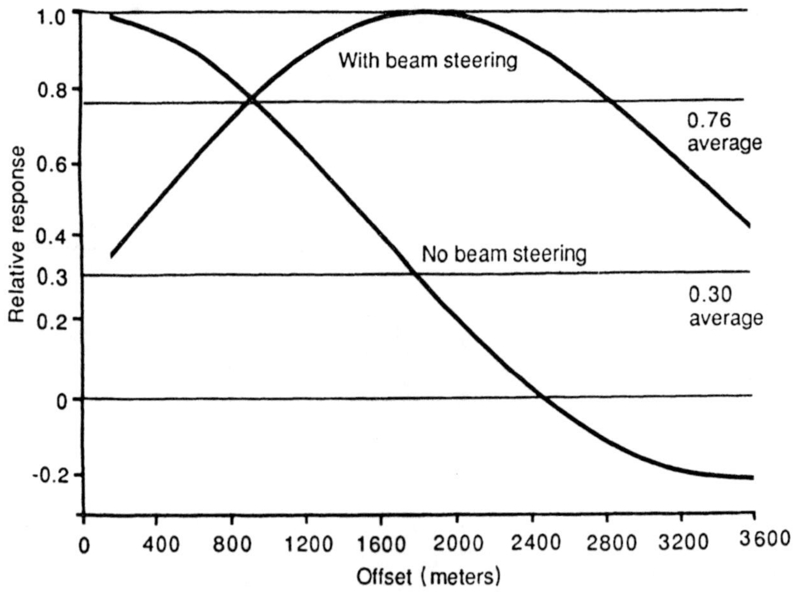

Fig. 8.7 Example of the relative response of the 55 Hz component of a reflection using a 300 m gun array both with and without beam steering. For this example the objective reflection time is 2.1 s and the moveout velocity for the objective is 4300 m/s. Without beam steering the average relative response for the 55 Hz component is only 0.303. With beam steering the average relative response for this component is increased to 0.762. Note that beam steering shifts the maximum response from the near end of the cable to near the center of the cable.

news too. The bad news is that we shall not suppress the multiples as well if we beam steer. Figure 8.8 shows the relative response of the 20 Hz component of the primary reflection both with and without beam steering. Again, beam steering helps, but the change is less dramatic at this lower frequency. Figure 8.8 shows also the relative response of the 20 Hz component of the multiple. The 20 Hz component of the multiple is suppressed twice as well without beam steering than it is with beam steering. In this case the author would tend to beam steer to preserve the higher frequencies of the primary reflection from the objective and to expect the processing to handle the multiple problem. But if multiples were an overwhelming problem and if low resolution data would suffice, he would either not beam steer or would use less time shift for a milder beam steer and better suppression of multiples.

How can we accomplish the time shift to achieve the desired beam steering? First of all, we do not want a full 14 ms of shift. The 14 ms was for the total length of an array of many, many equally spaced sources. We are approximating this response with five elements spaced 60 m apart for an overall length of 240 m for an array with an effective length of 300 m. The total time shift we

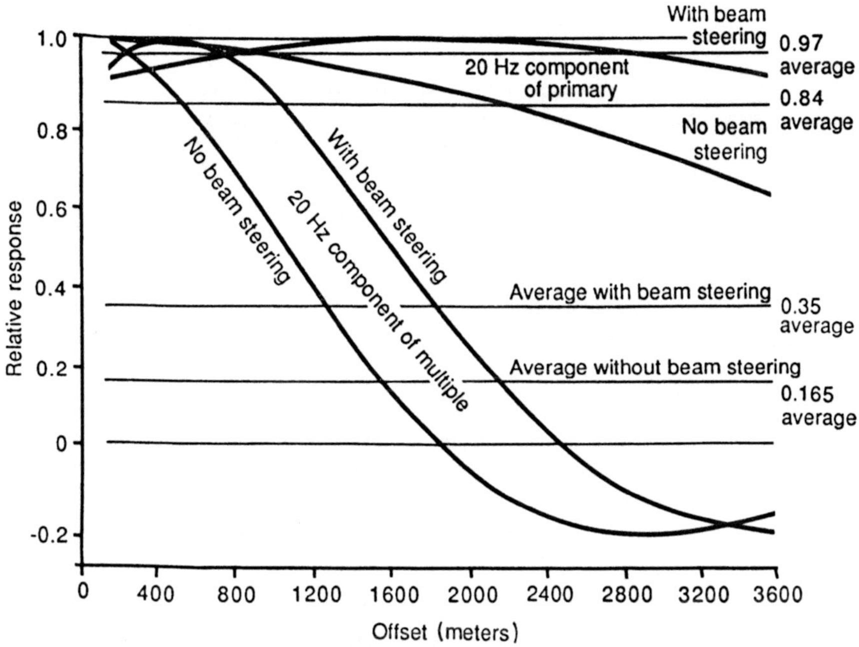

Fig. 8.8 Example of the relative response of the 20 Hz component of both the primary and a multiple using a 300 m gun array with and without beam steering. Again the objective reflection time is 2.1 s and moveout velocity is 4300 m/s for the primary and 2200 m/s for a troublesome multiple. Without beam steering the average relative response for the 20 Hz component of the primary is 0.844 which with beam steering is increased to 0.965. The average relative response for the 20 Hz component for the multiple is 0.165 without beam steering and increases to 0.348 with beam steering.

need is (4/5)(14) ms = 11.2 ms. Suppose that we provide a 2 ms time shift between adjacent gun groups. This would give us 8 ms of time shift leaving 3.2 ms to achieve with variation in the depth of the gun groups. The total differential depth needed is then 0.0032 s times 1500 m/s = 4.8 m (15.75 ft). If we put the gun farthest from the boat at a depth of 6.5 m, then succeeding gun groups would be at depths of 7.7, 8.9, 10.1 and 11.3 m respectively. (The gun depths in feet would be 21.3, 25.25, 29.2, 33.1, and 37.1 ft). The first notches for the source ghosts for individual gun groups would then be at frequencies of 115, 97, 84, 74, and 66.4 Hz. This variation in the notch frequencies would substantially diminish the spectral notches and peaks due to source ghosts in the composite response. This design is illustrated in Fig. 8.9.

An alternative design would use only 1 ms time shifts between adjacent gun groups leaving 7.2 ms of time shift to be achieved by variation in the depths of the guns. Now we need a differential depth of (0.0072 s) times (1500 m/s) = 10.8 m (35.4 ft). Again we might use a depth of 6.5 m for the shallowest gun

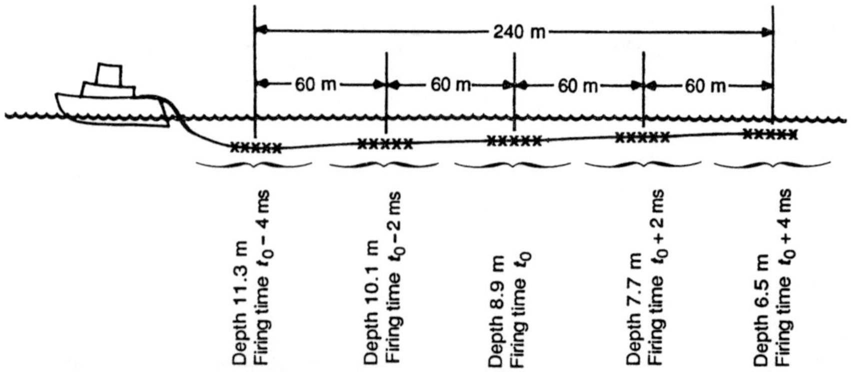

Fig. 8.9 An implementation of beam steering for the example case using variable depths and variable firing times for the gun groups. This figure illustrates one way to achieve the beam steering illustrated in Figs 8.7 and 8.8 using advances and delays in the firing times between gun groups and by using variations in the depths of the gun groups. These variations in the depths of the gun groups will also suppress notches and peaks in the spectra due to source ghosts.

group (farthest from the boat) and the deepest gun group nearest the boat would be at a depth of 17.3 m. The depth in feet would range from 21.3 to 56.8. It is probably not feasible to operate guns shallower than 6 m, and it may not be practical to operate guns as deep as 17 m, therefore this alternative design may not be attractive. For this alternative design the ghost notches for individual gun groups would be at frequencies of 115, 82, 63, 51, and 43 Hz. The spectra of the composite wavelets from the source should have only minor notches and peaks due to source ghosts (Chapter 9 gives a comprehensive treatment of ghost suppression).

8.11 MARINE OPERATIONS

We started this chapter discussing water depth for the guns and the hydrophones. We need to select depths which are operationally practical. We need to know if we are likely to encounter shallow water even if it is only on a turn to get into position for a line or to go to the next line. We must remember that with a long cable in tow we have to make wide, sweeping turns, and we need to know if there are obstacles which stick up from the bottom which may snag the cable or the guns if either is too deep.

For operations in very shallow water we may choose to use a weighted cable lying on the sea floor. In shallow water and in marshes we may choose to use geophones and to push them through soft sediment down into firmer sediment, along with 'spudded guns' that are imbedded in the soft sediment.

Individual guns should fire within one millisecond of time zero (unless you are purposely using delays and advances for beam steering. This requires a time break sensor for each gun and equipment that monitors the firing time of each gun and either advances or delays the firing signal as is required to keep the firing within specifications. Obviously in a low-frequency, low-resolution survey we could tolerate more drift in gun firing. But the contractor should be able to meet this specification.

We shall need to specify how many guns can be dead and still keep recording. If the normal limit of dead guns is exceeded slightly on a short line, we would probably allow the crew to finish the line before correcting the problem. With a tuned air gun array or with several, separately tuned air gun units the loss of even one gun is upsetting to the precisely tuned output. But we need to be practical in setting our allowable limits on dead guns.

For marine seismology, signature deconvolution is practical and is quite advantageous. This does require measurement of the wavelet injected into the water by the gun array. Ideally this measurement is made with a calibrated, high-fidelity hydrophone at a depth of several hundred feet (more than 100 m) where the water depth is more than 1000 ft (more than 300 m). (Fricke 1985; Reed 1987.)

We shall need to specify the sea conditions under which data can be recorded and to specify the boat speed for recording (usually 4.5–5.5 knots (8.3–10.2 km/h)).

Navigation must be planned and plotted carefully before shooting begins. We need to specify the tolerance for drift of the boat and the tail buoy relative to planned line locations. Far at sea there are no markers to return to. There is no point in finding a potential reservoir with seismic data if we cannot return to the line locations. Navigation data must be precise and complete and must be plotted and compared with the pre-plots. The post plots must include the location of the cable relative to the boat. Navigation is a highly specialized field which is beyond the scope of this book. Normally the field geophysicist discusses navigation problems with specialists and relies on these specialists to provide acceptable navigation and location control. The navigation precision required depends upon the type of survey. For regional lines in an early exploration phase, more error can be tolerated than in a tight, three-dimensional survey designed to define precise structure and faulting in a complex field. Particularly in three-dimensional surveys we need precise information as to the relative positions of the detectors and sources. Particularly when cross currents are strong, we shall need excellent means for providing this information.

Before drilling wells at sea we often need to make detailed seismic surveys of the sea floor and the shallow earth section. In such surveys it is particularly important to detect the presence of free gas (which could pose a hazard to drilling). For this purpose we want high-resolution data (Chapter 6) that is processed so as to preserve relative amplitude variations. The concepts involved are discussed at the beginning of Chapter 15.

REFERENCES

Larner, K., Chambers, R., Yang, M., Lynn, W. and Wai, W. (1983) 'Coherent noise in marine seismic data' *Geophysics* **V** 48, 854–84.

Fricke, J. R., Davis, J. M. and Reed, D. H. (1985). 'A standard quantitative calibration procedure for marine seismic sources'. *Geophysics* **V** 50, 1525–32.

Reed, D. H., Selsam, R. L. and Knox, A. E. (1987). 'SEG standards for marine seismic hydrophones and streamer cables'. *Geophysics*, **52**, 242–8.

9

Shot crews

In this chapter we discuss shot hole crews including ways to suppress ghosts (from reflections from above the shot). We also discuss the use of shallow minihole patterns of shots.

9.1 USE OF EXPLOSIVES

The use of explosives in any operation requires extra precautions for the safety of the crew and of the public. Care must also be exercised to prevent the theft of explosives (which might then be used in criminal activities). Not only must we be careful that we operate safely, we must also operate in compliance with all laws regarding the transportation, storage and use of our explosives.

We use electrically detonated caps to initiate explosions. It is imperative that we use high-grade caps that have a short and consistent delay time between excitation and detonation. Ideally the delay time would be zero. We had a crew that had been working for us quite successfully for several months when it was discovered that some caps they had recently received had variable detonation delay times ranging from about 3–5 ms. This, of course, was intolerable and the crew had to repeat about one day's work where these caps had been used. We need to use a reliable brand and type of cap and we need to keep an adequate supply on hand.

9.2 LESS THAN FULL-FOLD RECORDING

Particularly with deep-hole shot crew operations it is often too expensive to use a shot point interval as short as the receiver group interval. So we settle for a lower CMP fold and use a shot point interval that is equal to a multiple of the group interval. Can we do this and still achieve the desired, uniform sampling of the earth in each CMP gather? One might hope that if we used double group interval arrays that we would achieve this uniform sampling if the shot point interval were twice the group interval (yielding half the maximum CMP fold).

It does not work out as well as we might hope, but there are some positive features. First, we will consider an odd trace gap, achieved by locating shot points at alternate trace centers rather than halfway between alternate trace centers.

The first time the author did an analysis with an odd trace gap he thought that things would work out perfectly when the shot point interval was twice the group interval. The offsets from the front and back cables did interlace for the first gather that he happened to check, and a double group interval array would provide the desired uniform sampling in each gather. But then he

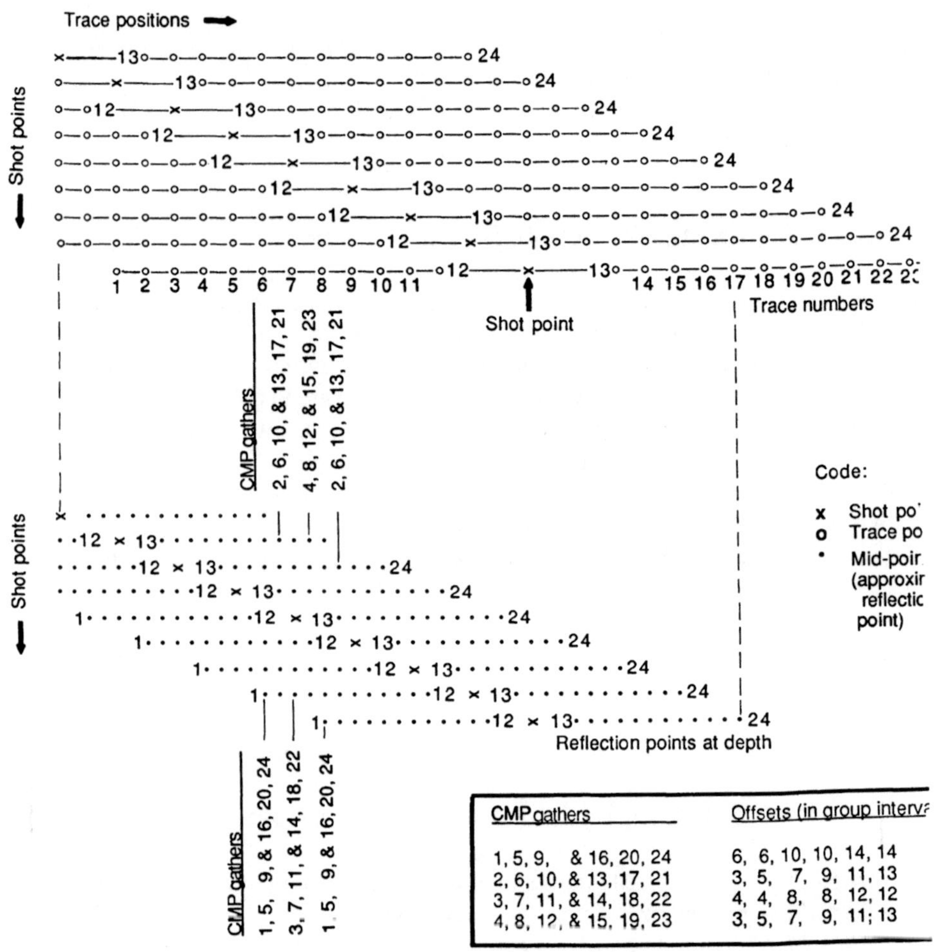

CMP gathers	Offsets (in group interv
1, 5, 9, & 16, 20, 24	6, 6, 10, 10, 14, 14
2, 6, 10, & 13, 17, 21	3, 5, 7, 9, 11, 13
3, 7, 11, & 14, 18, 22	4, 4, 8, 8, 12, 12
4, 8, 12, & 15, 19, 23	3, 5, 7, 9, 11; 13

Fig. 9.1 CMP gathers with an odd trace gap where the shot point interval is twice the group interval. If we record only half maximum CMP fold (source interval equal to twice the receiver group interval), we would have to use quadruple group interval arrays with an odd trace gap to achieve a stack-array.

Table 9.1 Traces and offsets in gathers where the shot point interval is twice the group interval—odd trace gap

Gather	*Traces*	*Offsets in group intervals (rearranged)*
First	1, 5, 9 & 16, 20, 24	6, 6, 10, 10, 14, 14
Second	2, 6, 10 & 13, 17, 21	3, 5, 7, 9, 11, 13
Third	3, 7, 11 & 14, 18, 22	4, 4, 8, 8, 12, 12
Fourth	4, 8, 12 & 15, 19, 23	3, 5, 7, 9, 11, 13
Fifth	1, 5, 9 & 16, 20, 24	6, 6, 10, 10, 14, 14
	(Fifth same as first; pattern repeats)	

checked a second gather and saw that offsets between the front and back cables duplicated rather than interlaced and this left gaps in the offsets of the gather. For the gathers where the offsets duplicate rather than interlace, the double group interval array would not provide the desired uniform sampling of the source-generated noise. Quadruple group interval arrays would work out nicely, but in most cases we would not want to use such long arrays.

Figure 9.1 will help to explain how this works out, where we consider a 24-trace system with a symmetrical split spread with a five-trace gap. The various gathers consist of four sets of six traces that then repeat down the line. The six traces in these four gathers and the offsets distances (in group intervals) are given in Table 9.1

In Table 9.1 the offsets included in the even numbered gathers interlace but the offsets in the odd numbered gathers duplicate. Suppose we recorded half-fold data with double group interval arrays. If we displayed a section using only the even gathers, noise suppression should be excellent. But a display of only the odd gathers might be contaminated with source-generated noise.

If the stacks of the odd gathers are noisy, might it be possible to suppress this noise in processing? For example, suppose we used noisy traces 3 and 5 to interpolate a noisy trace 4. Then suppose we subtracted the measured, low-noise trace 4 from the interpolated, noisy trace 4. By this technique we could define the coherent, source-generated noise field for all of the even traces. Could we then estimate the noise field for the odd traces and subtract the noise from each measured odd trace? It may or may not be like lifting ourselves with our bootstraps depending upon how we went about the process. We may or may not achieve a better result than simply interpolating odd traces from the adjacent even traces. At worst we could settle for a display of the even-numbered gathers. In such a display we would have the maximum CMP fold available from the data. But we would have lost the horizontal resolution and redundancy provided by the other half of the traces.

Table 9.2 Traces and offsets in gathers where the shot point interval is twice the group interval—even trace gap

Gather	*Traces*	*Offsets in group intervals (rearranged)*
First	3, 7, 11	2.5, 5.5, 6.5
Second	4, 8, 12	3.5, 4.5, 7.5
Third	1, 5, 9	3.5, 4.5, 7.5
Fourth	2, 6, 10	2.5, 5.5, 6.5
Fifth	3, 7, 11	2.5, 5.5, 6.5
	(Fifth Same as first; pattern repeats)	

Now to consider recording with an even trace gap with the shot point interval equal to twice the group interval. We consider Fig. 2.1 and assume that we delete alternate shot points B, D, F, H, J, etc. and shoot at A, C, E, G, I, etc. Now the shot point interval is equal to twice the group interval. The first stack gather will still have traces 3, 7 and 11. Table 9.2 lists the traces for the first several gathers. Notice the set of pairs of gathers: (first and second), (third and fourth), (fifth and sixth), etc.

Within each pair, as given above, the offsets do interlace between the front and the back cables. In these pairs of gathers we have all the offsets we would have if we had recorded full fold with a shot point interval equal to the group interval. Therefore for the gathers as paired, the sampling is uniform if the array length is an integral multiple of the group interval.

It is important to note, however, that if we were to drop the first gather and start gathering in pairs of second and third, fourth and fifth, etc., the offsets in the pairs of gathers would duplicate rather than interlace and the sampling would be non-uniform. Therefore if we use pairs of adjacent gathers in any of our data-processing procedures, we should be very careful as to how we do the pairing. One procedure would be to sum traces to begin with. The required sums involve traces in single shot gathers. For each shot gather we could simply sum traces 1 & 2, 3 & 4, 5 & 6, etc. This would increase the effective length of the geophone arrays which may or may not result in significant loss of higher-frequency components of reflections from dipping interfaces and of reflections on long offsets. At the other extreme we could process all the way to the final stack and then sum adjacent CMP gathers, but we would have to be careful about where we started the summing of pairs of stacked traces.

Alternatively, we could sum pairs of gathers, run velocity analyses and determine static corrections, repeat the velocity analyses and then apply a normal moveout correction to the individual traces. We may then sum them and proceed. This would take care of the previous concern about losing higher-frequency components of reflections. In addition, we could, at some point, solve for static corrections with unsummed data in order to make static corrections between traces before they are summed. There is a multitude of processing options; the best procedure is dependent upon the problems with the data and the requirements imposed by the exploration objectives.

So if we have data acquired in the manner described (with an even trace gap), we may be able to improve the final stack by taking the preceding into consideration. We could start by summing adjacent traces of the final stack. If in doubt about the right pairing, try both combinations (i.e. 1 and 2, 3 and 4, etc. as well as 2 and 3, 4 and 5, etc.) for a moderate sample section. This also lets us judge the importance of uniform noise sampling in our case.

The only valid justification for using a shot point interval greater than the group interval is to reduce costs by reducing the number of shot points per mile. If we double the shot point interval, you can maintain CMP fold by doubling the number of traces per shot. This is economic because adding traces is usually less expensive than the cost of the deleted shots. But the resulting seismic data are not likely to be as good (unless we can use a quadruple group interval array and an odd trace gap).

9.3 DEEP-HOLE SHOOTING

The advantages of deep-hole shooting include less ground roll and larger explosive charges without causing surface damage. We also can record the 'up-hole time' to aid in static corrections. Often we can drill deep enough to get into a more competent and more homogeneous material which will yield a better and more consistent seismic pulse with less static variations between shot points. One disadvantage is cost; another is possible problems with ghosts resulting from waves from the source traveling upward and being reflected back down at the base of the weathering, at the air–earth interface, or at some other interface above the shot. The ghost delay will usually vary from shot point to shot point due to variations in the near surface and due to some variation in the depth of the shot. In high CMP fold recording when the gathers from many shots are stacked, the ghost effects may tend to be suppressed. But where the near surface is rather uniform and the shot depth is kept nearly constant, the reflections from the ghosts may stack right along with the wanted reflections from the initially downward traveling wavefronts.

Under these circumstances the problem becomes similar to the ghost problem for marine work except that the ghost is only from the source and not also at the receivers. On land the shot depth is frequently much deeper than the gun and cable depths normally used in marine work. Therefore land ghosts are frequently delayed more in time. Like the marine case, land ghosts create notches in the spectra of the reflections from depth. There will be a notch at zero frequency and a first notch at a frequency where the distance from the shot depth up to the ghost interface is one-half of a wavelength (and the travel time is one-half of a period). For this frequency the total added ghost path is a full wavelength (and the total added time is a full period). Since the reflection coefficient at the ghost interface will be negative, the ghost and the primary will be out of phase and will tend to cancel. The author has seen cases in land shooting where cancellation was nearly complete, indicating a ghost reflection coefficient of very close to minus one for the frequencies of the first few notches. It would be desirable to at least partially fill in the deep notches so that the frequency bands in notches are not lost from our reflection data. This is particularly important for stratigraphic plays.

One solution is to vary the shot depth purposely in a systematic way, such that each CMP gather will have traces which were recorded from shots at a variety of depths. (We could shoot charges at a number of depths at each shot point, but this would be expensive.) We can think of the various shot depths as forming a source pattern and we want to design this source pattern to suppress a noise which is the ghost. The spacing for this design is twice the depth increment since the ghost travel path is two way.

Figure 9.2 shows a plot of 'uphole times' from shots at depths of 75–300 ft

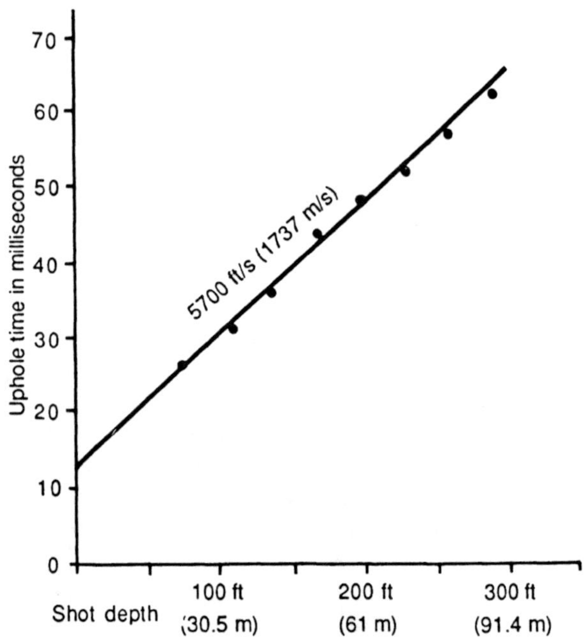

Fig. 9.2 Uphole survey in a bad ghost area. The earth in this area has a nearly uniform velocity of about 5700 ft/s (1737 m/s) over the depth range of 75–300 ft (23–91 m).

23–91 m) for a real earth case where ghosts were a severe problem. The data reasonably fit an earth model with a velocity of 5700 ft/s (1737 m/s) over the depth range of the data with one or more overlying layers with lower velocities (required for the uphole time at zero depth to be zero). In this immediate area we recorded reflection data with prominent and rather consistent notches in the spectra of reflections from many depths.

There was a prominent crest at a record time of 1.1 s which was followed by a trough. The time difference between the two events increased with an increase in the depth of the shot. The trough was obviously a ghost of the prominent crest. Figure 9.3 shows a plot of the time delay of the trough relative to the

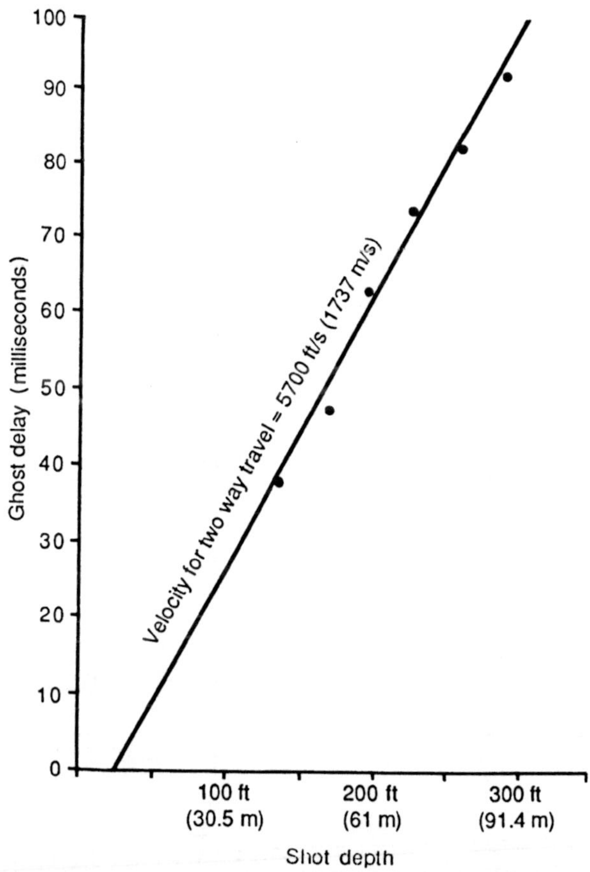

Fig. 9.3 Ghost delay as a function of shot depth. The ghost delay was measured from a primary peak near 1.1 s to a succeeding ghost trough. For shots at depths of less than 135 ft the ghost trough was not well resolved from the preceding primary peak. A velocity of 5700 ft/s (1737 m/s) fits the data quite well for a differential ghost travel distance equal to twice the differential change in shot depth.

peak as a function of the shot depth. Again the data fit reasonably well an earth model where the velocity in the depth range of the shots is 5700 ft/s (1737 m/s). The data imply that the ghost reflection interface is at a depth of about 25 ft (7.6 m). Figure 9.4 is a drawing of this simplified model of the real earth case. Based upon this earth model we can calculate the frequencies where we should expect notches in the spectra of our reflection events using the same formulae (Appendix B) as were used for the marine ghosts, but with a velocity of 5700 ft/s (1737 m/s) rather than 5000 ft/s (1524 m/s). These modified formulae are:

First notch $\quad {}_1f_{\text{notch}} = 1/(2h/5700) = 2850/h \text{ Hz}$

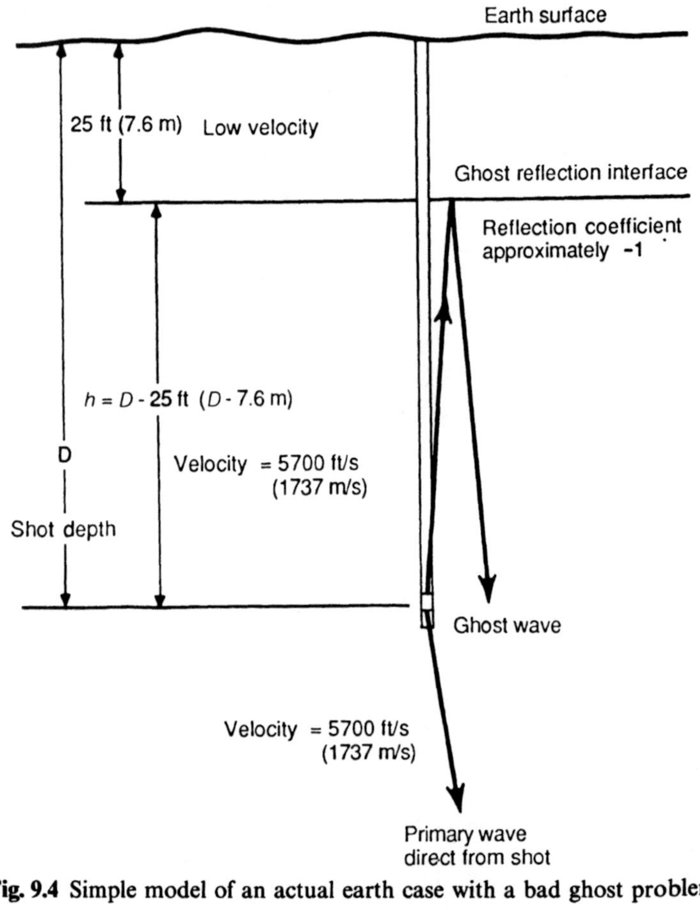

Fig. 9.4 Simple model of an actual earth case with a bad ghost problem.

Table 9.3 Expected notches and peaks in the spectra

Depth (*ft.*)	*Notches* $_1f_{notch}$ (*Hz*)	$_2f_{notch}$ (*Hz*)	*Peaks* $_1f_{peak}$ (*Hz*)	$_2f_{peak}$ (*Hz*)
75	57	114	28.5	85.5
100	38	76	19	57
125	28.5	57	14.3	42.8
150	22.5	45	11.3	33.8
175	19	38	9.5	28.5
200	16.3	32.6	8.2	24.5

Where $h =$ distance in feet from shot depth up to the ghost reflection interface.

First peak $\quad {}_1f_{peak} = 0.5({}_1f_{notch})$

Second notch $\quad {}_2f_{notch} = 2({}_1f_{notch})$

Second peak $\quad {}_2f_{peak} = 3({}_1f_{peak})$

If we assume that the ghost reflection interface is at a depth of 25 ft (7.6 m), then the distance h in the formulas is the depth less 25 ft. The frequencies for which we could expect notches and peaks in the spectra for shots at various depths are given in Table 9.3.

The spectra showed general agreement with Table 9.3, particularly for notches and peaks at the lower frequencies.

In this seismic program we did have stratigraphic objectives and we needed frequencies of 80 Hz or more. The ghost wavelengths (velocity divided by frequency) ranged from 380 ft (116 m) at 15 Hz to 71 ft (21.6 m) at 80 Hz. We needed to design a source pattern for our CMP gathers that would suppress the ghost over this range of wavelengths. If we shot at six depths with a depth increment of 25 ft (7.6 m) that ranged from 75–200 ft (23–61 m), we would have an *nd* product of $6(2 \times 25) = 300$ ft (91 m). (Remember the ghost path increment is twice the depth increment.) Table 9.4 gives the notches and peaks for the ghost wavelengths and the corresponding ghost frequencies. This table also gives the amplitude of the suppressed ghost as a function of ghost wavelength as well as the composite response of the primary plus the ghost. (We assumed the worst case reflection coefficient of -1 for the ghost.)

Figure 9.5 shows the ghost rejection response of this source pattern along with the composite response of the primary plus the suppressed ghost. The relative, composite response is one plus the relative ghost response, where the ghost response is negative at very long wavelengths (at low frequencies). Ghost frequencies corresponding to various ghost wavelengths are also given in

Table 9.4 Ghost suppression by shots at multiple depths (worst case)

$n = 6$ shot depths; d = two-way depth increment = (2)(25 ft) = 50 ft (15.2 m); nd = 300 ft (91.4 m)

Notches wavelengths and frequencies	*Peaks wavelengths and frequencies*	*Ghost amplitude*	*Primary + ghost amplitude*
300 ft (91.4) 19 Hz		0	1.00 (0 dB)
	200 ft (61 m) 28.5 Hz	−0.25 (−12 dB)	0.75 (−2.5 dB)
150 ft (45.7 m) 38 Hz		0	1.00 (0 dB)
	120 ft (36.6 m) 47.3 Hz	+0.20 (−14 dB)	1.20 (+1.58 dB)
100 ft (30.5 m) 57 Hz		0	1.00 (0 dB)
	85.7 ft (26.1 m) 66.5 Hz	−0.20 (−14 dB)	0.80 (−1.94 dB)
75 ft (22.9 m) 76 Hz		0	1.00 (0 dB)
	66.7 ft (20.3 m) 85.5 Hz	+0.25 (−12 dB)	1.25 (+1.94 dB)
60 ft (18.3 m) 95 Hz		0	1.00 (0 dB)
	50 ft (15.2 m) 114 Hz	−1.0	0.00

Fig. 9.5. The composite response is plotted on an expanded scale to show better how well the notches are filled in by the source pattern. Notches and peaks still exist but they no longer approach the former limits of nearly zero and nearly 2 respectively. This figure does not show wavelengths shorter than 100 ft (30.5 m), but the shortest notch prior to the first zero peak is at 60 ft (18 m) which corresponds to a ghost frequency of 95 Hz. Ghost suppression is good for frequencies from 15 Hz to more than 100 Hz. Excellent data were

Fig. 9.5 Ghost suppression by use of multiple shot depths. The use of multiple shot depths to suppress a ghost is similar to the use of a source pattern to suppress source-generated noise. To save time and money the various shot depths can be used at different shot points so long as the composite response of each CMP gather combines the responses to shots at the different depths. In this example the net response of the primary plus suppressed ghost varies between 75 and 125 per cent for frequency components between 15 and 100 Hz.

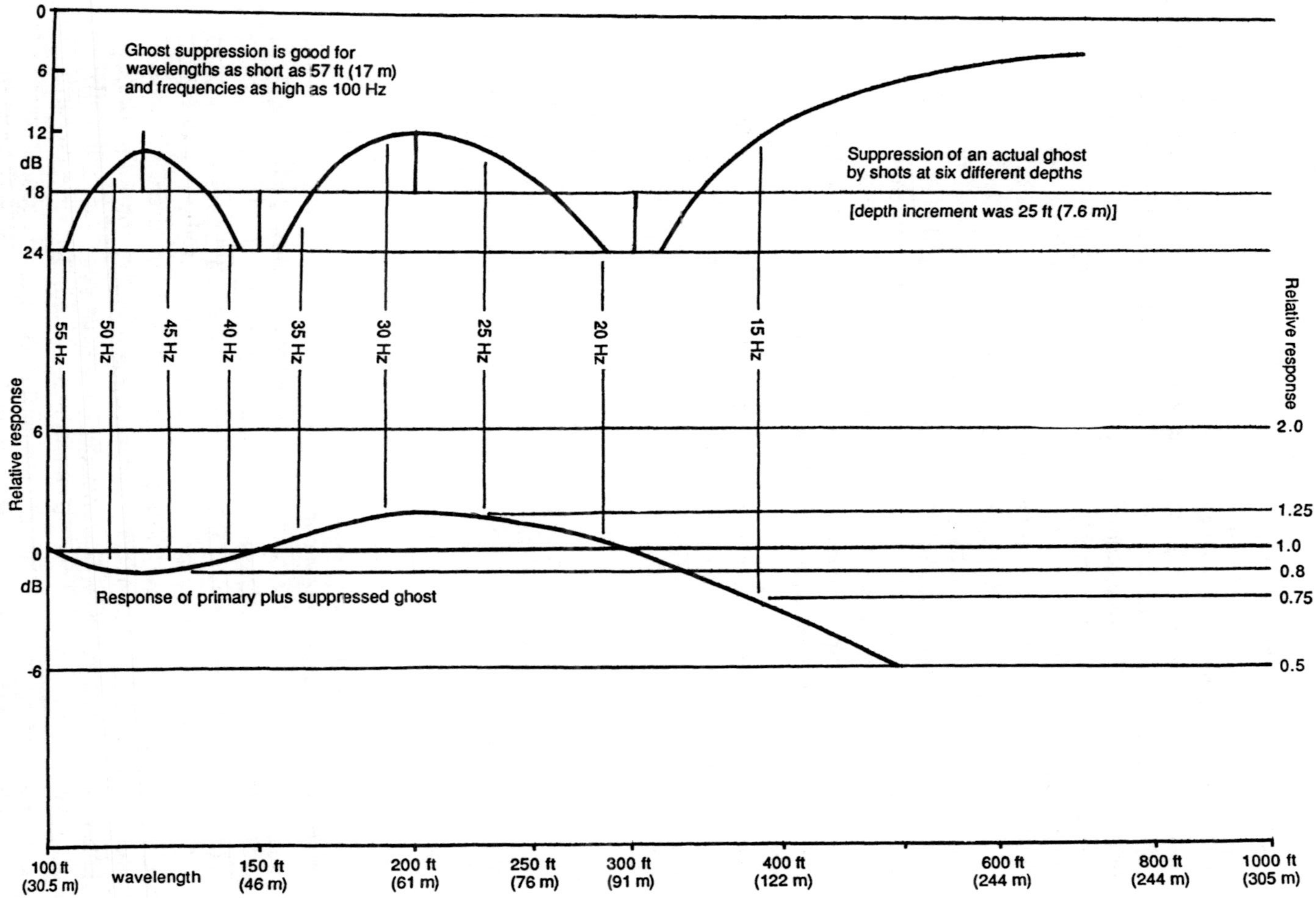

Ghost suppression is good for wavelengths as short as 57 ft (17 m) and frequencies as high as 100 Hz
Suppression of an actual ghost by shots at six different depths
[depth increment was 25 ft (7.6 m)]
Response of primary plus suppressed ghost
0
6
12
dB
18
24
Relative response
6
0
dB
-6
Relative response
2.0
1.25
1.0
0.8
0.75
0.5
55 Hz
50 Hz
45 Hz
40 Hz
35 Hz
30 Hz
25 Hz
20 Hz
15 Hz
wavelength
100 ft (30.5 m)
150 ft (46 m)
200 ft (61 m)
250 ft (76 m)
300 ft (91 m)
400 ft (122 m)
600 ft (244 m)
800 ft (244 m)
1000 ft (305 m)

recorded with this technique with a CMP fold of 24. In each CMP gather we had traces from four shots at each of the six depths.

The preceding is not meant to suggest that such an elaborate means must be employed whenever we encounter a ghost problem. Depending upon circumstance and need, we may get by with as few as two, well-chosen shot depths. Also we may be able to use multiple charges in the same hole, fired either separately (by reloading?) or nearly simultaneously by using primacord and fuse to achieve a precisely calculated delay time. We have to fire each deeper charge just as the wavefront from the shallower charges reaches the depth of each deeper charge. Not only is it difficult to achieve a desired delay time and desired charge separation, but the system will degenerate if the velocity in the shot range changes down the line. By contrast, the timing is readily adjusted in processing for the previously discussed technique of using different shot depths for shots in each CMP gather. In that technique the timing is accomplished using the measured uphole times and is then improved by the statistical, static correction program. (Data quality has to be good enough to achieve a good static correction solution.)

Having killed and buried the ghosts, we need to discuss other noise problems with deep-hole shooting. We sometimes use patterns of deep holes to suppress source-generated noise despite the high cost of drilling. The same design techniques are employed as are used for surface sources or shallow-hole patterns.

The author recalls one very tough area where his crew did a noise test and then designed an in-line pattern of three deep holes and a complementary geophone array to do an excellent job of suppressing the measured noises. It did a beautiful job at that location, but just a few shot points down the line something changed in the earth and it was no longer effective. In this case they had strong noises with very narrow frequency bands and hence very narrow wavelength bands. They managed to provide notches at these precise wavelengths and achieved a high rejection of the noises. But a slight change in frequency, a broadening of the frequency bandwidth of the noises, or a change in velocity moved strong noise components to other wavelengths where the suppression was only moderate. We need to design our source patterns and geophone arrays with rejection bands that are broad enough to allow moderate changes in the noises without serious loss of suppression.

One of the key advantages for deep-hole shooting is being able to measure uphole times and to use this information to control static corrections and prevent accumulation of error that can effect the apparent regional dip of key horizons. To be satisfactory, however, the holes must penetrate sufficiently deep to get into an earth section that is laterally consistent. Of course, the earth is never completely consistent along a seismic line, but often the shallow earth section is particularly variable. Compressional wave velocity is sensitive to changes in consolidation and in the fluid content of the pore space. These properties change markedly in the weathered zone. If the shot holes are not deep enough to fully penetrate the weathered zone (or the zone of rapid changes in acoustical properties), then a deeper hole may be required occasionally for an uphole survey to measure the velocity gradients in the shallow earth. Ideally these holes should reach an interface that can be identified. It is helpful if reflections from this interface can be extracted from the data acquired from shots in the shallower holes used for the bulk of the recordings.

9.4 PORTABLE CREWS

With portable crews we usually use explosives at or very near the air – earth interface. We use explosives as follows:

1. Small charges in patterns of shallow holes;
2. Small charges laid out on the surface in patterns;
3. Small charges on sticks slightly above the surface and in patterns;
4. Primacord plowed into the earth in long lines to form patterns;
5. Portable guns that shoot slugs into the earth.

With surface or near-surface sources we usually need a source pattern to help suppress ground roll. In some cases the air wave is tremendous and the only solution is to use offsets long enough so that the reflected data is recorded before the air wave arrives. Sources with strong air waves are objectionable where there are people nearby who would be disturbed. Explosions at the surface or in the air can be a fire hazard particularly if any metal foil is used for packaging. But multiple explosions of small charges in the air can be carried out safely and can result in almost no surface damage even to small trees very close to small charges—individual charges are usually 1 pound (0.45 kg). Usually surface or air charges are connected together with primacord which also explodes and serves to detonate the attached charges. The velocity of the explosion along primacord is about 23 000 ft/s (7000 m/s) and could be used to provide a desired delay time for some desired directional result. With an end-on cable we could use the finite detonation velocity to direct the energy (beam steer) toward the cable at approximately the angle of emission required for a reflection from the objective to be received near the center of the cable. More often, the delay is not wanted and care is taken to use enough caps located at enough positions over a large pattern such that the delay time is kept low enough to approximate a single explosion (i.e. the delay time is short compared with the period of the highest seismic frequency to be sought in the reflected wavelets).

Small charges of a fraction of a pound (less than 0.5 kg) in shallow holes

5–10 ft (1.5–3 m) deep can yield excellent results in many cases. If the holes can be tamped with dry cuttings and the charge size is small enough, the air wave can be modest and the disturbance of the public can be minimal.

Normally a pattern of four to twelve holes is used. Inline patterns are frequently used to help suppress source-generated noises that appear to move along the surface along the seismic line. In Chapter 12 we discuss the use of areal patterns of shallow holes to attack certain problems.

Guns that fire slugs into the ground may be an acceptable source for shallow earth reflections (up to a few thousand feet) for earths with a firm soil surface. But even a few inches of loose material on the surface will seriously impair results with these sources.

With portable crews working in rugged and relatively inaccessible areas, one should be careful not to burden the operation with arrays that require an excessive number of geophones to carry in, plant and carry out. Even if geophones are moved in by helicopter, they still have to be dispersed from the drop sites to each receiver position, planted, picked up, carried to bagging locations, bagged, and then attached to the line from the helicopter. But, it is even worse not to use enough receiver array and record unsatisfactory data. Short tests can frequently help us to develop a plan that will result in satisfactory data with moderate effort and acceptable costs.

Another use for portable operations is to fill in data for limited line sections where surface sources or deep hole drills cannot go. The author often had vibroseis crews carry and maintain some portable drilling and shooting equipment for this purpose. The data from the shallow hole shots were recorded with the same instruments, geophones and cables as were used to record data from the vibrator sources. Our data processors were quite successful in nearly every case in adjusting the phase and amplitude spectra of the shot hole data to match that of the Vibroseis data so that the two sets of data were compatible and could be summed in CMP gathers. As a result the line locations where the sources were switched were not at all obvious (these locations were listed conspicuously in the section labels). Chapter 17 provides more discussion of this technique (section 17.5).

10

Off-road coverage and three-dimensional surveys

In this chapter we consider off-road seismic coverage, including three-dimensional surveys. Techniques for simultaneously recording data from two or more source positions are discussed.

10.1 AREAL COVERAGE TECHNIQUES

Often we need to record data from depth points which underlie surface positions which are not readily accessible for drills, vibrators or other sources. In these cases we may or may not be able to use a portable crew such as one that uses portable drills and small charges in shallow holes. In other cases we need data over an area rather than along lines even if the lines are separated by as little as a few hundred feet. Figure 10.1 shows single-fold areal coverage that can be achieved with a fixed cable that is perpendicular to a line of source points. Here both cable and sources may be along section line roads at an intersection, or the cable and geophones might be laid through pastures, fields, woods, or swampy land and the sources might be along a road or trail. Figures 10.2 and 10.3 are similar except that there is more than one cable laid perpendicular to the source line, resulting in multiple-fold coverage over part of the area.

In some cases the data quality for single- or low-fold coverage may not be adequate. If the CMP points are closely spaced, we may be able to sum data from clusters of CMP points to enhance data quality by suppressing noise which is not consistent for each of the points of the summation. Figure 10.4 illustrates some possible summations. Obviously we may not be successful in summing data from points that are too widely separated. If we had reason to believe that there was dip in some direction for a given event, we could try various time shifts representing various dips before summing. The summation that enhanced the event the most would give us a measure of the dip. We could then check to see if the dip indicated was consistent with data from adjacent summations.

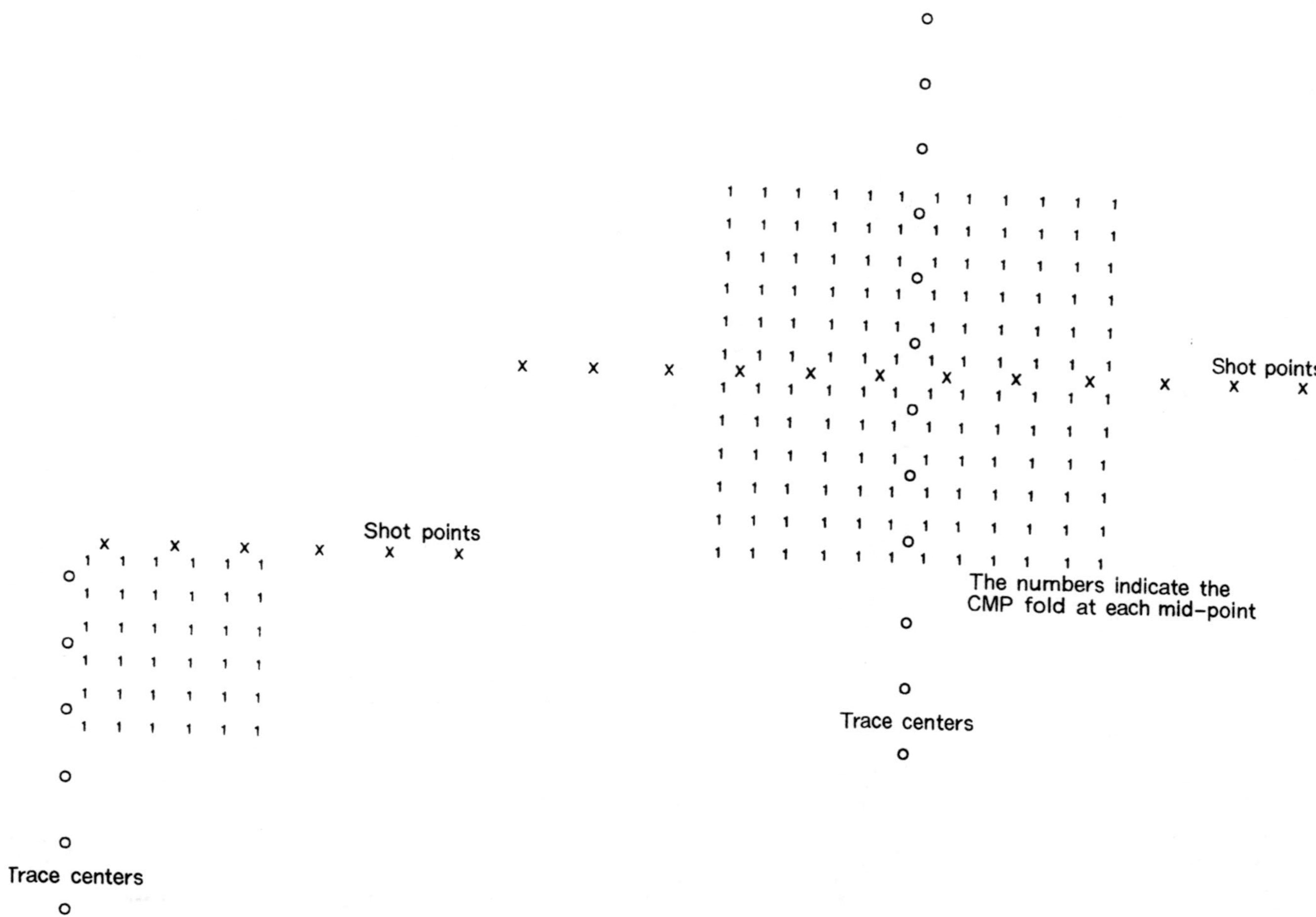

Fig. 10.1 Areal coverage with cable perpendicular to the line of shots. The reflection data are all single-fold.

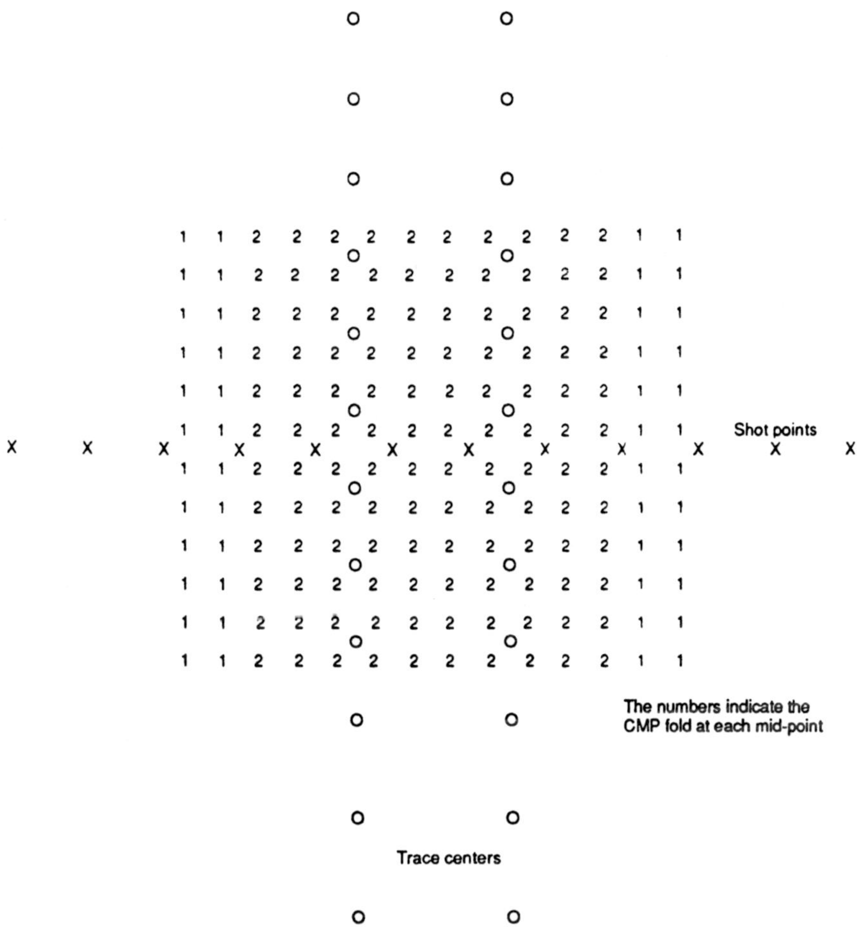

Fig. 10.2 Multi-fold and single-fold coverage with two cables perpendicular to the line of shot points. The CMP coverage is mostly two-fold that tapers to single-fold.

10.2 EXAMPLE OF AN AREAL SURVEY TO LOCATE A WELL SITE

We located some small anomalies on seismic lines that were recorded using vibrators along roads. We wanted to drill one of these anomalies, but we needed to define the anomalous area better in order to pick a well site. The anomaly was along an east–west road and was somewhat west of a fence line that ran north–south. We were granted permission to lay out cables and geophones along the north–south fence line and then vibrated along the east–

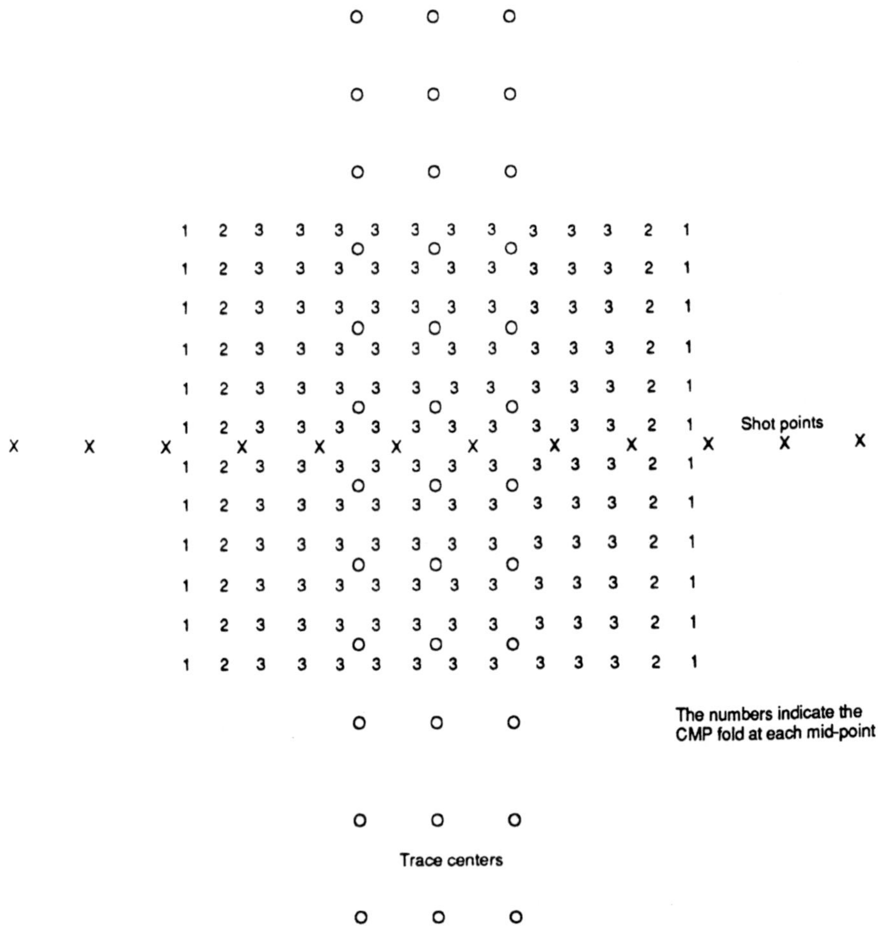

Fig. 10.3 Multi-fold and single-fold coverage with three cables perpendicular to the line of shot points. The CMP coverage is mostly three-fold that tapers to two-fold and single fold.

west road. The layout and CMP coverage was similar to that illustrated in the larger pattern in Fig. 10.1. There were 60 trace positions with a group interval of only 88 ft (27 m). The cable extended 2596 ft (791 m) to both sides of the road. We used 60 VPs and the VP interval along the road was also 88 ft (27 m). This gave us a 60 by 60 array of 3600 CMP points (for each reflection horizon) with adjacent points only 44 ft (13.4 m) apart. Think of this 60-by-60 array as 60 north–south lines with 60 traces per line and also as 60 east–west lines with 60 traces per line. The single-fold data were rather noisy. We summed adjacent east–west lines in groups of three and summed north–south lines in groups of three. This reduced the array of depth points from 60-by-60 to 20-by-20. The

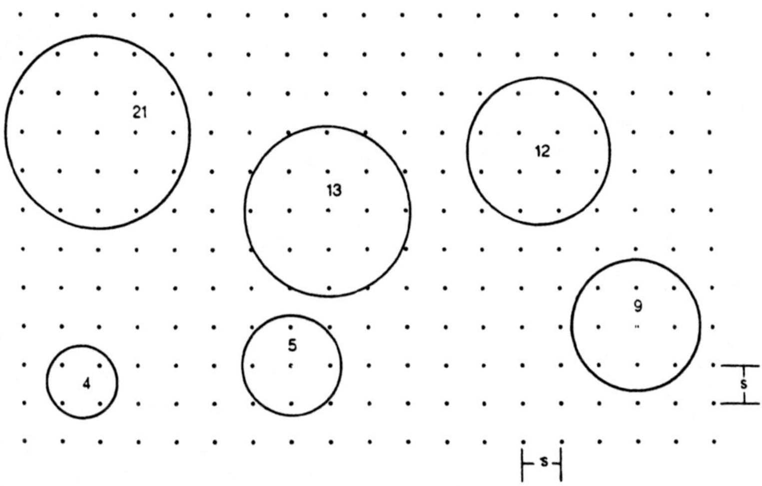

Number of samples	Maximum distance from sample center
4	0.707 s
5	1.000 s
9	1.414 s
12	1.581 s
13	2.000 s
21	2.236 s

Fig. 10.4 Possible summations of closely spaced depth points. If the areal data are noisy, then summation of data for closely spaced depth points may improve data quality.

summation that was used is the nine-trace summation illustrated in Fig. 10.4 and yielded quite satisfactory data for our objective. Figure 10.5 shows three of the 20 north–south lines after summation. A few data points near the center were still noisy because all the data were for very short offsets. In planning a program like this we should not have the main area of interest too near the intersection of the source line and the cable. The nine-trace summations resulted in composite depth points that were 132 ft (40 m) apart rather than 44 ft (13.4 m) apart. The data combined in any one nine-point summation were distributed as follows:

Distance from center (*ft*)	(*m*)	*Number of traces*	*Percentage of total*	*Cumulative percentage*
0	0	1	11.1	11.1
44	13.4	4	44.4	55.5
62	19	4	44.4	100.0
0–62	0–19	9	100.0	

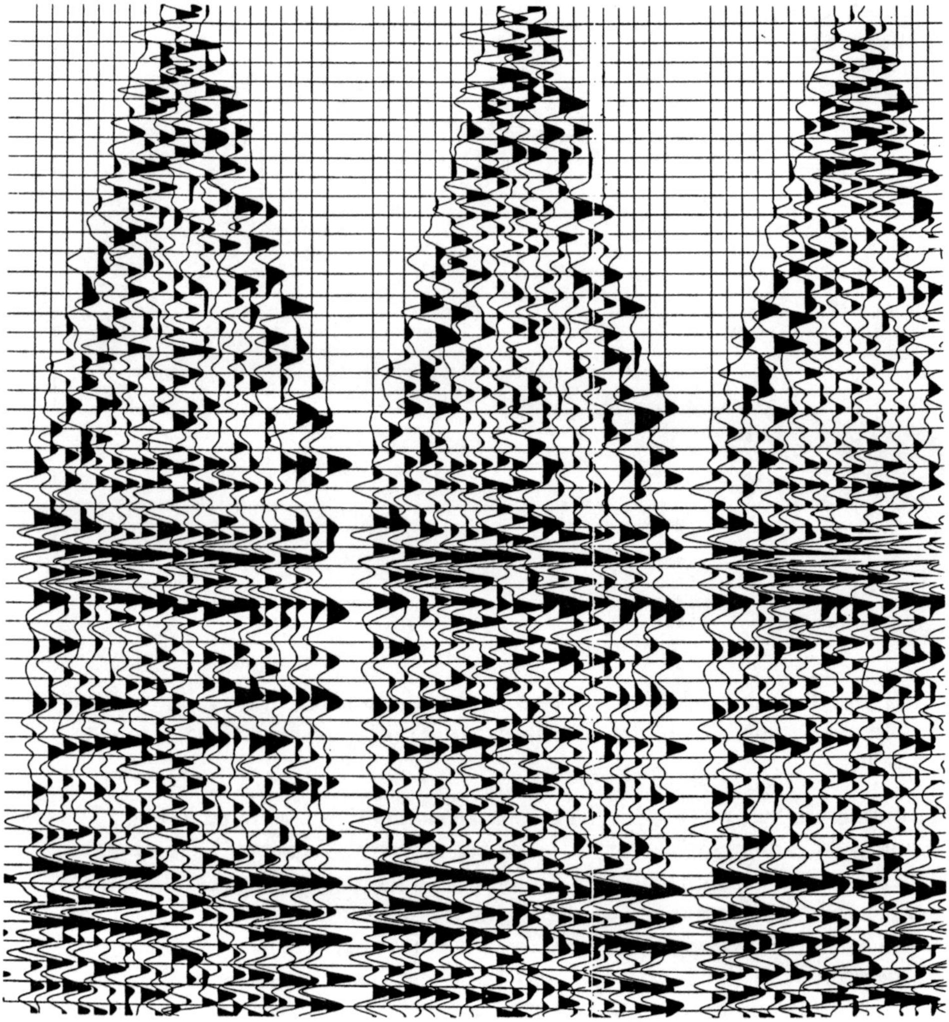

Fig. 10.5 Nine spot sums of single-fold data. The single-fold data were acquired with vibrators on a road and the cable laid out perpendicular to the road. The nine spot summations of traces for closely spaced depth points provided data of quite satisfactory quality to locate a well site. The separation between timing lines is 10 ms.

Even after the nine-trace summations, we still had a high density of data. Furthermore, the depth point area represented by each summation was small. If the data had not been satisfactory with nine-point summations, we could have summed a larger number of adjacent traces. For example, we could have used a 21-trace summation. For a 21-trace summation the data in any one composite trace would have been distributed as follows:

Distance from center (ft)	*(m)*	*Number of traces*	*Percentage of total*	*Cumulative percentage*
0	0	1	4.8	4.8
44	13.4	4	19.0	23.8
62	19	4	19.1	42.9
88	27	4	19.0	61.9
98	30	8	38.1	100.0
0–98	0–30	21	100.0	

The summation of 21 traces would have been acceptable had this been required to achieve a satisfactory result.

In recording the data the geophone arrays were aligned north–south (parallel to the cable) and the source patterns were aligned east–west (parallel to the source line). The geophone arrays were in-line with 24 geophones equally spaced over 100 ft (30.5 m). The source patterns were five vibrators bumper-to-bumper, 30 ft (9 m) pad-to-pad, with no move-up. The combination of north–south geophone arrays and east–west source patterns provided some suppression of surface noise from any azimuth. We used a logarithmic sweep of 20–120 Hz and we used equipment to cancel power-line pick-up without notch filters. We also conducted a full-fold 'walk-through' (Appendix E) with the vibrators run alongside the cable along the north–south fence line. All of the data were recorded in a single day. (The cables and geophones were laid out the previous afternoon.)

10.3 HIGH CMP FOLD COVERAGE ALONG LINES OFFSET FROM ROADS OR TRAILS

High CMP fold coverage can be recorded along lines offset from the line of the sources using essentially normal roll-along operations, but with the cables laterally offset by a significant distance. Figure 10.6 illustrates such an operation. CMP coverage is along a line half-way between the line of the sources and the cable. With a long lateral offset, there is no need for a center gap. (If we use a symmetrical split spread, then the source point centers may be offset half-way between the centers of the geophone arrays as shown, but this is not critical.)

There are some significant disadvantages to the use of long lateral offsets, which are:

1. The effective range of offsets is reduced and the difference in offsets is quite small for longitudinal offsets that are small compared to the lateral offset.

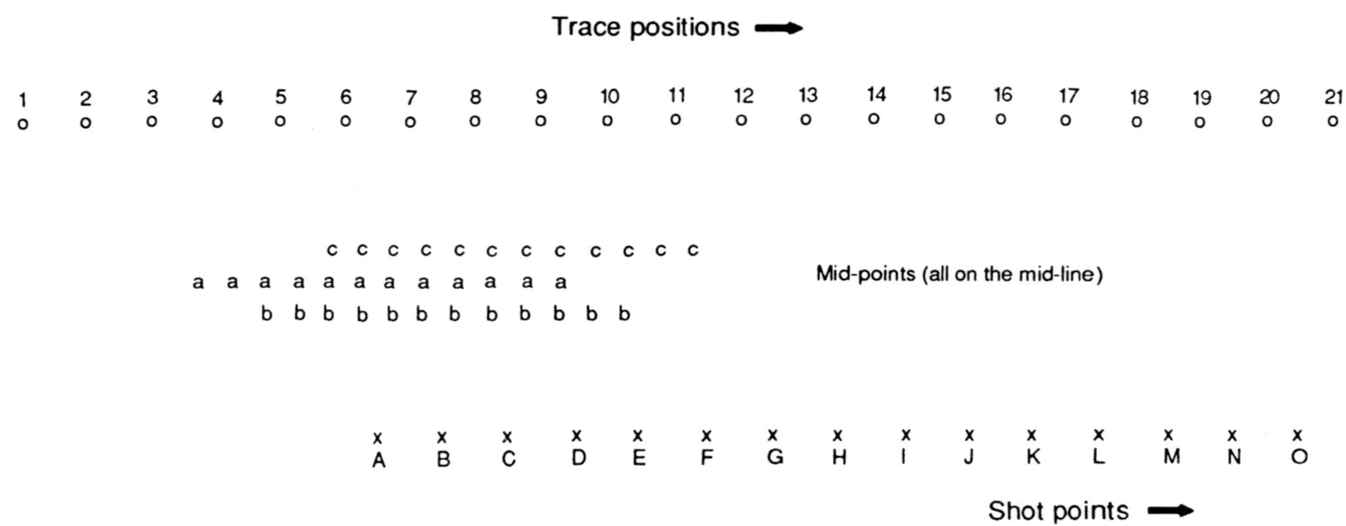

Fig. 10.6 Roll-along CMP recording with long lateral offset. In this sketch a 12-trace recording system illustrates CMP recording with a long lateral offset resulting in reflection coverage along a line half-way between the shot points and the locations of the seismic detectors (assuming that there is no cross dip). Shot point A is recorded by receivers at positions 1–12 yielding reflections from positions indicated by the letter 'a'. Similarly shot point B is recorded by receivers at positions 2–13, and shot point C is recorded by receivers at positions 3–14. Reflection points 'b' and 'c' all lie on the same line as points 'a' but are shown laterally displaced for clarity. With long lateral offsets no center gap is necessary.

These effects impair our capability for suppressing multiple reflections. Also our ability to measure the moveout velocity is reduced. Where multiples are not a problem and where we have good velocity measurements from other lines (without large lateral offsets), these disadvantages are insignificant.

2. The suppression of noise by geophone arrays and by source patterns is altered significantly. Indeed, we may even want to lay out linear geophone arrays at right angles to the cable and keep our source patterns parallel to the source line. Or we may want to use an areal geophone array such as a star array. Depending upon the severity of the noise problem this may or may not be a serious disadvantage.
3. The CMP stack itself is probably less effective in suppressing noise.

Despite the above disadvantages this technique can be used successfully in many areas and it is frequently economically advantageous over the use of recording methods and techniques that are totally off-road. The technique has the advantage that the air wave is less likely to be of concern since it may well arrive after we have recorded data from the objectives. Figure 10.7 illustrates this technique being used in sectionized country with the cable and geophones running down the centerline of sections and with sources along the roads to both sides of the centerline. Thus with one roll-along cable layout we can record two lines of data. Furthermore, if we used vibrators we could use polarity coding or some other type of coding to record data along both lines simultaneously with the same geophones (Fig. 10.8 and Chapter 14). In the simplest system sweeps are identical except for polarity, i.e. both upsweeps (or both downsweeps) are the same function (linear, logarithmic, etc.), are over the

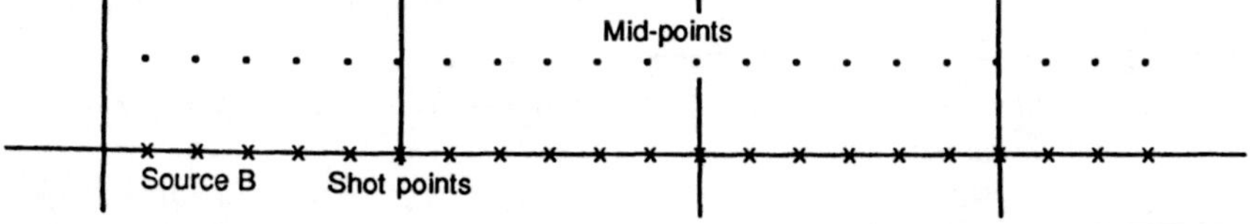

Fig. 10.7 Roll-along CMP recording with two sets of sources. Two sets of sources operate along parallel, section line roads with the cable and geophone line half-way between. With vibrators and appropriate coding, two sets of data for two seismic lines can be recorded simultaneously. For explosive sources, two shooting crews can be used to record data in quick succession for corresponding shot points on the two lines. Then the roll-along switch is advanced for the next shots.

same frequency range, and are of the same length. The only other requirement is that the offset distances for each receiver to each of the sources be the same (or nearly the same). The geometry illustrated in Fig. 10.7 meets this condition. Of course, we would need more vibrators than if we were recording only one line.

The technique illustrated in Fig. 10.7 could easily be carried out by a dynamite crew by having at least two shooting crews. In this case, data for the two lines would be recorded alternately in quick succession, but not simultaneously.

10.4 LARGE-SCALE, THREE-DIMENSIONAL SURVEYS

In this book a three-dimensional survey is one where data is recorded over some surface area in a manner such that a three-dimensional interpretation of the earth can be made (with time or depth being the third dimension). Large-scale, three-dimensional surveys are very expensive but may be undertaken for a variety of reasons where the stakes are sufficient to justify the costs. First, the complexity of the subsurface may require a high density of data over a large area in order to interpret properly the exploration objectives. Second, there may be large areas that are inaccessible for sources and some large areas that are also inaccessible for geophones. The problems may involve physical conditions such as water or may involve areas which cannot be permitted. Sometimes the only solution is to put detector arrays wherever they can be located over an area and then apply seismic energy at all locations where the sources can be used that are reasonably close. We frequently must then move the geophones to another available area, within the larger area to be surveyed, and again apply seismic energy wherever possible that is reasonably close.

Surveys of this latter type are 'one of a kind' and must be laid out with great care. A computer should be used to give a good picture of what the mid-point coverage would be and the multiplicity. Where coverage is weak within our area of interest, we should see what we can feasibly do to improve the fold. Also we need to know the range of offsets at each depth point and have a wide range of offsets to suppress multiples effectively. With only long offsets we cannot record shallow data.

In surveys of this type it is desirable to use receivers that do not have to be connected to a cable. There are several such systems; in all of them radio communication is required at least for a time break. Some systems record data on cassettes which are played back later after the cassettes are retrieved. Other systems send the data back to a central system by radio.

Large-scale, three-dimensional surveys of the type we first discussed (complex geology) can best be carried out either at sea or in relatively open country where sources and receivers can be positioned at almost any desired

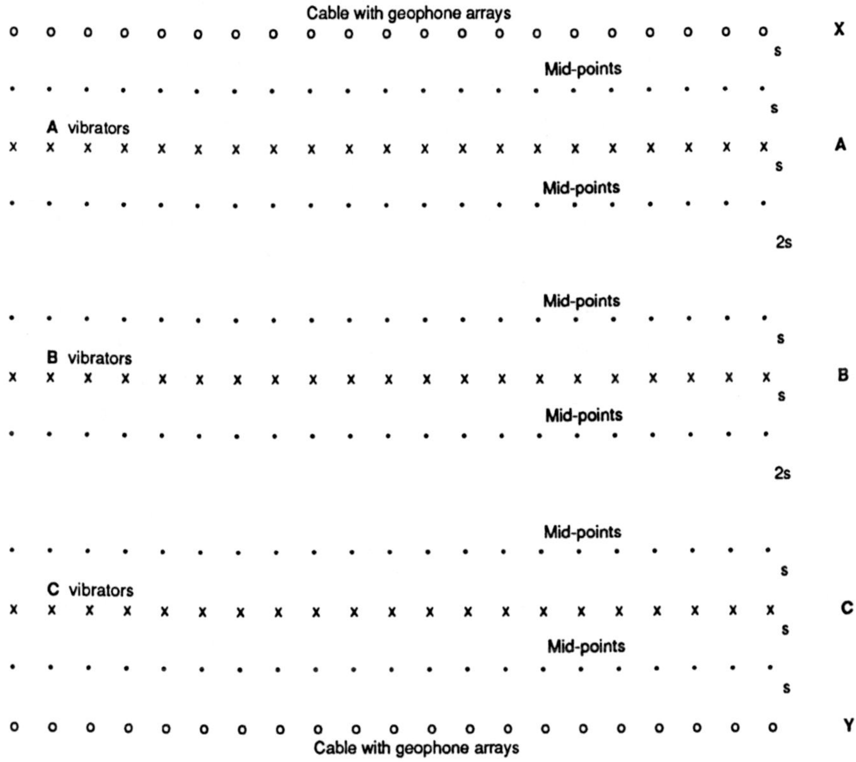

Sweep number	Polarity of Sweeps Vibrator Set A	Vibrator Set B	Vibrator Set C
1	+	+	+
2	+	-	+
3	+	+	+
4	+	-	+
5	+	+	-
6	+	-	-
7	+	+	-
8	+	-	-

Vibrators at: **A, B & C**
Cables at: **X & Y**

Fig. 10.8 Geometry and coding to record simultaneously six lines of data with three sets of vibrators and two cables. Vibrators operate simultaneously at corresponding points on lines A, B, and C with cables and receivers along lines at X and Y. This results in six lines of data as shown by the rows of dots. The coding shown will provide limited separation of the signals which may well be inadequate. The proposed system is presented for consideration and would probably need improvement (see text) to make it practical.

location. Generally it is more feasible to record data with a relatively short spacing between depth points along lines, as compared with the spacing between lines. It is therefore quite desirable to run such lines in a dip direction or a direction perpendicular to faults to be mapped, which will then give us a high density of data in the critical direction where changes occur more rapidly. Three-dimensional seismic projects require a lot of surveying of receiver and source locations. Where the initial plan for field layout must be changed, the changes in locations must be well documented and precisely surveyed.

In marine work it is frequently difficult to determine the precise, relative location of all of the detectors and sources and this often results in serious degradation of three-dimensional seismic sections. It is quite important to provide excellent means for determining the position of a feathered cable over its entire length. The simultaneous recording of multiple lines could be carried out in a manner similar to that shown in Fig. 10.7 and the results should be quite satisfactory. At sea or in open country, the separation between the lines could be whatever distance is wanted rather than what is dictated by existing roads. But it may be difficult to operate multiple boats in unison if there are strong currents. On land the technique shown in Fig. 10.7 should work quite well where the road spacing and the terrain between are satisfactory.

It is not difficult to devise a coding scheme to allow simultaneous recording from three sets of vibrators with subsequent separation of the three data sets recorded by each geophone array. Also a geometry can be worked out with three sets of sources and with two cables to record simultaneously six equally spaced lines, as illustrated in Fig. 10.8. But the problem is that the signal levels at the cables may be too different from the three different sources because of the differences in source to detector offsets and the signal separation may not be adequate to sort out the desired data sets. But the geometry for the deep reflections is not much different for the three sources. Therefore, one can argue that it is primarily noise and shallow reflection data that would be greatly different in magnitude from the three different sources. If the processing can extract one deep reflection signal from strong noise from a nearby source, it should be able to extract a nearly equal signal from the same reflector but from a different, more remote source. It may well be worth trying.

In the coding scheme illustrated in Fig. 10.8 we record eight sweeps for each VP and each sweep is recorded separately (i.e. no summing). Then we reverse the polarity of selected sweeps to make the polarities all positive for a particular signal we want to decode. For example, to extract signals from the B set of vibrators we must reverse the polarity of the data from sweeps 2, 4, 6, and 8. Then when we sum up the data from these eight sweeps we get the desired signal. Note that each of the other signals would still have four sweeps with positive polarity and four with negative polarity, so their signals would approximately cancel. Whereas this simple type of coding and decoding may or may not be satisfactory for the recording of six lines from three sets of vibrators and with two cables, it is satisfactory for recording two lines as shown in Fig. 10.7 and for other uses where the signal levels and source-generated noise levels are nearly equal in amplitude from all sources.

Chapter 14 provides a discussion of various simultaneous recording systems including some with more complex coding and decoding systems involving polarity reversal, phase shift, upsweeps and downsweeps, time separation of sweep frequencies, etc. These more complex systems can provide more effective separation of simultaneously recorded data.

If none of the coding and decoding systems discussed in Chapter 14 are adequate, this should be a challenge to the ingenuity of future geophysicists to find a satisfactory solution. A possible improvement with the simple polarity coding illustrated in Fig. 10.8 would be to sum the positive sweeps and separately sum the negative sweeps of a given code. Then, before combining the positive and negative sums, find a weighting such that for the longest offsets the energy prior to the first breaks (from nearer sources) is minimized. This should improve the separation of the sought record from the signals generated by the other sources. Another appraoch would be to make an initial separation of the several signals and then seek to minimize the cross-correlation of the signals by subtracting various small fractions of the various other signals. This approach may or may not work, or be practical, at the present. If it would work, but were not yet practical, it may become so with future computers and data processing techniques (including artificial intelligence?).

The planning and successful recording of large, three-dimensional surveys is a formidable task. But the real problems are in processing and interpreting the data. Tremendous efforts have been expended to develop programs and equipment to accomplish these tasks. Certainly we need to give the data processors and the interpreters all the help we can by recording data that is well suited to their needs. It is helpful if the data can be recorded with a consistent grid so that the need for interpolation is greatly reduced. Frequently it is simply not feasible to record data consistently with the desired grid, but we should do all we reasonably can to comply.

11

Steep dips and complex geology

Chapters 11–15 take up special exploration problems and practical solutions to these problems. This chapter deals with mapping steeply dipping beds and complex geology and the possible advantages of using beam steering and long offsets.

11.1 REFLECTION RAY PATHS AND ANGLES OF EMISSION AND ARRIVAL

In this chapter we consider problems in mapping steep dips and complex geology. We will deal mostly with apparent dip parallel to the seismic line as opposed to cross dip. Waves reflected from steeply dipping interfaces generally arrive at the surface at a significant angle to the vertical. At some updip offset the effect of normal moveout and the effect of the dip may nearly cancel and the rays may arrive nearly vertically. But for reflections from steeply dipping interfaces this fortuitous situation may occur well beyond the length of the cables normally used to record data. Even if we did record very long offsets, we might have difficulty in successfully combining data over a wide range of offsets.

The conventional approach, where dips are likely to be steep, is to acknowledge that the angle of arrival, from the vertical, will be significant and the apparent wavelengths of reflections will be relatively short (Chapter 4). Therefore, to prevent spatial aliasing, we would need very closely spaced sampling points, which means a short group interval and hence a short cable unless you record a huge number of channels. With the usual approach we may use a shorter cable even if we increased the number of channels moderately; therefore reflections even for the updip cable may arrive at sizeable angles to the vertical. Geophone arrays must be short or we would severely limit the recording of higher-frequency components of reflections.

Steeply dipping interfaces are not likely to persist for very far along our

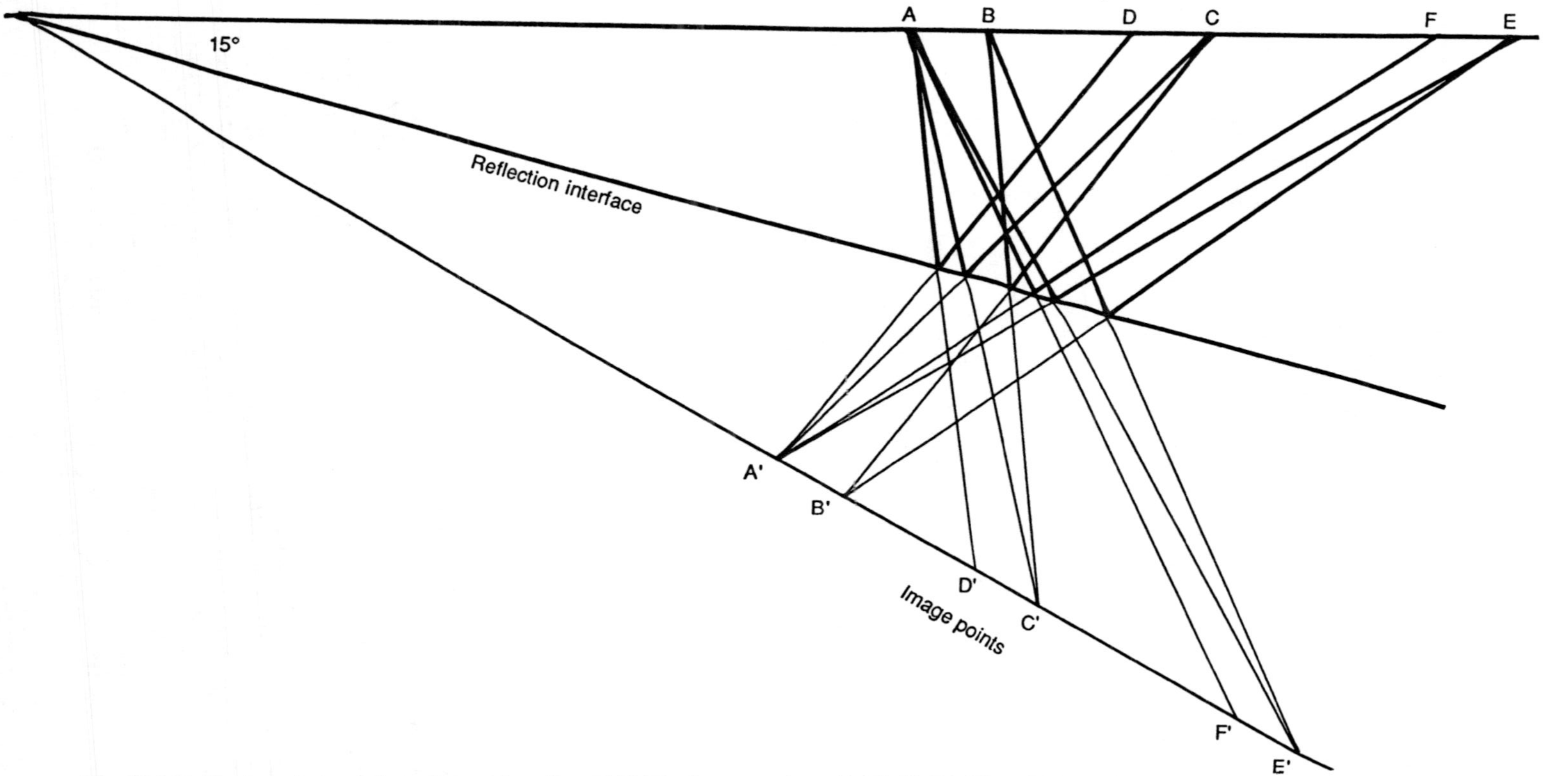

Fig. 11.1 Reflections from an interface with a dip of 15°. The image points A′, B′, C′, D′, E′, and F′ are shown to facilitate comparisons of relative path lengths. The figure illustrates that the moveout due to difference in path lengths is significantly more downdip than it is updip. For example, the path AC′ is considerably longer than path AD′ but the path CA′ is not much different than the path CB′. The angles of arrival or emission updip are more nearly vertical than the arrival angles downdip at comparable offsets.

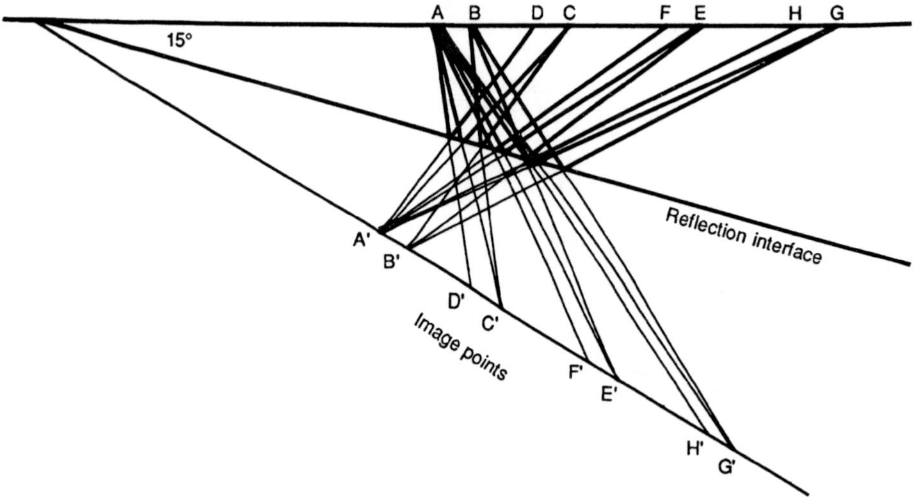

Fig. 11.2 Reflections from an interface with a dip of 15° at half scale. The scale used in this figure matches that used for Figs 11.3 and 11.4 to facilitate comparisons.

seismic line because the interface would soon either outcrop or else it would plunge to depths beyond our planned recording time. When we do get a reflection from a short segment of a steeply dipping interface, the reflection point is located somewhere far different from directly beneath the mid-point between our source and our receiver. Obviously there are many problems to be overcome in successfully mapping steeply dipping interfaces.

Figures 11.1–11.4 illustrate ray paths for reflections from interfaces with various dip angles. In these figures we show the mirror image points and rays as well as the actual reflection paths in order to more easily compare path lengths. The dip angle of 15° in Figs 11.1 and 11.2 presents little problem in that the angle of arrival for updip offsets is nearly vertical for offsets that are no longer than the depth of the reflector, at the reflection point. The offsets actually shown are longer than the depth of the reflector. Therefore, we could still use a rather long geophone array to suppress noise for the updip half of the cable for dips of 15°. If long geophone arrays are used, then the source pattern must be short since the angle of emission to the updip cable will be significantly displaced from vertical. Figures 11.2–11.4 are at half the scale of Fig. 11.1. (Fig. 11.2 is for comparison with the larger dip angles illustrated in Figs. 11.3 and 11.4.) For larger dip angles and with moderate length cables, the angles of both arrival and emission are significantly displaced from the vertical. Figure 11.3 is for a reflector dipping at an angle of 30° and Fig. 11.4 for a dip angle of 45°. All four of these figures represent ray paths for shot gathers. Note in Fig. 11.3 where the dip is 30°, that the angles of arrival, or of emission, at

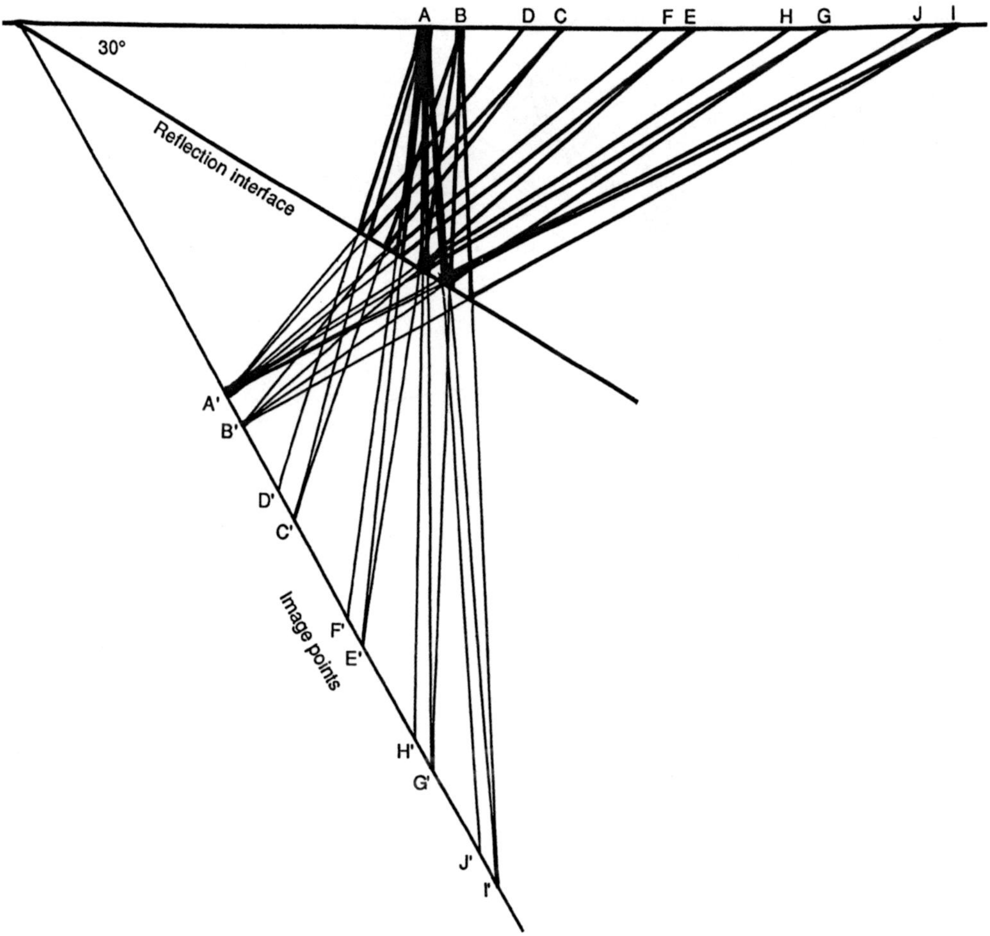

Fig. 11.3 Reflections from an interface with a dip of 30°. At a long enough offset, such as from I to A, updip reflections from an interface with a dip of 30° arrive at the receivers at a nearly vertical angle. This means that the apparent wavelength for the reflection is quite long and the apparent velocity is quite high.

updip points A and B are nearly vertical for offsets to points E and F. But these offsets are greater than the depth of the reflections and are not likely to be used in conventional recording. But the shorter offsets to points C and D still result in nearly vertical angles of arrival or emission at points A and B. Therefore for a dip angle of 30° we can still use a moderately long array updip from a short source pattern. (With a split spread, half of the traces would be updip with a favorable angle of arrival and half downdip with a very unfavorable angle of arrival.)

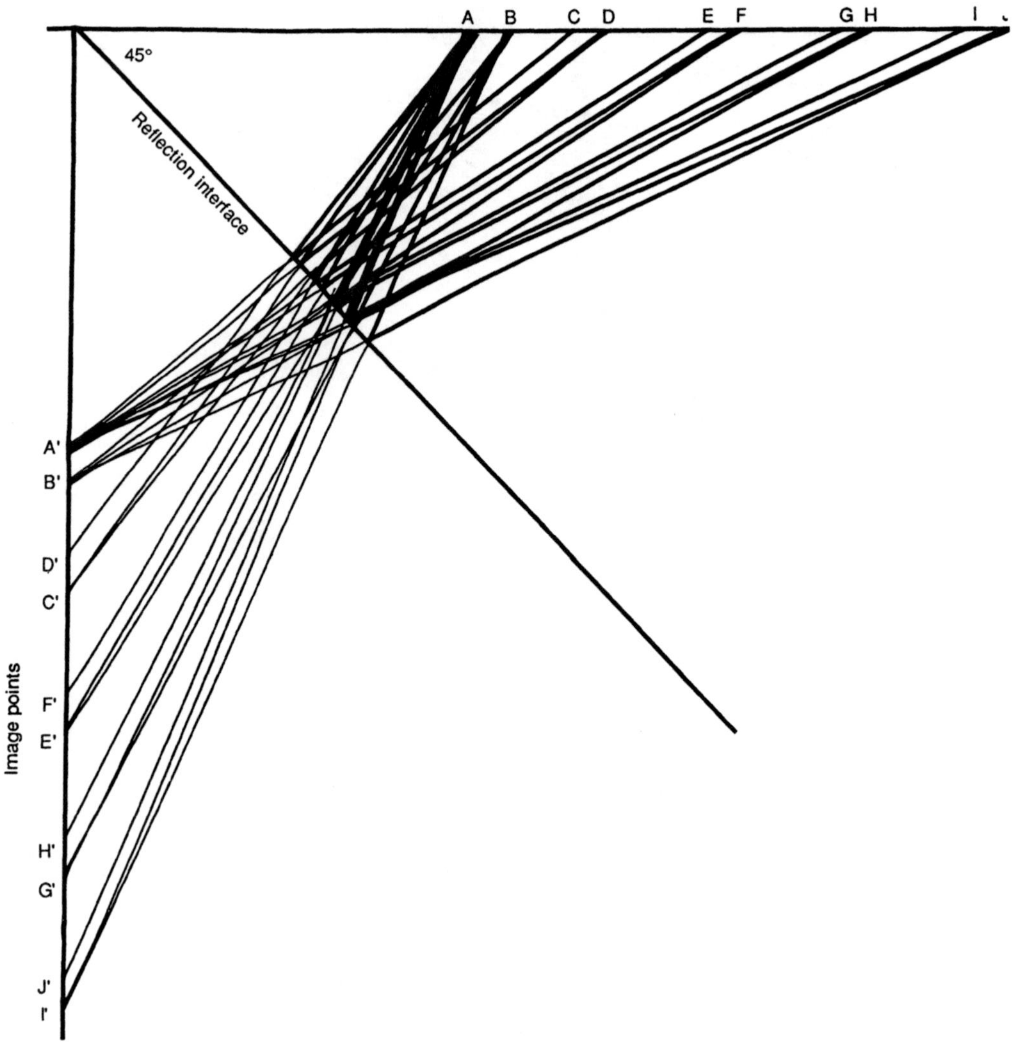

Fig. 11.4 Reflections from an interface with a dip of 45° arrive at a substantial angle to the vertical even for long, updip offsets.

Note in Fig. 11.4, with a dip angle of 45°, that the angles of arrival or emission at updip points are significantly displaced from vertical for offsets that would normally be used in recording and that these angles are still not very steep even for very long offsets. Figure 11.5 is similar to Fig. 11.4 except that it represents a CMP gather (rather than a shot gather) for a cable that is about as long as the reflector is deep. The reflector dips at a 45° angle. To simplify the drawing, we assumed that we used an odd gap so that ray paths for

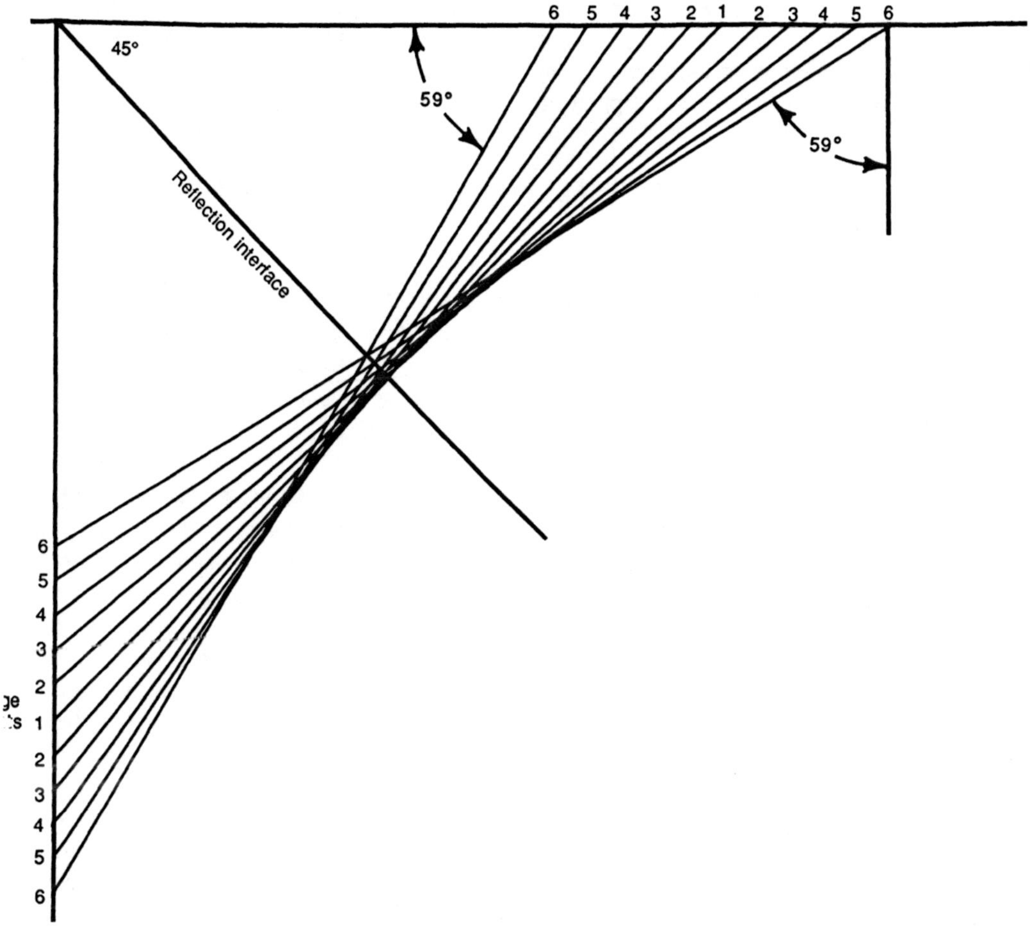

Fig. 11.5 A CMP gather of reflections from an interface with a dip of 45°. The ray paths shown (1–1, 2–2,... 6–6) represent rays to both a front and back cable (i.e. receivers both updip and downdip from the source). The cable length shown is approximately equal to the reflector depth. Note that these traces with common mid-points do not have a common depth point. Instead reflections lie along a line segment. The reflection points for the longest offsets are updip from the reflection points for the shorter offsets.

the front and back cables are identical. (Each ray path shown is for both the forward and back cable in the CMP gather and the rays travel both directions.) The common mid-point rays do not converge at a common depth point. The rays from the farthest offset reflect from farther updip than the rays from the shortest offset. Note that the angles of arrival and emission (measured from the vertical) range from 31° to 59° and the sine of these angles ranges from 0.51 to 0.86. Therefore reflections from the dipping interface would not have large apparent wavelengths and would be severely suppressed by either long geophone arrays or long source patterns even if they were updip. Note in Fig. 11.5 that the average angle of the rays from the vertical is 45° for reflection from an interface that dips at an angle of 45°.

11.2 BEAM STEERING IN THE FIELD AND IN PROCESSING

When we use a large source pattern with vibrators or explosive charges, we in effect direct most of the energy directly downward. Suppose that we would like to direct the energy at some other angle. For example suppose in Fig. 11.3 that we had a source pattern at point E and a large geophone array at A, we would like to direct our energy not directly downward, but instead at an angle of about 53° from vertical in an updip direction. To illustrate the technique of beam steering consider an exaggerated example. Let us suppose that we have a source that extends from E to F. The path length EA′ is longer than the path length FA′. If shots at points E and F are to arrive at A at the same time, then the shot at F will have to be delayed relative to the shot at E. Delay caps are available which can be used to provide time delays and accomplish beam steering in the field. Delays for beam steering can also be achieved with Primacord (or fuses and caps) which is particularly suitable for beam steering large areal patterns. With vibrators we would need to delay the start of the sweep for vibrators that are updip from the vibrator that is farthest down dip. Recently Sercel has introduced its VE 416 vibrator electronics which provides means for beam steering with vibrators.

Suppose that we had some delay caps that would provide delays in 2 ms steps and we wanted to beam steer to some desired angle. What should the spacing be between the shot holes in a pattern?

Let

θ = the desired beam angle relative to vertically downward.
d = the spacing between adjacent shot points.
v = the moveout velocity of the earth to some objective.

Then the difference in the path lengths from two adjacent shots along the beam direction is $d \sin \theta$ and the difference in time is $(d \sin \theta)/v$ (Fig. 11.6). If we make this time difference equal to the 2 ms increments of our delay caps then we could solve for the separation distance, d.

$$d = 0.002v/\sin \theta \qquad \text{(for a 2 ms delay cap)}$$

If the earth velocity were 15 000 ft/s (4572 m/s), then the spacing for various

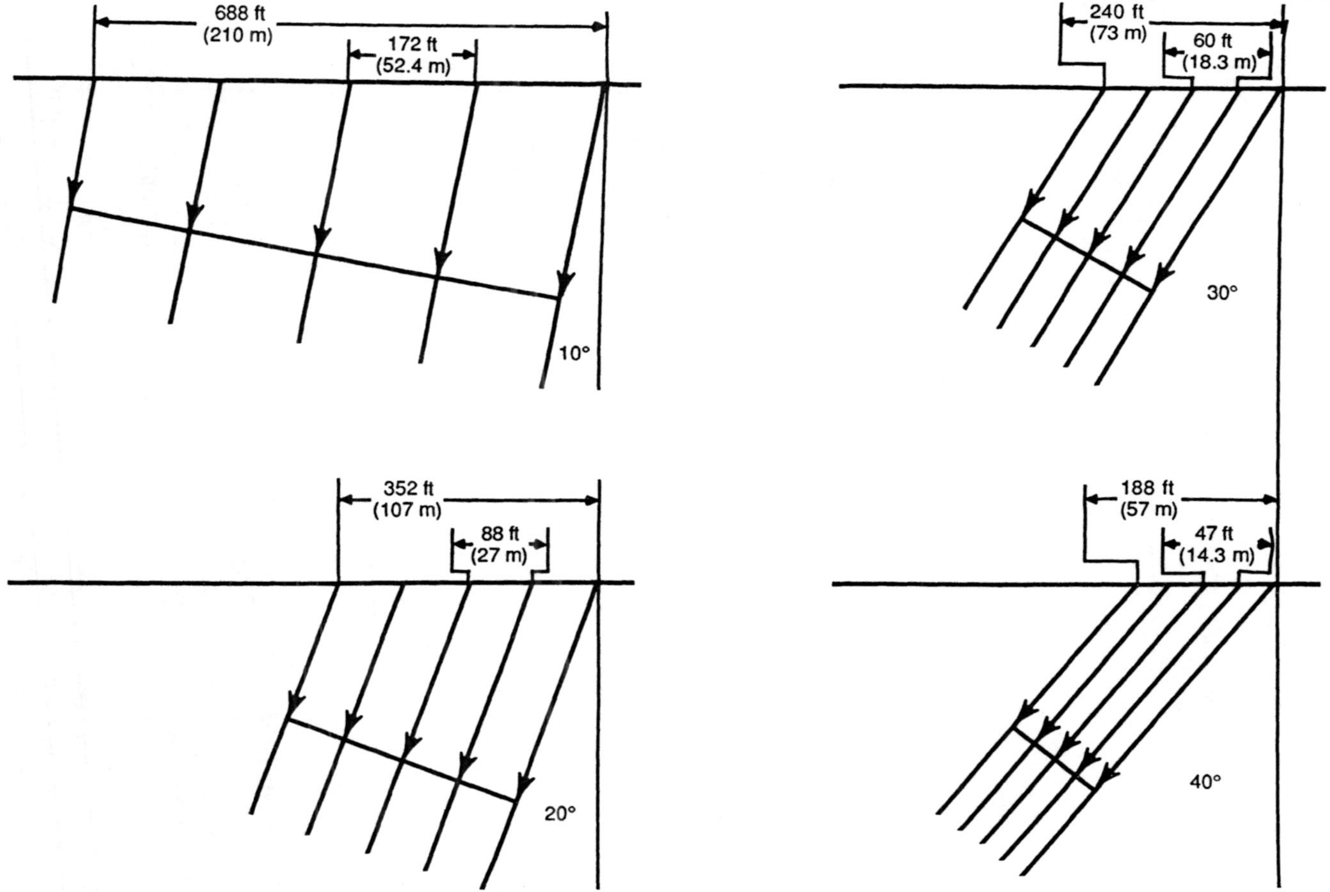

Fig. 11.6 Beam steering with a 2 ms delay between shot holes for an earth velocity of 15 000 ft/s (4572 m/s). For a given delay time and earth velocity, the beam angle is determined by the spacing between the shot holes.

beam angles would be:

Beam angle (degrees)	*Spacing d for 2 ms delay (ft)*	*(m)*
10	172.8	52.7
20	87.7	26.7
30	60.0	18.3
40	46.7	14.2
50	39.2	12.0
60	34.6	10.5

Figure 11.6 illustrates these spacings for beam angles of 10° to 40° for patterns with five holes with delays of 2, 4, 6, and 8 ms.

For Fig. 11.3 where the dip of the reflector is 30°, assume the moveout velocity is 15 000 ft/s (4572 m/s) and we are using long offsets with a source pattern at E and a long geophone array at A. With delay caps we could use a sizeable source pattern at E and direct the beam at an angle of 53° and our reflection would be in nearly perfect time phase from the source and across the array. Furthermore, source-generated noise with different emission and arrival angles would be suppressed. In this case we would need a spacing between adjacent holes of 37.6 ft (11.5 m). If we were using a five-hole pattern, the *nd* product for noise suppression would be only 188 ft (57 m). This may not be long enough to handle much of the source-generated noise. We could either increase the number of holes or choose to double the spacing to 75 ft (23 m) and double the time delays to 4 ms between adjacent holes.

If we had a number of geophone arrays laid out along a seismic line centered at A, then there would be some differential moveout from the source and across the arrays. But this differential moveout would not be nearly as great as the differential arrival time from a long source pattern without appropriate time delays. Likewise, if the dip of the reflector varied from 25° to 35° then the differential moveout from the source and across the arrays would increase but would hopefully remain small enough to avoid significant damage to the higher frequency components of our reflected signals. Beam steering either in the field or in processing can be used effectively for specific targets. The more we know about what we are trying to find, the better we can apply beam steering to the problem.

In processing we can apply the beam steering concept to both the sources and to the receivers. Another use for source beam steering in processing would be with a split cable to direct the beam toward the back cable and then toward the front cable. For beam steering in processing we would like to have closely spaced geophone arrays and closely spaced source points. We may also prefer short source patterns and short geophone arrays, but then we run the risk that unsuppressed noise will be too great and we may not have enough dynamic range to recover weak reflection signals. In addition to the concern about dynamic range, there is another problem with short patterns and short arrays. This other problem is that the data processor may become discouraged by the dominance of noise, and he may not pursue the project long enough to find just the right combination of processing steps and parameters to be successful.

Processors like to see massive amounts of good data streaming out of the computer, and its display peripherals. Processors are judged, to a large degree, by the amount of good data they produce. If a data set does not yield encouraging results reasonably soon, even a very good processor who enjoys a challenge may not continue to give it the thought, time, and effort that would be required for success. Many processors will disagree, but many other geophysicists will heartily agree. For this reason and because of concern about dynamic range, there needs to be a balance between noise suppression in the field and noise suppression in processing. We cannot allow the field techniques and parameters to damage the objective reflection signal badly in an effort to suppress noise. But we also need data that is not too discouraging to work with in processing.

For beam steering in processing we have some sample interval in our data and what we need to know is what time shifts we need to make between samples to result in a desired beam angle. So instead of starting with a time delay and solving for the spacing, we start with the spacing and solve for the appropriate time adjustment. In processing it is just as easy to advance as it is to delay so we generally leave the center element alone and advance or delay the other elements as required to align the wavefront at the desired beam angle. Beam steering is much more versatile in processing. We can beam steer at one angle to updip receivers and to a much different angle for downdip receivers; also we can vary the beam angle as a function of offset. Consider the CMP gather illustrated in Fig. 11.5. A beam steer angle of 45° for both source and receivers would work quite well to bring out the reflection from the interface with a slope of 45°. But if we processed a whole line that way, we would suppress reflections from low dip interfaces and from interfaces that dipped in the opposite direction. Beam steering is a powerful tool, but it needs to be used appropriately, which is not an easy task. Even after we have produced multiple data sets for multiple beam angles, how do we display all of the data for the interpreter? Or how can the interpreter assimilate the data if the processor does not combine it for him in some useful manner?

11.3 BEAM STEERING TO AVOID ALIASING

Beam steering can be used to prevent the aliasing of signals reflected from steeply dipping interfaces or reflections from horizontal interfaces at long offsets. Consider Fig. 11.6 and note that with beam steering it is as if the sources or receivers (or both) were aligned along an oblique line perpendicular to the beam rather than along a line on the surface. Apparent velocity and apparent wavelength are now determined by the angle between the beam angle and the angle of arrival or the angle of emission. This means that if we choose to steer the receiver beam such that it nearly parallels rays from a reflector, then the apparent angle of arrival is small and the apparent wavelength for this reflection is long. Then we would no longer need to worry about spatial aliasing our reflection if our group interval is not short. Likewise if we choose to steer the source beam so that it is nearly parallel to the angle of emission required for a reflection to arrive at a detector location, then the apparent angle of emission is small and the apparent wavelength of the reflection is long. This should reduce the need for closely spaced source points to avoid aliasing reflections from the subject interface. Of course, samples must be close enough together to allow beam steering and the reflector has to have enough continuity to interact with a beam of the composite source dimensions.

Beam steering does require some judgment just as interpreters years ago used judgment in successfully migrating sparse data. But we need the use of good judgment in all phases of exploration. It is not reasonable for geophysicists to insist upon very closely sampled data to map possible quite steep reflectors, if the exploration problems can be well handled at greatly lower cost with beam steering and good judgment. Very close sampling leads to huge amounts of data at high cost in terms of time and money. It also leads to huge processing costs. Routine processing is for routine exploration problems. Trying to map extremely steep dips and complex geology is not routine and calls for extraordinary processing and good judgment.

11.4 RECOMMENDATIONS FOR SURVEYS IN GOOD RECORD AREAS

For seismic surveys in good record areas where dips may be very steep (more than 35°), the conventional approach of using a short group interval, a short geophone array, a short source pattern, and a high CMP fold is recommended. We would depend upon the CMP stack for most of our noise suppression. But we could also use beam steering in processing, if the group and source intervals are small enough so that the loss in horizontal resolution is acceptable. (Again, the loss in horizontal resolution may be very little since we are dealing with waves that reflect from large areas rather than pencil-like rays that reflect from points.) If we know that dips will be steep only in one direction (or if we are concerned only with measuring steep dips in one direction), we may want to use an end-on cable and use a moderately short geophone array updip and a very short source pattern (or no pattern). Under some circumstances we may want to reverse this and use a moderate source pattern updip and little or no geophone array.

If only moderately steep dips (25–35°) are anticipated, we should consider using a moderate length geophone array and a short source pattern. We may consider also using somewhat longer offsets. Since we would still need a short group interval, we would need to use a recording system with many channels unless we used an end-on cable or a very long center gap. The author would not recommend the end-on cable unless it is certain that the steep dips would be only in one direction. The use of a long center gap eliminates recording shallow reflection data.

If dips are expected to be less than about 25°, we do not need to be too concerned about dip. We should tend toward using source patterns of moderate length and arrays only slightly shorter than we would select for horizontal objectives.

11.5 RECOMMENDATIONS FOR SURVEYS IN POOR RECORD AREAS

If record quality is poor, then even if steep dips are anticipated, we must first be reasonably sure that noise is under adequate control. If this forces us to use an array that is rather long with a short or moderate source pattern, then we should still be able to map lowdip events and perhaps to map some events with dips that become moderate. Remember the somewhat long offsets in Fig. 11.3, where rays to or from an interface with 30° dip are nearly vertical at updip locations? If we used longer offsets we could use a long geophone array (updip) to suppress noise and still record reflections from short source patterns.

In poor record areas we may do well to suppress noise and record what dips we can. This should let us pin-point the portions of the lines where steep dips may exist that we would still need to map. For a limited portion of a line we might plan a more intensive effort and one tailored to an inferred, specific target. For a specific target we might be able to say with some assurance what the dip direction should be and what the depth is likely to be. We could then model the anticipated dip and plan an acquisition program (and a processing plan) to bring out best the reflections from the anticipated dipping bed. For example, we may use some very long offsets to improve the angle of arrival to updip geophone arrays that are long enough to attack some of the noise. We may select a portion of the line of keen interest where we would acquire very high CMP fold data with a short sampling interval. (Appendix E gives high CMP fold methods that might be employed.)

11.6 RECOMMENDATIONS FOR FAIR RECORD AREAS

In fair record areas where steep dips are anticipated our approach should be intermediate in the sense that we still have concerns about suppressing noise, but this concern is moderate so we can direct more effort at trying to record reflections from steeply dipping interfaces. We can shorten arrays more and essentially eliminate a source pattern.

If the survey is of high interest and priority, it may be worthwhile to expend more effort not only to map steep dips, but to try to improve record quality from fair to good. We may then record many more channels of data to increase the fold and to record with longer offsets.

11.7 CROSS DIPS

On a strike line we deal primarily with cross dips. Reflection points for beds that have cross dip will be laterally displaced updip and the apparent depth of these beds (below a horizontal surface) will be less than the true depth directly below the seismic line. This does not cause problems in data acquisition, but it does cause problems in interpretation and in processing. By recording one or more dip lines, we can measure the cross dip. Alternatively we can record closely spaced lines and extract both components of dip (i.e. the apparent dip along the seismic lines and the cross dip). We may acquire some data with sources or receivers laterally displaced at various distances from the seismic line in order to get a measure of cross dip. Where we record data along crooked roads and where data quality is good, we may be able to extract a measurement of the cross dip from those CMP gathers where the depth points exhibit wide lateral scatter.

11.8 MODELING STEEP DIPS AND COMPLEX GEOLOGY

Interpreters use sophisticated computer models to see how the seismic sections should look for various geologic situations, and they compare actual seismic sections with seismic sections calculated for their models. Some computer models can be used to try out various seismic data acquisition ideas to map complex geology better including steeply dipping interfaces. For example, we could model an inferred steeply dipping interface and see what seismic data we could add by adding longer offsets. With our model we could trace ray paths for various offsets and for various CMP gathers. We could also generate synthetic seismic data which we could process and see what useful information would be contributed if we used longer offsets, increased fold, or decreased the group interval, etc.

For some complex earth cases, ray tracing for various acquisition techniques may be particularly important. Focusing and defocusing effects and progressively curved ray paths may effectively eliminate any chance to get reflections from some target with a planned acquisition technique. If a model shows this to be the case, then we should look for some change in the acquisition geometry that may improve our chance for success. Any promising technique could then be tried on the model. We should not rely totally on models, but we can use them to guide our efforts so that our field time and money yield better seismic data.

12

High-velocity surface problems and answers

In this chapter we discuss problems related to high-velocity surfaces both with and without heterogeneities that result in severe scattering. Use of star geophone arrays and star source patterns as well as large detector 'patches' is included. The use of shear waves is discussed, including both converted shear waves and all shear wave techniques using horizontally polarized shear waves.

12.1 ANGLES OF ARRIVAL AND EMISSION ON HIGH-VELOCITY SURFACES

Figure 12.1 illustrates ray paths in an earth section where velocity increases with depth, which is the usual trend. Only the reflection from the interface at 10 000 ft (3048 m) is shown for simplicity. Figure 12.2 illustrates similar ray paths in an earth with a high-velocity surface underlain by layers with lower velocity. Figure 12.3 illustrates similar ray paths in an earth with a thick permafrost layer underlain by clastics with a lower velocity. In this case the surface velocity is not very high, but it is high relative to the velocity of the underlying earth layers. In the high-velocity surface cases of Figs 12.2 and 12.3 note that the angles of arrival and emission deviate more from the vertical for a given offset than they do for the typical earth case shown in Fig. 12.1. Figure 12.4 is a plot of the angles of arrival and emission as a function of offset for the three cases of Figs 12.1, 12.2, and 12.3. For an offset of 10 000 ft (3048 m) these angles for the three cases are given in Table 12.1.

The reflection from a depth of 10 000 ft (3048 m) will have an apparent velocity at an offset of 10 000 ft (3048 m) that is equal to the moveout velocity times $1/\sin\alpha$. This means that the apparent velocity of the reflection and the apparent wavelength of each component frequency will be increased by a factor of 5.54 for the typical earth case but will be increased factors of only 1.92 and 1.71 for the two earths with high-velocity surfaces (see Table 12.1). For earths with high surface velocities, the apparent velocities and apparent

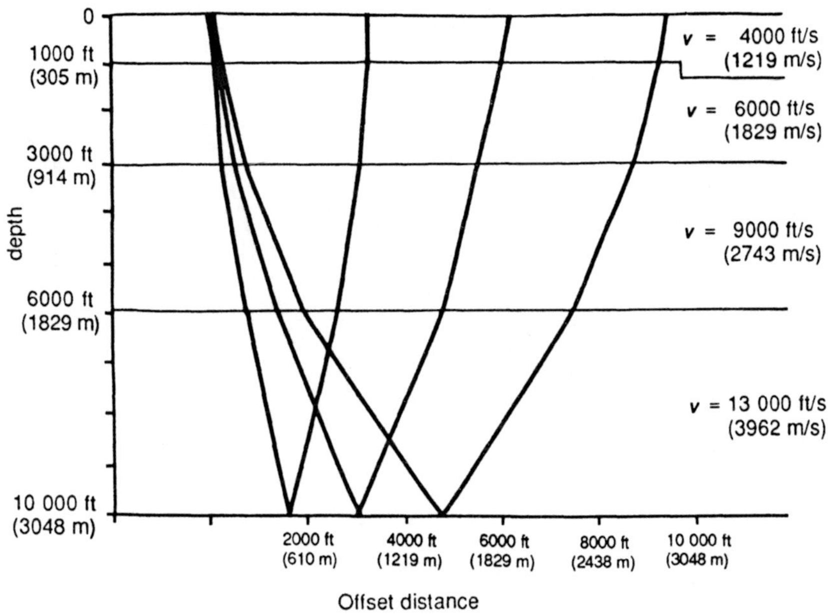

Fig. 12.1 Ray paths in a typical earth with increasing velocity with depth.

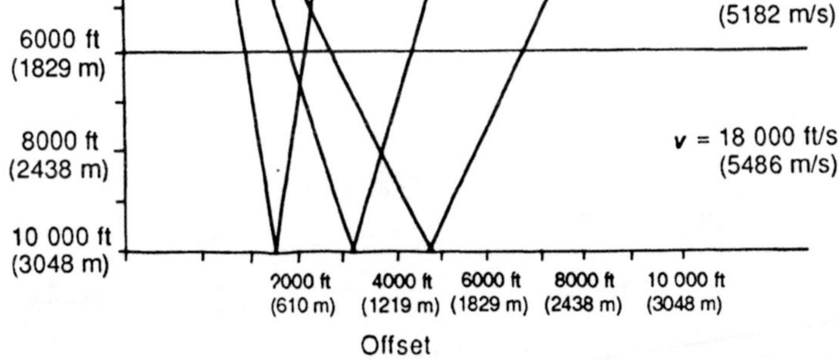

Fig. 12.2 Ray paths in an earth with a high-velocity surface. Note that the angles of emission and arrival for reflections at long offsets are more deviated from the vertical than for the typical earth case of Fig. 12.1.

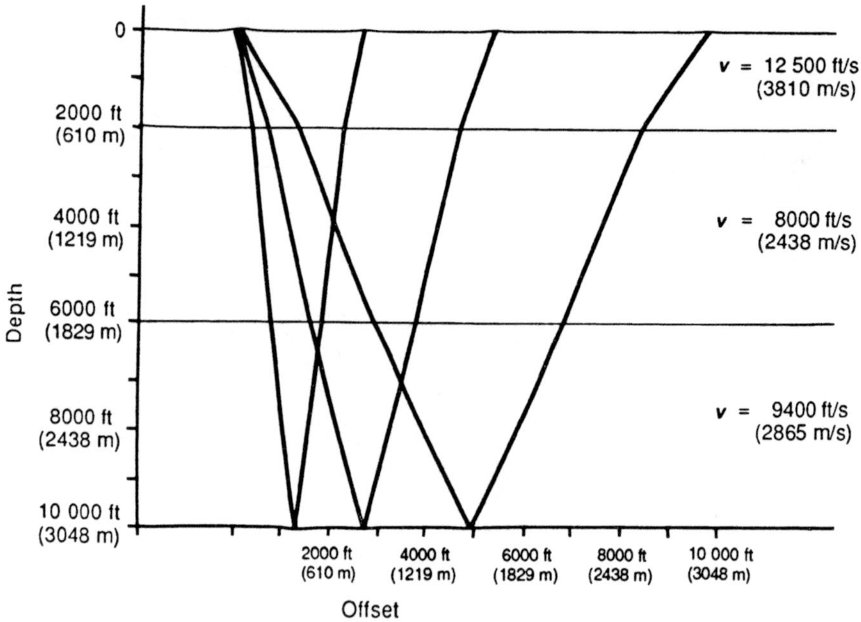

Fig. 12.3 Ray paths in an earth with thick permafrost. This case is patterned after the North Slope of Alaska where the section is dominantly clastics of low to moderate velocity with a thick permafrost section with much higher velocity. Again the angles of arrival of reflections are large at long offsets.

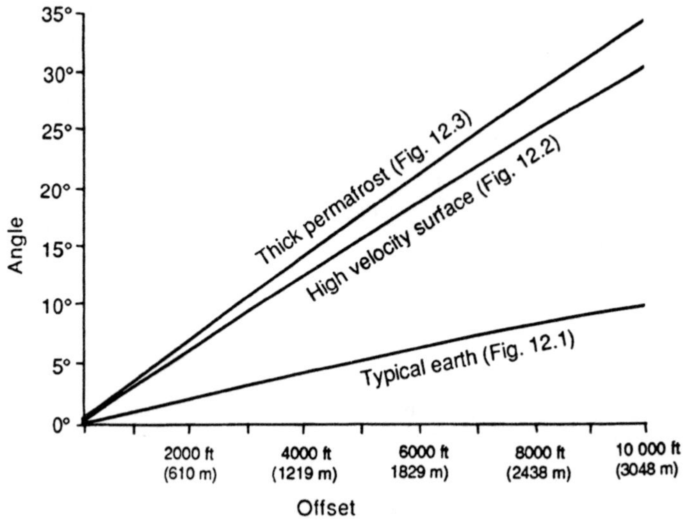

Fig. 12.4 Angles of arrival and emission for three earth cases. The angles of arrival and emission are relatively small for the typical earth case where velocity increases with depth. But these angles are much larger where the velocity near the surface is greater than the underlying velocity.

Table 12.1 Angles of emission and arrival for three earth cases for an offset of 10 000 ft (3048 m)

Case	*Angle α (degrees)*	*sin α*	*1/sin α*	*cos α*
Typical earth (Fig. 12.1)	10.40	0.1805	5.54	0.984
High-velocity surface (Fig. 12.2)	31.25	0.5188	1.92	0.855
Permafrost earth (Fig. 12.3)	35.7	0.5835	1.71	0.812

wavelengths of reflections at long offsets may not be much greater than the apparent velocities and apparent wavelengths of source-generated noises. Therefore it may not be possible to suppress noises effectively with geophone arrays and source patterns without suppressing the desired reflections. If we must use long arrays and patterns to suppress noise we may have to limit the maximum offset more than we would for typical earths. Also a relatively low apparent velocity with its attendent short apparent wavelengths means that we will need a relatively short group interval to avoid spatial aliasing of reflections at our longer offsets.

12.2 CONVERTED SHEAR WAVES ON HIGH-VELOCITY SURFACES

Figure 12.5 illustrates a compressional wave and a converted shear wave arriving at a relatively long offset in an earth with a high-velocity surface. At any given offset the converted shear wave will arrive at a steeper angle than will the compressional wave if it were reflected from the same interface. But if waves of the two types arrive at the same offset and at the same time, then the shear wave was reflected from a shallower interface. Thus the angle of arrival of a contemporaneous shear wave may be rather large as is illustrated in the figure.

Ordinarily if we wanted to detect the shear wave we would use a geophone that was sensitive to horizontal motion since the earth motion for a shear wave is perpendicular to the ray path. But if a converted shear wave arrives at the surface at an oblique angle, it will have a substantial vertical component of motion and will therefore appear on our normal seismic records. Therefore for earths with a high-velocity surface at offsets that are relatively long, our seismic records may display converted shear wave reflections. These converted shear waves are a noise to be suppressed in regard to the data we ordinarily want to display.

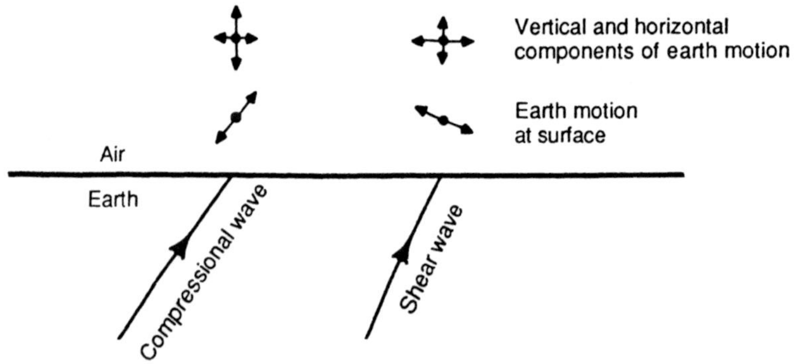

Fig. 12.5 Earth motion at the surface for compressional waves and shear waves with large angles of arrival. If the angle of arrival at the detectors is large, then a compressional-wave arrival will result in a radial, horizontal component of motion as well as the more familiar vertical component of motion. A converted shear wave with a large angle of arrival will result in a vertical component of motion as well as the more familiar radial, horizontal component of motion.

On the other hand, if we handle the data properly we can stack the converted shear waves and obtain additional information about the earth. In our compressional wave data recorded on high-velocity surfaces we may have valuable converted shear wave data that could be extracted and used (Chapter 15). Often, problems are also opportunities, look carefully behind every problem and seize any opportunities you find. But for compressional wave recording, converted shear waves are still a problem to be handled.

The vertical component of the earth displacement and velocity at the surface due to an arriving compressional wave will be proportional to the cosine of the angle of arrival. The vertical component of the earth displacement and velocity at the surface due to an arriving shear wave will be proportional to the sine of the angle of arrival. Both the cosine and sine of the angles of arrival are given in Table 12.1. But these angles are for reflected compressional waves. The angles for contemporaneous, converted shear waves for a real earth case would have to be determined using both the shear wave velocity and the compressional wave velocity of each earth layer.

For a typical earth where velocity tends to increase with depth, the vertically oriented geophones normally used record compressional waves effectively but not shear waves even at relatively long offsets.

12.3 KARST AND VOLCANIC SURFACES

A high-velocity karst or volcanic surface presents formidable problems in recording reflections from depth. We have all the problems of a high-velocity surface plus the problems associated with the heterogeneity of these rocks. Voids and pore space in leached limestone or in volcanic rocks may be filled with air or gas, water, or a fill material which may be only partially water-saturated. In almost any case, velocity and density will be greatly reduced which makes for a severe problem in static correction. But the biggest problem is likely to result from the scattering of seismic waves from the source and from scattering of reflections arriving from depth.

If we record a noise spread on a karst or volcanic surface, we may well find little noise that is coherent. This does not mean that we do not have much source-generated noise; what it means is that the noise is scattered. This is double trouble. We cannot systematically attack incoherent noise, and, if the noise is scattered and therefore incoherent, we can expect the reflections also to be scattered and made incoherent. The only solution is to find some way to average out the static variations and the effects of scattering.

First let us consider how we can do the most averaging of the response to reflections at the receiver end. Not only do we want to average the reflection response, we would like our averaging to suppress noise. Therefore we want to average over some distance an area such that:

1. The reflection response should be consistent except for the effects of the heterogeneities.
2. The noise response should vary even if there were no heterogeneities.

Obviously we can do more averaging that will meet these objectives if we average over a circular area around the receiver position, rather than along a line centered on this position. (One may well argue that the area should be elliptical rather than circular, but the ellipticity desired would vary with the depth of the reflector and the offset.) So a star geophone array seems appropriate. To meet the first objective, we need to keep the radius of the star array short enough not to suppress our reflections. But to try to meet the second objective we want to use a star array that is big enough to suppress the longer wavelength components of the noise. We absolutely must meet the first objective. We may be able to handle the noise problem with the CMP stack. So we use a star geophone array that is about as big as you can and still meet the first objective.

A scattering layer such as karst at the surface is bad, but this is not as bad as a scattering layer at depth and above an objective interface. Chapter 21 includes a discussion of why scattering near the source or near the receivers is less serious than scattering at greater depths.

From both a theoretical standpoint and from field experience in recording seismic reflections of karst surfaces, the author recommends the use of both star geophone arrays and star source patterns. This recommendation also applies to volcanic surfaces and other surfaces where severe scattering is a

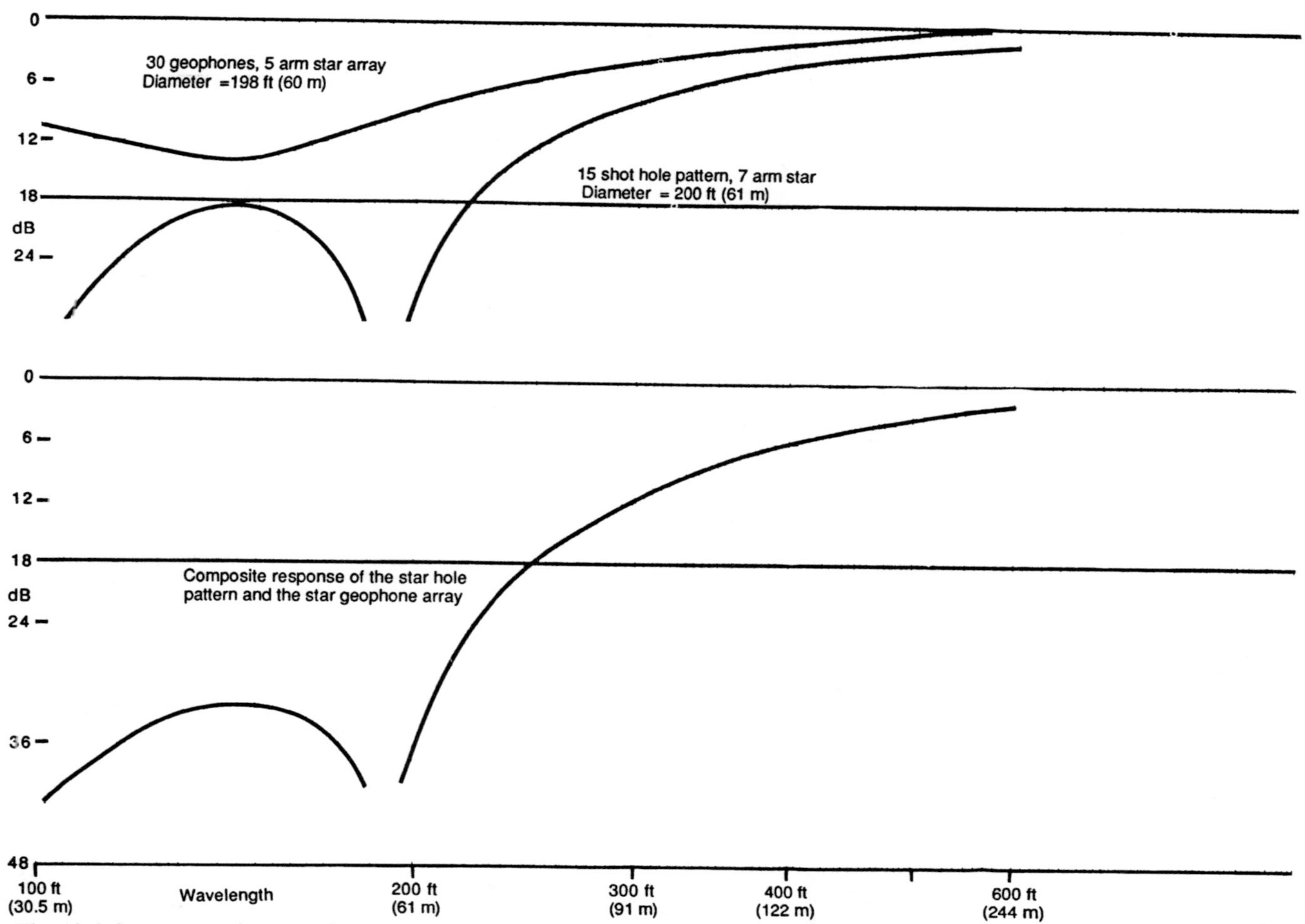

Fig. 12.6 Separate and composite responses of a star source pattern and a star geophone array. Star patterns and arrays are less effective than in-line arrays in suppressing noise traveling parallel to the seismic line, but they are more omni-directional in the suppression of noise.

problem. This method can be carried out in a regular CMP roll-along operation. An effective star source pattern can consist of mini-holes about 5 ft (1.5 m) deep with $\frac{1}{2}$ lb (0.25 kg) charges. The star geophone array and the star source pattern should be selected so as to provide reasonably good suppression of coherent noises over a wide band of wavelengths without also suppressing the wanted reflections. Figure 12.6 illustrates the response of a star geophone array and a star source pattern and the composite response of both together. The star source pattern had seven arms and a total of fifteen holes with a diameter of 200 ft (61 m). The star geophone array had five arms and a total of thirty geophones with a diameter of 198 ft (60 m). These star patterns are shown in Figs D-10 and D-11 (Appendix D) along with a discussion of stars.

One may well argue that there is no need to be concerned about the composite response if scattering from heterogeneities has completely randomized the noise so that there is no coherent noise. But the noise is not completely randomized and one may as well provide some suppression for

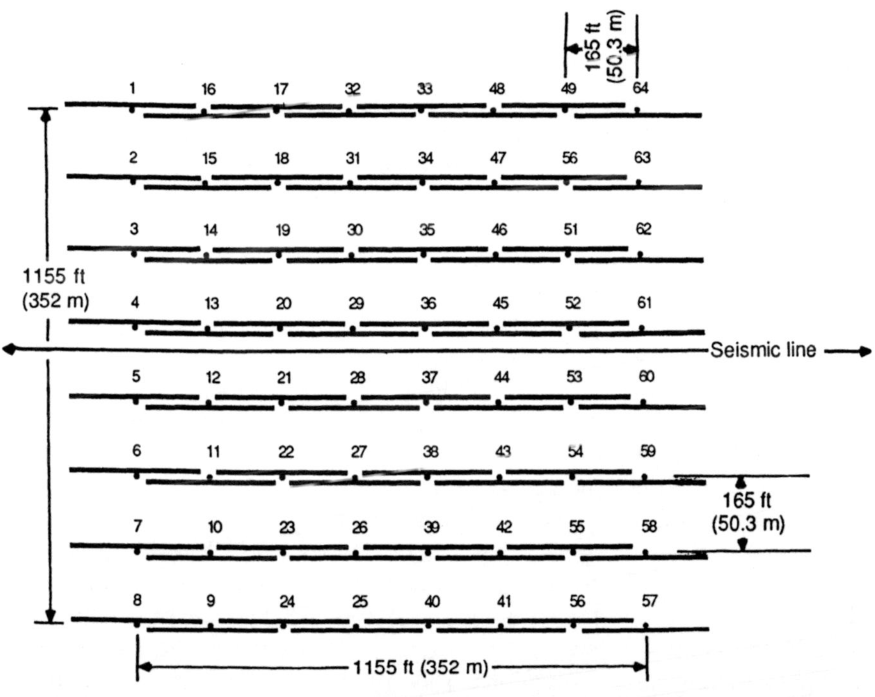

Fig. 12.7 Example of a receiver patch. The receiver patch illustrated consists of 64 geophone arrays laid out in an 8 × 8 matrix. The 64 traces recorded for the patch are summed together to yield one trace. Before summing, relative NMO corrections and possibly static corrections can be applied.

coherent noise. Indeed, we shall probably see some coherent noise in both noise spread and production data and there is certain to be more coherent noise that is hidden by the random noise.

If we use vibrators for the seismic source on a karst or volcanic surface, it will likely not be feasible to use anything like a star source pattern. In this case we can use a star geophone array and a feasible source pattern that complements the geophone array in terms of adding notches where the array is less effective.

If the star–star, CMP roll-along technique does not provide acceptable data on a karst or volcanic surface, then we need even more averaging. In this case we should try a large receiver 'patch', which is illustrated in Fig. 12.7. In the patch technique we lay out receiver positions over a large area, with dimensions of about 1000 ft (305 m) or more, and record the responses on separate channels. At each position use a small array, which may be inline or it may be an areal pattern such as a star. In processing the data we treat all data recorded from one source position as if it were the response at the center of the patch. Ultimately we shall sum all of the patch data into one trace, but first we shall want to make NMO corrections and possibly make some static corrections as well. These NMO corrections will not be large since the dimensions of the patch will be short compared with the range of offsets in a CMP gather. Therefore, the moveout velocity to be used is not very critical, which is fortunate since we may have to estimate it initially and refine it later.

In addition to NMO corrections, before the traces of a patch are summed, we would like to make corrections for static variations within the patch. The unsummed data will probably be too poor to extract reflection information that would allow us to make a good static correction, but we may find that we can see first arrivals at some moderate offsets. Figure 12.9 illustrates a possible method for recording first arrivals and using these to make relative static corrections within the patch. The offset should preferably be considerably longer than the length of a side of the patch. Then at that offset we need to locate shot points both forward and back along the seismic line and also at right angles to this line and to both sides (locations A, B, C, and D in Fig. 12.9). At each of these four positions we drill a single, deep hole or a small pattern. The holes in a pattern should be not more than 5–10 ft (1.5–3 m) apart.

From the arrival time at the nearest traces we calculate a velocity to use for a moveout correction for the first break arrival times only. This will not be the velocity we shall use for reflection data. Consider the record made from a shot at position A. The path length from A to traces 1 and 8 is greater than that from A to traces 3 and 4. So we need to correct for this difference in path length before extracting static corrections. Then the relative static variations for a column of traces such as 1–8 can be determined from the NMO corrected arrival times from sources positions A and B and then averaged. Static corrections for all of the columns of traces (such as 48–55) can be determined from the arrivals from source positions A and B. The relative static variations

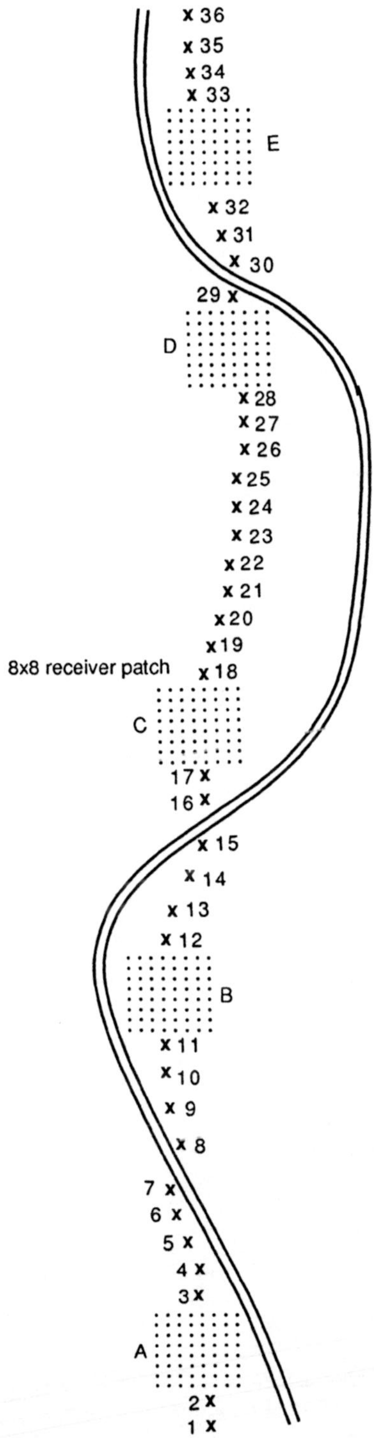

Shot points	Patches recorded
1-14	A, B & C
7-31	B, C & D
15-35	C, D & E
24-?	D, E & F

Fig. 12.8 Illustration of layout and recording of patches in groups of three. The illustrated CMP fold is approximately two.

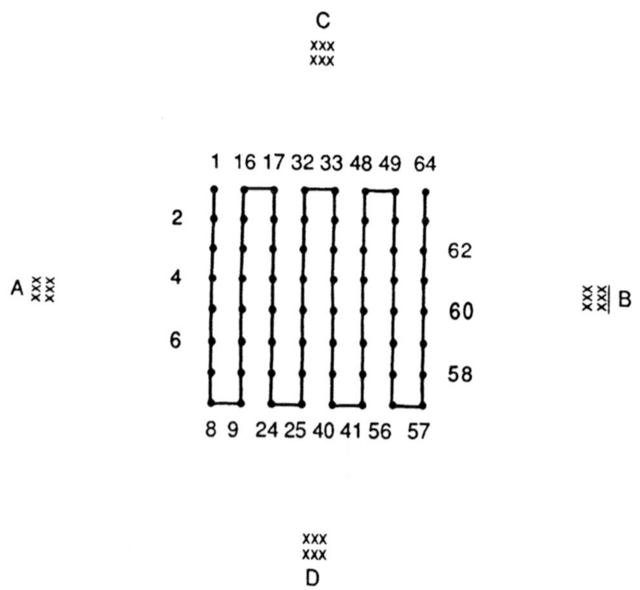

Fig. 12.9 Shots to record first breaks for static corrections within a receiver patch. Shot patterns at A, B, C, and D should be located at an offset where good 'first breaks' can be recorded. Preferably this offset should be long compared with the length of the side of the patch (to minimize the moveout correction that will need to be applied). The shot patterns should be closely spaced to approximate a point source. A deep, single hole shot would be even better. A relative moveout correction needs to be made on a row-by-row basis from each source position using a 'first break velocity' determined from the measurements. An average static correction is then determined on a relative basis for each receiver position within the patch.

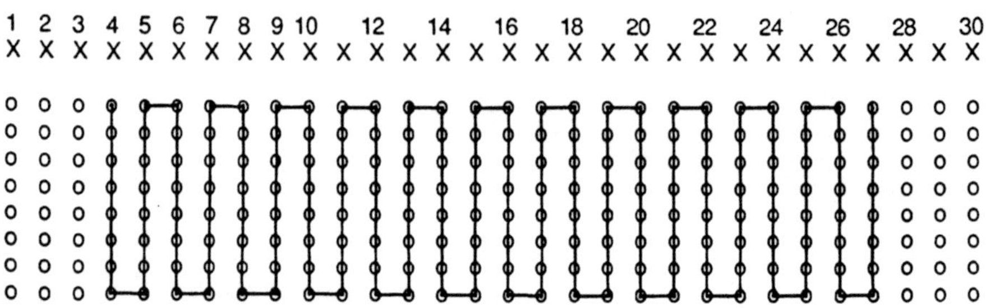

Shot location	Record columns (8 traces/column)	Process as 8x8 patches for columns
1	4 - 27	4-11, 5-12, 6-13, ... 20-27
2	5 - 28	5-12, 6-13, 7-14, ... 21-28
3	6 - 29	6-13, 7-14, 8-15, ... 22-29
etc.	etc.	etc.

Fig. 12.10 Illustration of a roll-along patch. After processing 8 × 8 matrices of traces as patches, the data can be stacked to achieve a CMP fold of 12 with a maximum overlap of the patches or a lower fold with less overlap.

for any row of traces such as the row 8, 9, 24, 25, 40, 41, 56, and 57 can be determined from the NMO corrected arrivals from source positions C and D. The row and column corrections are combined to yield a relative static correction for each trace in the patch.

Figure 12.8 illustrates how we might layout receiver patches along a winding road and how we might layout the shot points to use with the patches. The shot points may be single holes or mini-hole patterns. If we have a road available that is not too crooked, we might use vibrators for a source. We try to find locations for patches which are relatively flat and open. However, surprisingly good results have been obtained in cases where the slopes were steep and the surface was quite irregular within patches.

With a recording system with many channels, we may find it feasible to record a roll-along patch as is illustrated in Fig. 12.10, with 192 channels. After each record the source point advances one group interval and the receivers are moved one column of eight traces. This results in a great deal of overlap between the successive patches. It may be preferable to double or quadruple these move-ups even though that would reduce the CMP fold. But remember, we are trying to do a lot of averaging to offset the randomizing effect of the scatterers. The author never had a recording system with enough channels to try a roll-along patch. But such recording systems are available and it is a technique that may be the key to acquiring satisfactory data in some really tough areas where scattering is a major problem and the exploration stakes are high.

In some cases we can record good reflection data through a karst or volcanic surface with horizontally polarized shear waves. This requires good shear-wave sources such as horizontal vibrators. (Chapter 15 gives a discussion of an inclined impactor source called ARIS*.) Where the first sizeable layer is underlain by a thick layer with a lower shear wave velocity, there is no Love wave generated and unusually good shear-wave data may be recorded (Fix 1983; Danbom 1987). Shear waves have an additional advantage as well. On a karst or volcanic rock surface, shear-wave static variations may be significantly less severe than compressional-wave static variations. Usually the reverse is true and static variations are much worse for shear waves than for compressional waves, because shear-wave velocity changes radically with slight changes in the degree of consolidation of soil and poorly consolidated rocks. But in karst and volcanic rocks the rock is well consolidated. It is true that the shear-wave velocity is affected by the voids and pore space, but this velocity is insensitive to whether or not these pores and voids contain free gas or air. By contrast the compressional-wave velocity is very sensitive to even small amounts of free gas or air in these pores and voids. Therefore on karst and volcanic rock surfaces, shear-wave static variations and even scattering may not be severe, whereas compressional-wave static variations and scattering will be.

*Registered trademark of the Atlantic Richfield Company.

12.4 HOMOGENEOUS HIGH-VELOCITY SURFACE LAYER

Where the earth has a high-velocity surface and the surface layer is relatively homogeneous, we do not have the scattering problem, but we do have the other problems previously discussed. In these cases we can probably acquire reasonably good seismic data with a rather conventional approach. We are likely to do better if we limit the offsets to less than the depth of the deepest objective. But a well-planned test should be conducted to see what works best in that area. If we see some low-velocity events on long offsets these may be shear waves (converted shear waves from shallow interfaces may appear at rather modest offsets).

REFERENCES

Fix, J. E., Robertson, J. D. and Pritchett, W. C. (1983). 'Shear-wave reflections in three West Texas basins with high-velocity surface rocks'. Presented at the 53rd Annual International Meeting. (Also in Danbom 1987, Chapter 4.)

Danbom, S. and Domenico, S. N. (eds), (1987) 'Shear-wave Exploration,' SEG.

13 Other problem areas and answers

In this chapter we discuss poor record areas in general and then take up the problems associated with the special case of a low velocity near a surface layer overlying a much higher velocity section.

13.1 POOR RECORD AREAS IN GENERAL

There is a number of reasons why some areas tend to yield poor seismic records. In the preceding two chapters we dealt with problems resulting from steep dips, complex geology, karst and volcanic surfaces, and other high velocity surfaces both with or without scattering. Later in this chapter we shall deal with the problem of a low velocity surface underlain by a high-velocity reflector and refractor. In this section we shall discuss poor record areas in general and conditions that cause them.

In some cases there are no good reflectors. A good reflector must have a reasonably smooth interface, a good acoustic contrast, and enough thickness such that the reflections from the upper and lower interfaces do not largely cancel. If the problem is that the reflectors are too thin, then high-resolution techniques may be a solution. If there is too little acoustic contrast for compressional waves, we may consider the possibility that the acoustic contrast might be better for shear waves.

With some poor record areas the problem may be high background noise. Can the background noise be reduced or circumvented? If wind noise is a problem, can you bury the geophones or work at night and in the early morning if the wind is strongest in the afternoon? If pumping wells are the problem, can arrangements be made to shut down the pumps? If boat and streamer noise is a problem, slow down, use a quieter boat, or a quieter streamer?

When the background noise level cannot be controlled, can you increase the effective source power enough to overcome the problem? More source power, directional source patterns, and source beam steering can be used to boost the reflection signal relative to background noise. Can you beat the noise by a huge multiplicity of measurements for which reflections would be in phase and additive and the noise would not be in phase? More geophones, more traces, more CMP fold, and receiver patches all tend to favor reflections over background noise if properly used. But there is a practical limit and increasing the number of geophones, the CMP fold, etc. produces diminishing returns as these numbers get large.

Often the problem in poor record areas is too much source-generated noise. More source power will not help, but better source patterns may do so. Chapter 7 deals with ways to test to see how source-generated noises can best be suppressed.

Sometimes the earth may be too heterogeneous, so that the seismic waves are scattered somewhere between the surface and the reflectors of interest. The problem could be at an intervening unconformity. Big receiver patches may help in this case just as patches often help on hetrogeneous surfaces, as described in Chapter 12.

Sometimes there just does not seem to be an answer as to why data quality is poor. This is particularly perplexing when data quality varies abruptly from rather good to very poor and there is no reasonable explanation for this abrupt change in quality. In the deep Black Warrior Basin (Mississippi and Alabama) reflectors from the Paleozoic section appear and disappear abruptly, yet reflections from the Cretaceous and from the unconformity at the base of the Cretaceous are continuous. This is not an ideal seismic area as it suffers to some degree from the problem discussed in the next section. But if we can handle this problem and get good data why does it abruptly go bad with no apparent change in the noise? The Paleozoic beds are thick with lowdip and must be generally continuous. It is unlikely that the acoustic contrast changes abruptly.

13.2 LOW-VELOCITY SURFACE PROBLEM AND ANSWERS

Reflection and refraction from the interface between a high-velocity layer underlying a low-velocity surface layer can result in a serious problem with multiple reflections and refractions. In one such case the shallow earth section was composed of Cretaceous clastics with a velocity of about 6000 ft/s (1829 m/s). At a depth of about 2000 ft (610 m) there was a major unconformity with an abrupt increase in velocity to about 16 000 ft/s (4877 m/s). If we calculate the reflection coefficient (considering both the density change and the velocity change) and the critical angle we get:

Reflection coefficient = 0.52; Critical angle = 22°

A reflection coefficient of 0.5 means the amplitude of the reflected wave is 50 percent of that of the incident wave. This means that less than 75 per cent of the vertically (downward) traveling energy would penetrate the refractor just below the unconformity and none of the energy outside a cone of 22° (the critical angle) from the vertical would penetrate. The bulk of the energy would be reflected and refracted back to the surface where it would be reflected back down to be reflected or refracted again at the unconformity. Is it surprising then that the raw field records clearly showed the primary reflection and refraction from the unconformity plus three multiples down the record of the reflection and of the refraction? The whole record beyond the primary reflection time of about 0.67 s was dominated by multiple reflections, and multiple refractions.

What can we do in a case like this to get more energy through the unconformity and to reduce the noise later in the record? One solution is to use a source pattern that is highly directional and will emit most of its energy vertically downward within an angle of less than 22° from the vertical. Likewise, we want a geophone array that also discriminates against angles of arrival that are as much as 22° from the vertical. (The geophone array may not have to discriminate against the lowest frequencies if the source pattern is effective for them.) If we used a 14–56 Hz bandwidth, the longest refraction wavelength would be 16 000/14 = 1143 ft (348 m). If we cut our bandwidth back to a lowest frequency of 16 Hz, we would still calculate a noise wavelength of 1000 ft (305 m).

We recognized our basic problem at the time, but we just did not go far enough in our tests to find the solution. We went to the trouble of arranging for eight vibrators but we tried spacings of only as long as 85 ft (26 m) between the vibrators. This gave us notches at wavelengths of 680 ft (207 m) and less (Chapter 3 and Appendix D) which was not sufficient to suppress adequately the refraction and its multiples. Refraction frequencies of less than about 20 Hz were not well suppressed. At the time we did not have a nonlinear sweep generator which could have helped us by spending less time at frequencies below 20 Hz. As a result, our big effort with eight vibrators yielded results that were not much better than previous efforts with only four vibrators.

What we should have done (since we did not have a nonlinear sweep generator) was to space the vibrators 125 ft (38 m) apart to give us notches at wavelengths of 1000, 500, 333, 250, 200, 167, and 143 ft (305, 152, 101.5, 76, 61, 51, and 43.5 m). We would then have had low suppression peaks at wavelengths of 667, 400, 285, etc. ft (203, 122, 87, etc. m). We should have used a geophone array of 36 over 642 ft (196 m) to provide notches at wavelengths of 660, 330, 220, 165, 132, 110, etc. ft (201, 101, 67, 50, 40, 33.5, etc. m). We could have used eight sweeps with a move-up of 50 ft (15 m) between sweeps to provide additional notches at wavelengths of 400, 200, 133, 100, etc. ft (122, 61, 40.5, 30.5, etc. m). If we used a sweep of 16–56 Hz we would have strongly suppressed the refraction and its multiples which was our main problem.

Would these long source patterns and long arrays have killed our reflection data as well? Suppose we had used only 48 traces and a group interval of 165 ft (50 m). Then the geophone array specified above would have been a quadruple group interval array. With such long patterns and arrays we might have gone beyond our listing in the tables of cables in Chapter 4 and used a gap of fourteen traces with minimum and maximum offsets of 1217.5 and 5032.5 ft (371 and 1534 m) respectively. For an objective at a depth of 10 000 ft (3048 m) with a moveout velocity of 16 000 ft/s (4877 m/s), the long source pattern would result in a differential arrival time of 0.0152 s at the longest offset. The frequency response for our objective reflection would have been:

$f_0 = 66$ Hz (frequency for zero response)
$f_{0.3} = 49$ Hz (frequency for 30% response)
$f_{0.64} = 33$ Hz (frequency for 64% response)

Response for the higher frequencies is bad. We need another trick. Suppose we used upsweeps of 16–56 Hz for 10 s. The rate of change of frequency would have been 4 Hz per second of sweep. Suppose we swept only 6 s with the first and last vibrators and only 8 s with the second and seventh vibrators. Then the sweeps would have been as shown in Table 13.1.

Table 13.1 Illustration of a shortened sweep time for the outer vibrators of a source pattern

Time (s)	*Frequency* (*H*)	*End of sweep for vibrators*	*nd product of source* (*ft*)	(*m*)
0	16		1000	305
			8 × 125	
1	20		1000	305
2	24		1000	305
3	28		1000	305
4	32		1000	305
5	36		1000	305
6	40	1 & 8	1000	305
7	44		750	229
			6 × 125	
8	48	2 & 7	750	229
9	52		500	152.4
			4 × 125	
10	56	3, 4, 5 & 6	500	152.4

As shown in Table 13.1, the *nd* products for the source pattern would have been 1000 ft (305 m) for 16–40 Hz, 750 ft (229 m) for 40–48 Hz, and 500 ft (152 m) for 48–56 Hz. One may choose to adjust this somewhat, but the point is that we could have come up with a recording system to suppress the refraction with a long source pattern at the lower frequencies and still get the data we needed with shorter source patterns at intermediate and higher frequencies.

It is true that by limiting the maximum offset we would be less able to suppress any other multiple reflections. The multiple reflection from the unconformity would have a much lower velocity than the primary reflections at depths of about 10 000 ft (3048 m) and should not be a problem. For this short cable the total moveout for the unconformity multiple at objective time would have been 255 ms compared with the moveout of the objective of only 39 ms.

If we had a nonlinear sweep generator and spent less time at the low frequencies, we probably could have used significantly shorter source patterns and shorter geophone arrays. With the Pelton nonlinear sweep generators we can use the trick of reducing the sweep time of some of the vibrators to shorten the pattern at the higher frequencies, but it is difficult. The problem is that the sweeps must be identical for the time periods when all are sweeping and both the ending times and ending frequencies must be integers. Therefore we have to use a sweep that results in an integer frequency at one or more integer times between the start and end of the sweep. In Table J-2 of Appendix J the Pelton 12 dB/octave sweep of 20–120 Hz over 16 s yields a frequency of 75.94 Hz at 4 s and 105.99 Hz at 11 s. These frequencies are very nearly integer numbers so that we could combine sweeps of 20–76 Hz over 4 s, 20–106 Hz over 11 s, and 20–120 Hz over 16 s.

The author knows of no way to use this trick with the Texas Instruments logarithmic sweep generators, but we could do it with EPROMS (Appendix J) with their sweep generators.

Unfortunately we usually cannot go back and do now what we wish we had done the first time. 'Ve get too soon old und too late schmart!'.

14
Shallow objectives and simultaneous recording

In this chapter we tackle the problem of recording very shallow seismic data and reducing the cost by recording data from two or more sets of sources simultaneously.

14.1 ECONOMIC SEISMOLOGY FOR SHALLOW OBJECTIVES

For shallow objectives the usable maximum offset is quite limited. If we want to record enough traces to have a high CMP fold, then group intervals and source intervals become very short. This means many recordings per mile and lots of data to process. This costs money. In 1984 we encountered this problem when we needed to record some regional lines with a Vibroseis crew. To reduce field time and cost we decided to record pairs of VPs simultaneously to get twice as many records in a given time. Figure 14.1 illustrates the technique we used quite successfully. With 60 data channels we recorded 60 traces per VP or 120 traces at a time. We used two vibrators at each A position and two more at each B position (a total of four vibrators). The coding used to allow separation of the data from the two sets of sources was designed to be compatible with the use of the system to cancel power line pickup. This power line cancellation equipment and technique are discussed in Chapter 16. The coding used is given in Table 14.1.

After this sequence of sweeps with no move-up, we moved both sets of vibrators up two positions and advanced the roll-along switch two positions.

The coding shown in Table 14.1 was very easy to implement and was simple for the crew to utilize. The B vibrators used commercially available equipment that automatically reversed polarity after each sweep.* The A vibrators did

*Polarity reversing was developed originally to minimize second-harmonic ghosts when using downsweeps. This interesting idea was conceived by Sam Sorkin who was with Teledyne. Now there are better ways to handle second-harmonic ghosts (Chapter 5).

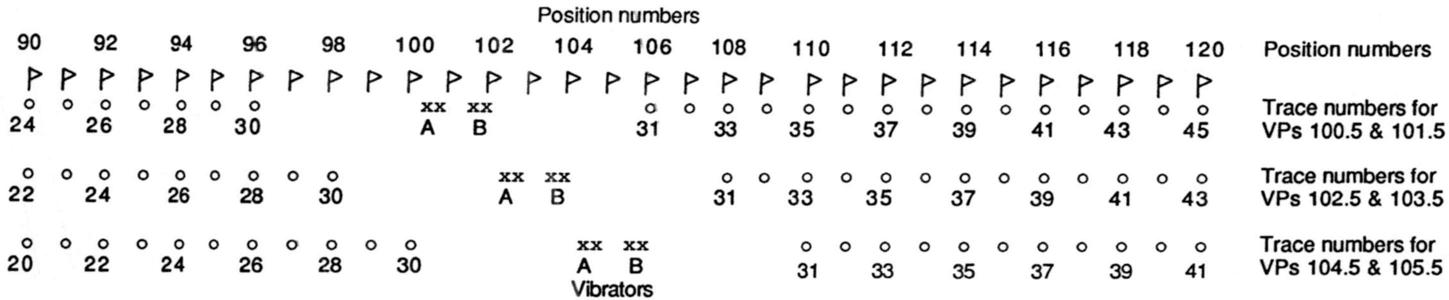

VP numbers	Trace numbers	Receiver positions
100 & 101	1-30 & 31-60	67- 96 & 106-135
102 & 103	1-30 & 31-60	69- 98 & 108-137
104 & 105	1-30 & 31-60	71-100 & 110-139

Fig. 14.1 Illustration of the method used to record pairs of VPs simultaneously
1 Use two vibrators at positions A and two more at positions B;
2 Do not reverse polarity for vibrators at positions A;
3. Reverse sweep polarity of vibrators at positions B—see text;
4. The figure shows operations with a 60-trace crew with nine traces in the gap. The cable is not quite symmetrical for either VPA or VPB, but would be symmetrical if the two were composited subsequently in processing;
5. Following each simultaneous recording of VPs A and B, the observer advances the roll-along switch two positions and both sets of vibrators move up two positions;
6. Both VPs are centered half-way between receiver positions.

Table 14.1 Coding used to record and separate the earth response to two sets of vibrators

	Polarity of vibrators		*Polarity for summing in processing*	
Sweep No.	*Vibrs A*	*Vibrs B*	*For A record*	*For B record*
1	+	+	+	+
2	+	−	+	−
3	+	+	+	+
4	+	−	+	−

Then we changed the polarity of the power line cancelation box without triggering any sweeps prior to the next sweep.

5	+	+	+	+
6	+	−	+	−
7	+	+	+	+
8	+	−	+	−

To extract the B record we reversed the polarity of the signals recorded for sweeps 2, 4, 6, and 8 which made all the B records have positive polarity at the same time that half of the A records were made to have negative polarity. Thus the A record was suppressed and the B record was extracted.

not use polarity reversing. We had an unusually good crew with an excellent observer and this was an important ingredient in our success.

In this technique we have to record the separate sweeps so that we can make the polarity reversals (in processing) for the summation to get the B records. But we can sum and correlate the A records in the field for quality control purposes. Ordinarily we edit and sum individual sweeps in the field. In this case we did the editing and summing in processing. Initially we were concerned that in editing-out noise bursts (not source-generated) we would upset the cancellation of power-line noise. We had a solution that proved to be unnecessary in our case but might be required in some future work. The idea was to first extract one of the two signals and then sum sweep responses in pairs and edit these summed pairs to suppress noise bursts. The traces in each pair must have opposite polarities for the unwanted source.

The simultaneous recording technique worked well and saved time and money. We also wanted to save money in processing. What we tried to do was to make an approximate normal moveout correction and then sum the A and B records and back out the approximate normal moveout correction. In this way we would have half as many records for the bulk of the processing. The sections with the summed data were good, but full-fold processing of all the separate data always gave better results. We tried extracting static corrections and applying these before summing the A and B records, but the full processing was still superior.

In our case in 1984 we used a 20–120 Hz logarithmic sweep, a group and VP interval of 66 ft (20 m) and recorded 3698 VPs over 46.23 miles (74.4 km) in 154 hours of 14 working days including a move. We recorded 80 VPs per mile (50 VPs per km) with an overall acquisition cost of $2872 per mile ($1785 per km) and averaged 264 VPs per day. Data quality including resolution, continuity and S/N ratio was superb. Without simultaneous recording of pairs of VPs our acquisition cost would have been about $5000 per mile ($3100 per km). The work was done by an excellent contract crew we had used continuously for about five years. The low costs were not the result of a low bid by a desperate contractor.

Several papers presented at technical meetings in 1984 reported disappointing results with simultaneous recording. Why were we so successful? In devising a solution to our problem we provided ourselves with two advantages. First, we saw that our geometry and recording system were such that the amplitude of the combined signal plus the source-generated noises were nearly the same from both sources at all receivers. This was true at early record times when the energy level was relatively high and still true at late record times when the energy levels were quite low. Therefore the separation of the two sets of traces (of signal plus noise) was less difficult than for cases where one signal plus noise is much larger than the other signal plus noise at some time during the recording. Second, our geometry was such that the two signals were very much alike in that the centers of the reflection points were only 33 ft (10 m) apart. Consequently there was substantial overlap between the two areas from which the two signals were reflected from the various interfaces of interest. Therefore the separation of the two signals did not have to be perfect. If the A signal was slightly contaminated with the B signal, it would make very little difference.

Use a good tool the right way and it works. Use it carelessly and you may get hurt.

14.2 EXTENSION TO SEXTUPLE SIMULTANEOUS RECORDING

With as few as eight sweeps we can set up codes to allow the recording of data from as many as six different groups of vibrators. As envisioned here all vibrators would use the same sweep (but some could cease sweeping before others, as was discussed in chapter 13). The method for coding and decoding six different data sets is given in Table 14.2. Possibly there are more codes that could be used, but six is probably more than is sufficient. Certainly if we

Table 14.2 Codes for recording up to six data sets simultaneously

As recorded & decoded for code 1

		codes				
1	2	3	4	5	6	
+	+	+	+	+	+	
+	+	+	+	-	-	
+	+	-	-	+	+	
+	+	-	-	-	-	
+	-	+	-	+	-	
+	-	+	-	-	+	
+	-	-	+	+	-	
+	-	-	+	-	+	
+8	0	0	0	0	0	sum

Decoded for code 4

			codes		
1	2	3	4	5	6
+	+	+	+	+	+
+	+	+	+	-	-
⊖	⊖	⊕	⊕	⊖	⊖
⊖	⊖	⊕	⊕	⊕	⊕
⊖	⊕	⊖	⊕	⊖	⊕
⊖	⊕	⊖	⊕	⊕	⊖
+	-	-	+	+	-
+	-	-	+	-	+
0	0	0	+8	0	0

Decoded for code 2

		codes				
1	2	3	4	5	6	
+	+	+	+	+	+	
+	+	+	+	-	-	
+	+	-	-	+	+	
+	+	-	-	-	-	
⊖	⊕	⊖	⊕	⊖	⊕	
⊖	⊕	⊖	⊕	⊕	⊖	
⊖	⊕	⊕	⊖	⊖	⊕	
⊖	⊕	⊕	⊖	⊕	⊖	
0	+8	0	0	0	0	sum

Decoded for code 5

			codes		
1	2	3	4	5	6
+	+	+	+	+	+
⊖	⊖	⊖	⊖	⊕	⊕
+	+	-	-	+	+
⊖	⊖	⊕	⊕	⊕	⊕
+	-	+	-	+	-
⊖	⊕	⊖	⊕	⊕	⊖
+	-	-	+	+	-
⊖	⊕	⊕	⊖	⊕	⊖
0	0	0	0	+8	0

Decoded for code 3

		codes				
1	2	3	4	5	6	
+	+	+	+	+	+	
+	+	+	+	-	-	
⊖	⊖	⊕	⊕	⊖	⊖	
⊖	⊖	⊕	⊕	⊕	⊕	
+	-	+	-	+	-	
+	-	+	-	-	+	
⊖	⊕	⊕	⊖	⊖	⊕	
⊖	⊕	⊕	⊖	⊕	⊖	
0	0	+8	0	0	0	sum

Decoded for code 6

			codes		
1	2	3	4	5	6
+	+	+	+	+	+
⊖	⊖	⊖	⊖	⊕	⊕
+	+	-	-	+	+
⊖	⊖	⊕	⊕	⊕	⊕
⊖	⊕	⊖	⊕	⊖	⊕
+	-	+	-	-	+
⊖	⊕	⊕	⊖	⊖	⊕
+	-	-	+	-	+
0	0	0	0	0	+8

Notes:

1. Uncircled signs in the above table refer to the polarity of the sweep;
2. Circled signs mean that the polarity of the data from that sweep is to be reversed in processing before summing in order to decode a particular data set. The sign in the circle is reversed from the original sweep polarity to show the net effect;
3. All codes except number 6 are compatible with the power-line pickup cancellation system if this system is triggered once after the fourth sweep.

expanded the number of sweeps to twelve we could handle more different data sets. But remember that the signals cannot vary very much in amplitude at any given time if we are to extract the weaker signal or signals. The first five of these six codes are amenable for use with the power-line cancellation system if this system is triggered once after the fourth sweep to reverse its polarity.

To use some of these codes we would definitely have to have some special circuitry that would make them easy for the crew to implement. At the very least, the use of some of these codes opens up possibilities for error that should be carefully considered. Any added complexity needs ample justification and complexity increases the level of competence required of the crew and of its management (including contractor and client).

In Chapter 10 in the discussion of Fig. 10.8 we pointed out that signal separation with codes like these might well be inadequate for a particular set of measurements. But we also briefly discussed possible ways that signal separation might be improved. At the SEG international convention in 1987 several papers were presented which dealth with the problems and opportunities for simultaneous recording of seismic reflection data and VSP data. In addition, literature was available describing various coding and decoding systems for better separation of multiple data sets. In one case polarity reversal was combined with upsweeps and downsweeps to provide better data separation. In that system there was need to be concerned about harmonic ghosts from the upsweep sources in decoding the signal from the downsweep sources and vice versa. This problem was overcome for a two-source system by polarity reversal of both sets of sources with a quarternary code (Deluchi, 1987).

Coding using polarity reversal is a special case of the more general phase coding (the phase difference being 180° for polarity reversal). The Sercel VE 416 vibrator electronics system allows phase changes in 1° steps from 0–360° to allow more flexibility in developing and using phase coding and decoding systems. They also provide means in this system for alternate sweeps of say 10–30 Hz and 30–60 Hz for one set of vibrators and 30–60 Hz and 10–30 Hz for a second set of vibrators. In addition they provide some pseudo random sweep capability. All of these capabilities provide ways to code and decode for simultaneous recording of seismic reflection and VSP data. Some of these techniques can be combined to improve signal separation capability. Current information about the Sercel equipment is limited, but more complete information will be forthcoming. Other manufacturers provide some similar capabilities.

The author is confident that there are important applications for simultaneous recording which you can successfully implement both with currentlly available equipment and with equipment that will be available in the near future.

REFERENCES

Deluchi, L., Marschall, R. and Werner, H. (1987) '3-D, seismic survey of the Gaggiano Oil Field with quaternary encoding dual-source Vibroseis technique.' *Expanded Abstracts with Biographies 1987 Technical Program*, SEG, S8.2, 570–3.

Khan, T. A. and Olson, E. L. (1987) 'Field productivity: a challenge to manufacturers of field systems.' *Expanded Abstracts with Biographies 1987 Technical Program*, SEG, S8.3, 574–6.

15 Mapping porosity and free gas

Often a key element in exploring for hydrocarbons involves detecting the development of porosity or the presence of free gas in a potential target layer. In this chapter we consider methods for detecting and mapping porosity development and mapping the presence of hydrocarbons (as evidenced by the presence of free gas). In addition to compressional wave reflection techniques we discuss the use of comparative compressional and shear-wave data or the use of compressional-wave refraction data.

15.1 COMPRESSIONAL-WAVE REFLECTION DATA

An increase in porosity reduces the compressional-wave velocity of rocks and of unconsolidated layers. If the pore space is filled with a relatively incompressible substance such as water or oil, then the decrease in velocity is significant, but not dramatic. The reduction in velocity may be vastly increased if even a small amount of free gas is present in the pore space. Detecting these more dramatic changes in velocity amounts to detecting the presence of free gas more than it does to detecting the development of porosity. Even so, we may be able to detect porosity changes where there is no free gas by the effects of the decrease in velocity due only to porosity. Our chances for doing so are greatly improved if we can record high-frequency components of reflections from the zone of interest (to give us more detailed information). It is also helpful if the spectra of the reflections are relatively flat except for effects from the zone of interest. Good bandwidth is advantageous, too. Chapter 6 discusses how to record high-resolution data with either explosive sources or with vibrators, while Chapter 9 deals with spectral degradation due to ghosts and ways to minimize this degradation. Before planning a program to detect and map porosity it is almost mandatory to model the anticipated development and calculate synthetic seismograms for various porosities for each model. This will help to define the resolution required and to aid materially in interpreting the results of the survey.

To detect free gas we may not necessarily need high-resolution data. What we do need is data with a good S/N ratio and data that have been processed to preserve differences in relative amplitude. Relative amplitude displays can be used successfully to detect free gas only where the near surface is sufficiently uniform so that there are minimal variations in earth coupling of the source and of the receivers. Furthermore, the formation that contains the gas must have a high porosity and preferably should not be well cemented. In such cases the large decrease in velocity coupled with the decrease in density due to the gas results in a very large reflection coefficient above and below the zone containing the gas. This results in larger than usual amplitudes for reflections from these two interfaces, the so-called 'bright spot effect'. The most notable successes for this technique have been primarily with marine data (where the coupling of source and receivers is more uniform than on land). Lithologic changes such as lignite zones can also result in very large reflection coefficients and produce events that can easily be misinterpreted as bright spots due to gas.

High-resolution data offers the advantage that the interpreter may be able to tell the difference between a lignite zone and a gas-bearing sand by measuring the interval velocity and by estimating the thickness of the anomalous zone.

15.2 COMPARATIVE COMPRESSIONAL AND SHEAR-WAVE DATA

The development of porosity reduces the velocity of both compressional and shear waves. If the pore space is filled with a relatively incompressible substance such as oil or water, then the percentage change in velocity is not greatly different for compressional waves and shear waves. But generally the percentage decrease in velocity will be somewhat greater for shear waves. If we can acquire good shear-wave data in an area, we may be able to detect porosity development by carefully comparing shear wave and compressional-wave data. But this approach is expensive and may or may not prove to be satisfactory even if good shear-wave data can be acquired.

Shear-wave velocity is insensitive to the presence of free gas. Therefore a compressional-wave bright spot due to free gas will not show an anomalously large response on a shear-wave section. False bright spots on compressional wave sections are the result of lithologic changes which typically do result in high-amplitude responses on comparative shear-wave sections. Thus with both compressional-wave sections and comparative shear-wave sections we can discriminate between bright spots due to gas and false bright spots due to lithologic changes (Robertson, 1985).

Shear waves propagate rather well through well-consolidated rocks but are attenuated rapidly by loose materials. They will not propagate in water or air.

The velocity of a shear wave decreases dramatically as a material grades from consolidated to poorly consolidated to loose. In the surface and near surface, shear waves are often subject to high attenuation and significant and variable delay in traversing unfavorable, loose material. But as they penetrate deeper into the earth the weight of the overburden alone provides an added degree of consolidation even in the absence of cementation. Most of the problems with shear-wave seismology involve problems in the surface and near surface. If the surface and near surface appear favorable, it is probable that we can record good shear-wave data. If they appear unfavorable, then it is unlikely. But we have misjudged both ways. Good shear-wave data have been obtained where the horizontal vibrators literally plowed through small dunes of loose sand. Probably the sand firmed-up rather quickly with depth. In other places with firm surface material and no obvious problems, the results were quite disappointing. Even so, from an examination of the surface and shallow earth we can frequently guess correctly whether or not good shear-wave data could be acquired.

Shear-wave velocities are normally only about half the compressional-wave velocities. If we have an event on a compressional-wave section at a time of 1.1 s we do not know at what time the same event should be on a shear-wave section. We can guess that it should be somewhere near 2.2 s, but there will likely be several events near 2.2 s which may be the event we are looking for. We have to be able to make a detailed correlation between compressional-wave and shear-wave sections before we can compare the two responses to a target zone. Sometimes this is possible and sometimes it is not. A particular interface may have a rather strong reflection coefficient for one and not for the other. Sometimes the reflection coefficients will be different in polarity. Often our reflections are from sequences of thin beds and the two types of waves may respond quite differently to such sequences. Where we can record good shear-wave data, the upper effective frequency limit is usually about half that for compressional waves. This combination of lower velocity but lower upper-frequency limit means that we cannot achieve a significant increase in vertical resolution by using shear waves.

If we can get good, comparative compressional and shear-wave sections that correlate well, we can detect changes in porosity and can better detect the presence of free gas. We look for changes in various ratios between the two sections, and calculate the ratio of the travel time between correlated events and note changes in this ratio. The shorter the time difference between the two correlated events the better the sensitivity to a change in porosity of some intervening layer. With data processed to preserve relative amplitudes we calculate the ratio of the amplitudes of reflections from correlated events. By this method we can not only identify false bright spots on the compressional-wave record, but can also detect amplitude anomalies that are only 'half bright'. Free gas in a rather low porosity formation (particularly one that is also well cemented) will not result in a strong bright spot on a compressional-wave record. But the ratio of the amplitudes of the same event on the compressional and shear-wave sections can be used both to detect half-bright anomalies and to differentiate between half-bright spots due to free gas and other changes that cause half-bright amplitude anomalies.

We can record shear-wave data using a compressional-wave source and measuring converted shear waves. Conversion takes place when a compressional wave encounters a reflecting interface at an oblique angle. The angle of reflection of the converted shear-wave is always less than the angle of incidence of the compressional wave. For horizontal reflectors the converted shear-wave returns to the surface at a steeper angle than the downward going ray path of the compressional wave that created the shear wave. We cannot get converted shear waves (from horizontal interfaces) for short offsets because of the need for an olbique angle of incidence. At offsets where we do record converted shear waves, the reflection point at depth is not directly beneath a point halfway between source and receiver. Instead, the projection of the reflection point on the surface is much closer to the receiver because of the steeper ray path for the converted shear wave. The moveout of a converted shear wave is approximately hyperbolic. It is not a simple task to stack converted shear-wave data, but it can be done successfully. Converted shear waves are vertically polarized in the sense that at the wavefront there is at least a small component of earth vibration that is vertical. The horizontal component of the motion is radially in and out relative to the position of the compressional-wave source.

Whereas it is sometimes possible to record good shear-wave data using converted shear waves, it is generally better to use a shear-wave source and an all-shear-wave system. With a shear-wave source it is desirable to use horizontally polarized shear waves which are generated by applying a force on the surface of the earth at right angles to the seismic line. The source can be a horizontal vibrator or it can be an impact device like the Marthor* Hammer or an inclined ARIS* impactor. Since we will want to compare shear-wave data with compressional-wave data, it is very important to use a similar system for both. It would be asking for unnecessary trouble to compare a shear-wave section recorded with vibrators, with a compressional-wave section recorded with explosives.

There are advantages for systems where both data sets are recorded simultaneously from the same source. The ARIS impactor can be inclined at an angle to the vertical to generate both shear waves and compressional waves. The inclination is in a vertical plane that is perpendicular to the seismic line. By reversing the direction of inclination, the transverse, horizontally polarized shear waves are reversed in polarity without reversing the polarity of

*Marthor is a registered trademark of Institut Français du Pétrole.
ARIS is a registered trademark of the Atlantic Richfield Company.

the compressional waves and of the vertically polarized shear waves (and converted shear waves). Summation of the two records yields a record with compressional waves and vertically polarized shear waves, but the transverse, horizontally polarized shear waves are largely canceled. Subtraction of one record from the other yields a record with transverse, horizontally polarized shear-waves with the compressional waves and vertically polarized shear-waves largely canceled. We will call this record obtained by subtracting records with reversed, transverse inclinations the 'horizontally polarized shear-wave record'. The transverse, horizontally polarized shear wave data are recorded primarily by horizontal (transverse) geophones whereas the compressional waves are recorded primarily by the normally used vertical geophones. The use of these different geophones is significant in providing additional separation of the transverse, horizontally polarized shear-wave data from the compressional-wave data and vice versa.

It is desirable to use horizontal geophones with axes both perpendicular to the seismic line (transverse) and parallel to the seismic line (radial). If the earth was horizontally isotropic, then the horizontally polarized shear-wave record should be zero for the radial geophones. But horizontal anisotropy (due to stress, oriented vertical fractures, or other causes) can result in rotation of the direction of horizontal polarization (Willis, 1986). Indeed, the rotation can be so much that the main response may be on the radial geophones rather than on the transverse geophones.

Also it is possible to record both shear and compressional-wave data using horizontal vibrators for the common source. Cleats with sloping faces are used to couple horizontal vibrators to the earth and these cleats exert a vertically downward force as well as a horizontal force on the earth. This vertically downward force is exerted each time the vibrator pad moves to either side. Therefore the frequency of the compressional wave that is injected into the earth is twice that of the shear wave. So we must double the sweep frequency before correlating to extract the compressional-wave data. In some cases the horizontal vibrator is a satisfactory source for compressional waves. Again, we should use horizontal geophones both perpendicular and parallel to the seismic line to record shear waves.

15.3 SHEAR-WAVE DETECTION OF ORIENTED FRACTURES

Open fractures are a special type of porosity and one that is particularly favorable for hydrocarbon reservoirs because of the high permeability of open fractures. Open fractures at considerable depth tend to be aligned nearly vertically because the weight of the overburden acts to close horizontal fractures. Vertical fractures that have a dominant azimuthal orientation may be detectable with shear waves without using comparative compressional wave measurements. For a vertically traveling shear wave, the particle motion is horizontal. The shear modulus, and the shear-wave velocity, will be reduced no matter what is the azimuthal angle between a vertical fracture and the horizontal partical motion. But the effect on fluid flow in open fractures differs markedly depending upon the relative orientation between a fracture and the particle motion. If shear-wave particle motion is perpendicular to a fracture, then the motion tends to close and open the fractures and hence to increase and decrease fluid compression and to pump pore fluids.

The effects of a fracture and of the pore fluids in the fractures will be significant. These effects relating to fluid flow will be affected by the fluid properties (viscosity and compressibility) and by the relative permeability of the fracture system. The effects are complicated if several different fluids are present, such as water, gas, and oil. Conversely, if shear-wave particle motion is parallel to a fracture, there is little tendency to open or close a fracture and little compression or pumping of pore fluids. Therefore a fracture that is parallel to particle motion will have less effect on a shear wave. Shear-wave measurements at different azimuthal orientations will respond differently to a zone with oriented, vertical fractures.

Shear-wave seismology can yield excellent data, but the technique is not nearly as robust as compressional-wave seismology. It should be used in those cases where it has a good chance of success and where the data it may provide is worth the cost and worth the risk that the data will not be satisfactory. Horizontal vibrators are the best-proved source for shear waves. Record with horizontal geophones both perpendicular and parallel to the seismic line. The ARIS source may or may not prove to be effective. One of the main disadvantages of the Marthor Hammer is that multiple units cannot be used because of problems in synchronization.

15.4 REFRACTION SEISMOLOGY TO DETECT POROSITY AND FREE GAS

Where the target horizon is also a good refractor, compressional-wave refraction data may be quite sensitive to the development of porosity and to the presence of free gas. This is most likely to be true if the expected porosity development and free gas is at or near the top of the refractor. If there is an unconformity at the top of the refractor, then the expected porosity development may well be associated with weathering just below the unconformity which is where its presence would be most detectable from refraction data. The reason that the refraction data would be so sensitive to porosity development and to free gas is that the critical angle is related to the velocity ratio at the interface. Any change in the velocity of the refractor due to porosity or free gas would change the angle of refraction and this would significantly alter the response at the surface. It would cause skips and overlaps in the offsets where refractions reach the surface. Figure 15.1 illustrates these effects, the example shown being based upon a simple model of the Lisburne Limestone on the North Slope of Alaska. To receive the refraction arrival before the direct arrival we would need to use offsets of more than about 11 miles (17.7 km) for this particular example. It is not certain if the direct arrival through the permafrost would cause much of a problem at shorter offsets. But at any offset range the patterns of refraction arrivals at the surface would be the same as are illustrated in the Figure.

Table 15.1 shows some calculated angles, offsets, path lengths and travel times for the refraction from both a low-porosity Lisburne Lime and for this formation where the porosity is higher. If the normal Lisburne Limestone velocity were 22 000 ft/s (6706 m/s), then an increase in porosity of three per cent should reduce the velocity to about 19 960 ft/s (6084 m/s). (The change in velocity with porosity is based upon the 'time-average equation' often used in well log analysis.) An increase in porosity of six per cent should reduce the velocity to about 18 270 ft/s (5569 m/s). The refraction angle should change from 23.93° to about 25.84° and 28.44° respectively. The angle of arrival of the refraction at the surface should change from 34.62° to about 38.77° and 43.17° respectively. Figure 15.1 shows surface skip and overlap zones resulting from changes in the porosity of the refractor. The widths of the surface skip zones and the overlap zones are about 618 ft (188 mm) for a three per cent increase in porosity of the Lisburne Limestone and about 1306 ft (398 m) for a six per cent increase in porosity.

In the surface overlap zones there should be interference between the overlapping refraction arrivals. For the two refractions we can calculate the arrival times at the surface at a common offset in the surface overlap zone. Table 15.1 gives the travel times for refractions up to the surface. The refraction from the porous interval at the far edge of the first porous refractor interval reaches the surface at an offset that is 5142 ft (1567 m) farther out. The refraction from the low-porosity refractor must travel 618 ft (188 m) farther along the low-porosity refractor in order to reach the same surface offset. This added path length in the refractor adds 0.0281 s of travel time so that the total time after the split is $1.1080 + 0.0281 = 1.1361$ s compared with a travel time from the porous interval of 1.1375 s. The ray from the porous zone then arrives 1.4 ms later. The interference would be constructive at seismic frequencies.

Now consider the refraction from the low-porosity interval just beyond the first porous interval. This refraction reaches the surface at a point 4524 feet farther out. For a ray from the first porous interval to reach the surface at the same offset it would have to leave the refractor at a point 618 ft (188 m) closer in. This time the longer travel along the refractor would be at a lower refraction velocity. The added travel time would be 0.0301 s which would

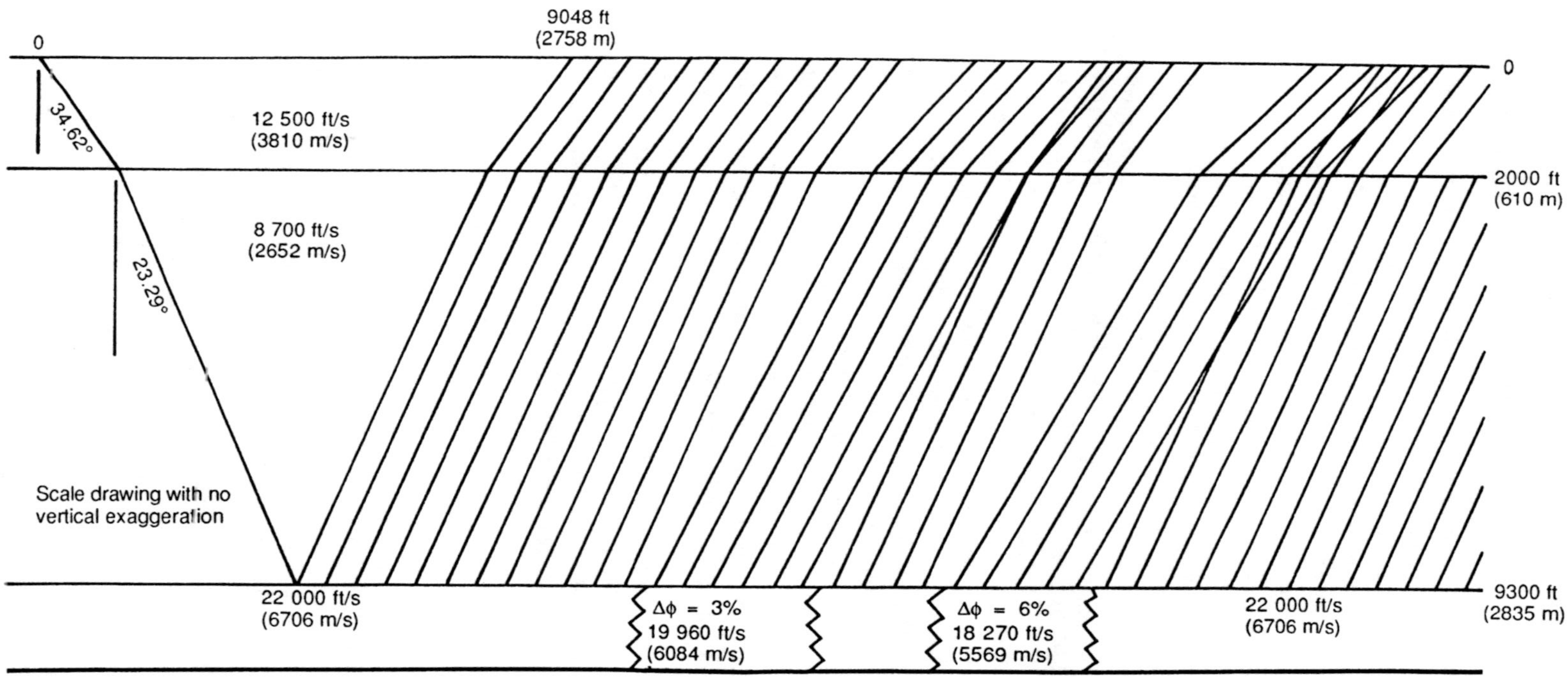

Refractor velocity	Refraction angle	Surface angle	One way offset (from refractor up)	One way path length (from refractor up)	Travel time up in seconds
22 000 ft/s (6706 m/s)	23.3°	34.6°	4524 ft (1379 m)	10 378 ft (3163 m)	1.1080
19 960 ft/s (6084 m/s)	25.84°	38.77°	5142 ft (1567 m)	10 676 ft (3254 m)	1.1375
18 270 ft/s (5569 m/s)	28.44°	43.17°	5829 ft (1777 m)	11 044 ft (3366 m)	1.1736

Table 15.1 Paths for refractions from the Lisburne Limestone to the surface for the model illustrated in Fig. 15.1

Refractor velocity (ft/s)	(m/s)	*Porosity increase* (%)	*Refraction angle* (*degrees*)	*Surface angle* (*degrees*)	*One-way offset (from refractor up)* (ft)	(m)	*One-way pathlength (from refractor up)* (ft)	(m)	*Travel time up* (s)
22 000	6706	0	23.29	34.62	4524	1379	10 378	3163	1.1080
19 960	6084	3	25.84	38.77	5142	1567	10 676	3254	1.1375
18 270	5569	6	28.44	43.17	5829	1777	11 044	3366	1.1736

make the total travel time after the split 1.1390 or 1.5 ms more than the refraction to the same offset from the nonporous refractor interval. Again the interference would be constructive.

The same calculations for the second porous interval with a porosity increase of six per cent show corresponding time differences of 6.3 and 5.9 ms. Again the interference would be constructive for the low frequencies that we would record in a refraction survey at long offsets.

Would we really have surface skip zones devoid of any refraction arrival from the Lisburne Limestone? No, rays from a low-porosity interval just ahead of a porous interval would diffract at shallower angles and fill in arrivals in the surface skip zones. Also rays from the beginning of the porous intervals would diffract at steeper angles and also fill in the surface skip zones. The arrival times in surface skip zones would be the same for both refractions and diffracted refractions. But these simultaneous arrivals in the skip zone are delayed relative to what the arrival times would have been in the absence of a porous refractor zone. From Table 15.1 this delay would be 29.5 ms for a porosity increase of three per cent or 65.6 ms for a porosity increase of six per cent. If the delays increased linearly along the length of the surface skip zones, the delays would be 4.77 and 5.02 ms per 100 ft for porosity increases of three per cent and six per cent respectively (15.65 and 16.47 ms per 100 m respectively). Along the surface the apparent velocity preceding each surface skip zone would be 22 000 ft/s (6706 m/s). The apparent velocity averaged over a surface skip zones would be 20 950 ft/s (6386 m/s) and 19 909 ft/s (6068 m/s)

Fig. 15.1 Refraction survey to detect porosity development in a refractor. The effects of an increase in the porosity of the refractor are:

1. A skip zone and an overlap zone for refracted waves at the surface. The length of these zones for the case illustrated are 618 ft (188 m) for a porosity increase of 3% and 1306 ft (398 m) for a porosity increase of 6%;
2. The refractions from the more porous refractor zones will be delayed in arrival due to the lower refraction velocity. There will be interference in the overlap zones, and the difference in arrival time is likely to be small so that the interference will probably be constructive.

for refractor porosity increases of three per cent and six per cent respectively. Between each surface skip zone and the next overlap zone the apparent velocity would be the porous refractor velocity of 19 960 or 18 270 ft/s (6083 or 5569 m/s) for refractor porosity increases of three per cent and six per cent respectively.

The amplitudes of the diffracted refractions in the surface skip zones would be reduced relative to the amplitudes of refractions prior to surface skip zones. In the surface overlap zones the amplitudes of the composite refraction would increase, and the incremental apparent velocity would undergo a transition from the refraction velocity of the porous interval to the refraction velocity of the low porosity refractor.

These changes in the first break arrival times and changes in apparent velocities are illustrated in Table 15.2 and Fig. 15.2. This example assumes a low-porosity Lisburne Limestone with a velocity of 22 000 ft/s (6706 m/s) from the shot point to an offset of 57 976 ft (16 671 m) where the refractor porosity increases abruptly by three per cent for a distance of 3000 ft (914 m). At an offset of 60 976 ft (18 585 m) the porosity decreases abruptly for a distance of 2000 ft (610 m) where it abruptly increases by six per cent for a distance of 2500 ft (762 m) and then decreases abruptly. The first skip zone in Fig. 15.2 is not too noticeable because this skip zone of 618 ft (188 m) is not much wider than the 500 ft (152 m) increment used in this illustration between points along the refractor from which ray paths have been calculated and drawn. Figure 15.2 shows the ray paths in only the upper 3000 ft (914 m) of the earth. The first skip zone begins just beyond an offset of 62 500 ft (19 238 m). As is indicated in Table 15.2 there are two rays that arrive at an offset of 63 118 ft (19 238 m). One is the refraction from the beginning of the first porosity interval and the other is a diffraction of the refraction from the end of the low porosity immediately prior to the first porosity interval. These two rays have the same path and the same arrival times (since the porosity increase is assumed to be abrupt).

The first overlap zone begins at an offset of 65 500 ft (19964 m) where there are two different ray paths to the same offset. One is a refraction from just beyond the first porosity interval but it arrives 1.5 ms after a second ray which is a refraction from the first low-porosity interval. This first overlap zone on the surface extends to an offset of 66 118 ft (20 153 m) where the refraction from just before the end of the first porosity interval arrives 1.4 ms after an arrival from the low-porosity interval beyond the first porosity zone.

A refraction from the low-porosity interval just before the second porosity interval reaches the surface at an offset of 67 500 ft (20 574 m). The second skip zone begins just beyond 67 500 but diffractions of the refractions from the

Fig. 15.2 Refractions and diffracted refractions arriving at the surface from a modeled Lisburne Limestone. The figure shows refracted rays near the surface from rays emitted at intervals of 500 ft (152.4 m) along the refractor. The two porosity zones modify the pattern of the refractions resulting in skip and overlap zones at the surface.

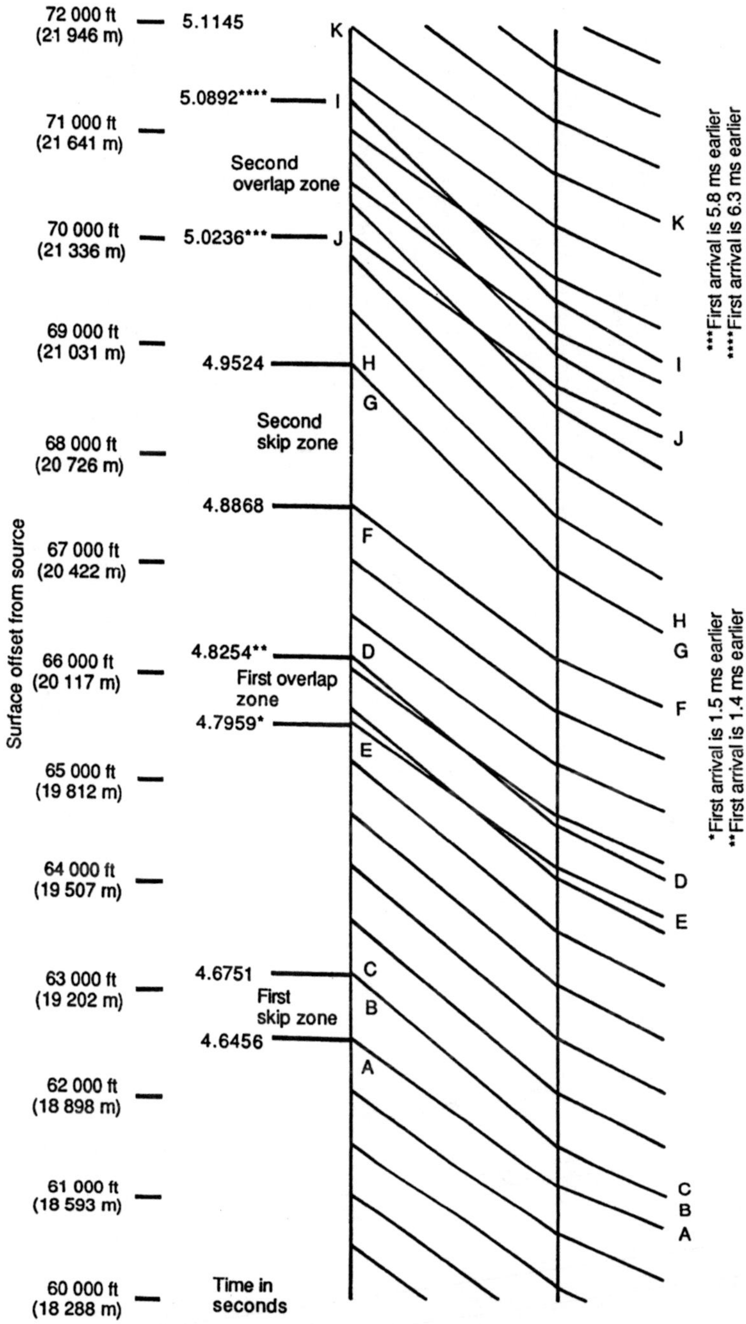

Porosity zone #	Porosity increase	Offset of zone from source from	Offset of zone from source to	Refractor velocity
1	3%	57 976 ft (16 671 m)	60 976 ft (17 585 m)	20 950 ft/s (6386 m)
2	6%	62 976 ft (19 195 m)	63 476 ft (19 348 m)	19 909 ft/s (6068 m)

Table 15.2 Refractions and diffracted refractions from low-porosity and increased-porosity intervals of the Lisburne Limestone model of Fig. 15.2

At refractor *Offset L* (ft)	*Porosity*	*Time* (s)	*At Surface* *Offset L* (ft)	*Time* (s)	ΔL (ft)	Δt (s)	$\Delta L/\Delta t$ (ft/s)	*Remarks*
55 476	ϕ	3.4240	60 000	4.5320				
					2500	0.011364	22 000	$v_a = v_r$ of low-porosity Lisburne Limestone.
57 976 −	ϕ	3.5376	62 500	4.6456				1st skip zone, but diffracted refraction from end of low-porosity interval fills in.
57 976 −	ϕ	3.5376	63 118*	4.6751				End of 1st skip zone.
57 976 +	$\phi + 3\%$	3.5376	63 118	4.6751				Refraction from start of 1st porosity interval.
					2382	0.11938	19 960	$v_a = v_r$ of medium-porosity Lisburne Limestone.
58 594	$\phi + 3\%$	3.6570	65 500	4.7945				1st arrival here (by 1.5 ms) and begin 1st overlap.
60 975 −	$\phi + 3\%$	3.6879	66 118	4.886				2nd arrival here (by 1.4 ms) and end 1st overlap.
60 976 +	ϕ	3.6879	65 500	4.7959				2nd arrival (by 1.5 ms) and begin 1st overlap zone.
61 594	ϕ	3.7106	66 118	4.8240				1st arrival here (by 1.4 ms) and end 1st overlap.
					1382	0.06281	22 000	$v_a = v_r$ of low-porosity Lisburne Limestone.
62 976 −	ϕ	3.7788	67 500	4.8868				Begin 2nd skip, but diffracted refraction from end of low-porosity interval fills in.
62 976 −	ϕ	3.7788	68 805*	4.9524				End of 2nd skip zone.
62 976 +	$\phi + 6\%$	3.7788	68.805	4.9524				Refraction from start of 2nd porosity interval.
					1195	0.0654	18 270	$v_a = v_r$ of high-porosity Lisburne Limestone.
64 171	$\phi + 6\%$	3.8442	70 000	5.0178				1st arrival here (by 5.8 ms) and begin 2nd overlap.
65 476 −	$\phi + 6\%$	3.9156	71 305	5.0892				2nd arrival here (by 6.3 ms) and end 2nd overlap.
65 476 +	ϕ	3.9156	70 000	5.0236				2nd arrival here (by 5.8 ms) and begin 2nd overlap.
66 781	ϕ	3.9749	71 305	5.0829				1st arrival here (by 6.3 ms) and end 2nd overlap.
					695	0.03161	22 000	$v_a = v_r$ of low-porosity Lisburne Limestone.
67 476	ϕ	4.0065	72 000	5.1145				

*Diffracted refraction filling in a skip zone

Table 15.3 Arrival times along surface and apparent velocities for refractions from Lisburne Limestone model of Fig. 15.2 and Table 15.2

Offset at refractor (ft)	(m)	*Offset at surface* (ft)	(m)	*Arrival time* (s)	*Apparent velocity* (ft/s)	(m/s)	*Remarks*
55 476	16 909	60 000	18 288	4.532			
					22 000	6708	$v_a = v_r$ of low porosity limestone.
57 976 –	17 671 –	62 500	19 050	4.6456			
					20 949*	6385	Skip zone with intermediate velocity.
57 976	17 671	63 118	19 238	4.6751			
					19 950	6081	$v_a = v_r$ of limestone with 3% more porosity.
60 358	18 397	65 500	19 964	4.7945			
					20 949*	6385	Overlap zone with intermediate velocity.
61 594	18 774	66 118	20 153	4.8240			
					22 000	6708	$v_a = v_r$ of low porosity limestone.
62 976 –	19 195 –	67 500	20574	4.8868			
					19 893	6063	Skip zone with intermediate velocity.
62 976	19 195	68 805	20 972	4.9524			
					18 270	5570	$v_a = v_r$ of limestone with 6% more porosity.
64 171	19 559	70 000	21 336	5.0178			
					20 046	6110	Overlap zone with intermediate velocity.
66 781	20 355	71 305	21 734	5.0829			
					22 000	6708	$v_a = v_r$ of low porosity limestone.
67 476	20 567	72 000	21 946	5.1145			

*Apparently this is a coincidence

termination of the low-porosity interval arrive at the surface to fill in beyond 67 500 ft. At an offset of 68 805 ft (20 972 m) we have simultaneous arrivals along the same raypath of a diffracted refraction from the high-porosity interval just prior to the second porosity interval and a refraction from the beginning of the second porosity interval.

The second surface overlap zone begins at an offset of 70 000 ft (21 336 m) where a refraction from just beyond the second porosity interval arrives 5.8 ms after a refraction from the second porosity interval. The second overlap zone ends at a surface offset of 71 305 ft (21 734 m) where a refraction from the second porosity interval arrives 6.3 ms after a refraction from the low-porosity interval beyond the second porosity interval.

Table 15.3 summarizes these arrival times and apparent velocities along the surface. The decrease in the apparent velocity (between skip zones and overlap zones) would be a measure of the increase in porosity. Of course, other changes could cause a change in the apparent velocity including the depth to the refractor, static variations, changes in the velocities of beds above the refractor, etc. But we could use reflection seismic data to measure many of these changes, and we could correct the refraction data for these effects.

By shooting refractions in the reverse direction we could locate accurately any porosity developments that were detected. If there were free gas in the developed porosity of the Lisburne Limestone, the reduction in velocity would be greater and the effects measured at the surface would be more dramatic.

If the porosity development were too far below the top of the refractor, then the effects of the porosity would differ from those illustrated in Fig. 15.1. Certainly the effects would be small if the porosity were as much as one wavelength below the top. But for the Lisburne Limestone a wavelength would be 2200 ft (671 m) for a frequency of 10 Hz, or 4400 ft (1341 m) for a frequency of 5 Hz. Since we would utilize rather low frequencies for refraction surveys at long offsets, it seems probable that porosity development in the Lisburne Limestone could be detected at whatever depth it occurred below the top. This example illustrates a possible method for mapping porosity development and free gas in a refractor. It may or may not be effective in this case or in other cases, but it is a method that should be considered where the stakes are sufficient to justify the measurements.

REFERENCES

Robertson, J. D. and Pritchett, W. C. (1985). 'Direct hydrocarbon detection using comparative P-wave and S-wave seismic sections'. *Geophysics* **50**, 383–93.

Willis, H. A. Rethford, G. L. and Bielanski, E. (1986). 'Azimuthal anisotropy occurrence and effect on shear-wave data quality.' *Expanded Abstracts with Biographies 1986 Technical Program* SEG, S9.7, 479–81.

16

Selecting and operating equipment

The next three chapters are involved with practical field operations. In this chapter we discuss the selection and operation of equipment including seismic sources (drills and vibrators), geophones and strings of geophones, cable or data link to the recorder, recording instruments, power-line noise suppression and cancellation, communication equipment, and miscellaneous equipment.

We need to select equipment which is well suited for the job to be done and to the terrain and weather in the work area. We need the right mix of equipment and personnel so that the recording crew loses very little time while waiting on cables and geophones to be laid out or on drillers to finish and load holes. Also we do not want the drillers waiting on the surveyor or the surveyor waiting on permits. Such delays will occur, but the frequency should be rare and the time lost should be minimal.

16.1 SEISMIC SOURCES

If we plan to use explosives in deep holes, we need drills which are effective for the rocks to be drilled and for the depths anticipated. Drills must be able to negotiate the terrain and sometimes more than one kind of drill is needed to cover varied requirements. If we are drilling shallow holes that are close together, we may want a short mast so that the drill can move from one location to the next without having to lower the mast. But for deeper holes we need a taller mast for efficient operation. We may need one or more drills that can negotiate steep, narrow, and crooked trails and one or more that can be easily dismantled into two or three components for transport by helicopters. For swampy areas we may need drills on marsh buggies. For mini-holes in rough terrain we may need small backpack drills and slightly bigger drills mounted on wheelbarrow frames. There are many types of drill in many different sizes. Where water flow is not a problem and we can effectively drill with air, this is usually more economical. In heavily populated areas the dust from air drilling may not be acceptable.

The crew needs to know if there are any cables, lines or drains that they might drill into. We and the crew need to know the local regulations regarding drilling holes and plugging holes and the local regulations regarding storage and use of explosives. The author had a portable crew using mini-holes that had been working in Virginia, Maryland and West Virginia. They were using an explosive that consisted of two liquids which were mixed and poured into plastic containers in the field just prior to loading. Since neither liquid alone is explosive, the operation was safer. But when the crew worked its way into Pennsylvania they could no longer do this. Why? Because in mixing the two liquids they would manufacture explosives, and in Pennsylvania this required a license that was not feasible for the crew to obtain. So the crew had to do their mixing outside the State of Pennsylvania and then transport the explosive material into the state for use. This was not as safe, but was required to comply with the law. We need to seek to have such laws changed or reinterpreted, but this takes time and concerted effort.

For either drills or vibrators we need to know gross weight and axle weight limits on the roads and bridges that we plan to use. Even if these limits are high, there are advantages for smaller vibrators. We may find it easier to get permits to work on roads or on the shoulders of the roads if the vibrators are not too big. But a small vibrator takes an operator and just as much electronic controls as a large vibrator. Where large vibrators are suitable and where more source energy is desirable, then large units are more economical.

With bigger vibrators or more vibrators it takes less recording time to achieve an adequate signal-to-background-noise level, therefore we can save recording time and lower the unit cost. The required recording time is inversely proportional to the square of the effective power, but there are practical limits. Total power may be limited by concern for possible damage to houses or other property. With too many vibrators per VP the source pattern may be too long, in which case we might consider simultaneous recording (Chapter 14) to reduce the unit cost.

We may want the vibrators mounted on buggies for mainly off-road use. This makes moves complicated because it is impractical to drive buggies more than about 50 mi (80 km) on highways. They are too slow and they are dangerous if driven faster than about 15–20 mi/h (24–32 km/h). Another risk is damage to the vehicles if an operator pushes them too hard on the road. The Vibroseis technique loses effectiveness if the vibrators cannot operate in a nearly continuous manner and in many areas buggies can make a big difference in this regard.

In Chapter 6 we briefly discussed the conflicting design requirements for high and low frequencies. If our main interest is in using higher frequencies we select a vibrator with a high ratio of force to reaction mass (greater than 6) and preferably a short stroke for less elastic loss. For high frequencies we also need the best electronics to meet the more precise timing requirements at higher frequencies (Chapter 5 and Appendix H). We should definitely require good nonlinear sweep capability.

If we require good performance at very low frequencies, we need a large reaction mass of at least 4400 pounds (2000 kg) and a big pump capacity in relation to the volume of the cylinder or cylinders. At very low frequencies we need a long stroke of at least 4 in (10 cm) or the reaction mass will not be able to impart much energy to the earth without reaching its stroke limit. There is a frequency below which vibrator output must be reduced because the peak force is sufficient to drive the reaction mass beyond the available stroke length. This frequency is called the displacement limit. The equation for the displacement limit (derived in Appendix H) is:

$$f_{dl} = \sqrt{\frac{19.56\,F}{s_{in}\,W}}$$

or

$$f_{dl} = \sqrt{\frac{49.65\,F}{s_{cm}\,W}}$$

Where

f_{dl} = the displacement limit in Hertz,
F = the peak force,
W = the reaction mass in the same units as the peak force,
s_{in} = the maximum displacement in inches,
s_{cm} = the maximum displacement in centimeters.

The pump volume needed for maximum displacement at the displacement limit is:

$$Q = 0.52\,A_{sin}(f_{dl})s_{in} \text{ in gallons per minute,}$$

or

$$Q = 0.12\,A_{scm}(f_{dl})s_{cm} \text{ in liters per minute}$$

Where

Q = the pump capacity required at the displacement limit (f_{dl}) in the units given above,
A_{sin} = the cross-sectional area of the piston in square inches,
A_{scm} = the cross-sectional area of the piston in square centimeters.

We would prefer the vibrator to have a displacement limit somewhat below the lowest frequency we plan to sweep. For very low frequencies we would prefer a pump capacity that exceeded the volume required at the displacement limit. But if we plan to use a nonlinear sweep that spends very little time at the lowest frequencies, high-pressure hydraulic fluid stored in accumulators may greatly reduce the need for high pump capacity. Even for low-frequency work, the author recommends the use of nonlinear sweeps to use the field time more effectively and to use the vibrator equipment to best advantage.

We need to tabulate comparative data on commercially available vibrators. We should calculate and include these two ratios:

1. Peak force divided by reaction mass.
2. Baseplate weight divided by peak force.

For high-resolution data the first ratio should be 6 or greater. If the second ratio is more than 1.5, then the vibrator is not likely to decouple and ground force amplitude control (Appendix H) may not be justified. Ground force phase lock is recommended for all vibrators. If the second ratio is less than 1.1, then the vibrator may well decouple if a high drive level is used; therefore the vibrator should definitely be equipped with ground force amplitude control as well as ground force phase lock. One might think that a vibrator could not decouple if the second ratio exceeded 1.0. But vibrations that build up in the earth cause the earth to kick back near resonances thereby adding to the instantaneous acceleration produced by the vibrator alone.

For categorizing vibrators it is helpful to plot each type of commercially available vibrator on a graph with the first ratio as the ordinate and the second ratio as the abscissa. If we are concerned about low-frequency performance, we may also want to plot each type of vibrator on a graph with stroke length as one axis and displacement limit as the other axis. These graphs can help us to select several vibrators which might be suitable. We should then consider other factors such as reaction mass, baseplate weight and overall vehicle weight.

If we anticipate a need for source beam steering or simultaneous recording with vibrators, we need vibrator electronics that provide these capabilities (such as the Sercel VE 416).

When using vibrators on paved roads, we need boards or a neoprene pad attached to the underside of each vibrator pad. Particularly with the boards, all edges that contact the road surface should be well chamfered. Neoprene might be better in terms of transferring the force to the road more uniformly over the surface area of the pad. But good results have been obtained with boards as well as with neoprene pads.

How do we convince the road commissioner that the vibrators will not damage his paved roads? For one thing we point out that the tires on a big heavy truck load the surface of the road with a pressure equal to the air pressure in their tires, which exceeds 90 lbs/in^2 (4654 mm Hg, 620 kPa or six times atmospheric pressure). Then we take the baseplate weight, or even the total weight of the vibrator, divide this by the area of your pad, and you will have a much smaller number. It is true that the vibrator pressure on the road is not a static load, but neither is the pressure on the road applied by the tires of a heavy truck—particularly when the brakes are applied.

A problem can arise if the road surface is so irregular or is so curved that only a small part of the pad is in contact with the pavement. For this reason, it is generally better to put the pads down near the center of a lane rather than near the edge of the pavement or partly across the edge.

For work on roads or alongside roads we need adequate signs and enough flag people wearing conspicuous clothing. This is a rather dangerous job.

Every Vibroseis crew needs excellent equipment to check vibrator similarities. For high-resolution work this equipment should give digital values for the phase errors as a function of frequency. Each crew should have equipment that allows the observer to check each vibrator by radio a few times a day during routine operations.

16.2 GEOPHONES AND STRINGS OF GEOPHONES

In Chapters 5 and 6 we discussed the resonant frequency of geophones and the advantages and disadvantages of using geophones with higher or lower resonant frequencies. Preferably the geophones should be damped individually at 60–70 per cent of critical. With modern instruments the input impedance is usually high enough so that it does not materially affect the damping.

It is desirable to be able to tie strings of geophones together, either end-to-end to form long strings, or at one end to form stars. This flexibility is best met by connecting the geophones in each string in series and then connecting strings of geophones together in parallel. We need to be able to connect to either end of any string of geophones. This requires two unbroken wires plus the wire that connects the geophones in series. In addition at each end we need hermaphroditic connectors (or separate male and female connectors that are polarized). For flexibility in the geophone arrays and the number of geophones used per array, six geophones are commonly used per string. If we have strings of twelve geophones, a center takeout for each string increases the flexibility. This makes it convenient to split the center string so that we can lay out the array symmetrically from a receiver position with 12, 36, or 60 geophones per string (as well as 24 or 48). It is not good field practice to try to skid the cable takeouts one half of a string from the receiver positions. This does not work well even with a good crew and it is disastrous with one that only claims to be good.

The spacing between geophones on each string needs to be long enough to layout any spread we are likely to want and still leave some slack. Slack is needed to minimize wind noise. Frequently the spacing required is greater for star arrays and other areal arrays.

Fancy recording equipment and a super computer with grand processing programs may be to no avail if the signal from the geophone is bad. It starts with the coupling to the earth. In soft earth we need long spikes that can reach firm earth. In rocky soil long spikes may not penetrate so that the geophone sits up on a swinging pedestal; so use shorter spikes. The geophones must be in firm contact with the earth. The crew may need to brush away pine needles, leaves, or possibly loose soil before planting the geophones. They may need a short shovel. In snow or ice we may need flat bases that are rotated in firm contact with the surface as the geophone is planted. In this case we want to melt snow or ice and have it freeze back to make a good coupling.

All geophones must be upright and active. Years ago it was fairly common practice to lay a geophone on its side if it could not easily be planted because of bare rock, for example. The idea was that gravity would cause the coil to lie against the magnet and the geophone would be rendered inactive. But if a geophone coil is restrained, this changes the electrical characteristics of the geophone. Once we were doing an experiment on a hard, bedded sandstone. We struggled with a hammer and a hand-held star drill to chisel holes where we could insert the spikes of our geophones. We needed only one geophone per location, but we had only strings of geophones with six to the string. So we tried planting only two geophones per string and laid the others on their sides. This did not work. We received a poor signal until we planted all six geophones at each location.

Is it windy? If so, we may need to cover the geophones with dirt or snow. This sounds like a formidable task, but it need not take too long if everybody pitches dirt.

16.3 GEOPHONE ARRAY AND SOURCE PATTERN LAYOUTS

We shall have occasion to worry about changes in the terrain within a geophone array or within a source pattern. A geophone array may go from a valley underlain by fill material up the side of a steep hill. But in this case the difference in statics (vertical travel time from a shallow common datum) may not be great. The low velocity of the valley fill may approximately compensate for the increased path length of the geophones planted well up on the steep hillside. The hillside is nearly sure to be a resistant rock with a high velocity. The probable differential statics within an array should not exceed one-half the period of the highest frequency components we expect to record. If the slope is too steep along the line such that the static difference would be too great, we may find it desirable to rotate the direction of a linear array sufficiently to reduce the difference in elevation to an acceptable amount. But if the steep slope is underlain by high-velocity rock, then we can tolerate quite a change in elevation. For example, half of a period at 80 Hz is 6.25 ms. If the rock velocity were 16000 ft/s (4877 m/s), then we could tolerate a difference in elevation of 100 ft (30.5 m).

Certainly we do not want geophones in the same array at the top and bottom of a steep, high cliff. If the geophone array is halved by being laid up to a steep cliff (or for any other reason), or if it is rotated substantially; this should be noted on the observer's sheets. The effect of terrain changes within a geophone array or source pattern are more likely to be serious for shear waves than for compressional waves.

16.4 CABLE OR DATA LINK TO THE RECORDER

Multi-conductor cables carrying analog signals were the dominant means to transfer the signal to a central recording truck. Today there are many options. Groups of signals or individual signals may be sent by wires to boxes out on the line where the signals are digitized, coded, and sent to the recording truck via a pair of wires or by radio. In other systems the coded electrical signals are changed to light and are sent via a fiberoptics cable to the recording truck. Or the data are recorded on magnetic tape cassettes at the remote receiver locations. Some advantages of these systems over the multi-conductor cable are:

1. With the increase in the number of traces recorded the multi-conductor cables are heavy and burdensome and other methods reduce the weight;
2. Power-line noise picked up by the cables is a problem that may be eliminated or greatly reduced by these other systems. But power-line noise picked up by the geophones and any connecting wires and connectors ahead of the digitization will not be reduced and often this is the main source of the problem;
3. Atmospheric interference (static) is sometimes a severe problem that may be eliminated or reduced by these other methods. Digitization and coding before transmission is a key reason for this advantage in some of these systems;
4. Certain obstacles are easier to work around with other methods of data transmission. Obstacles can be physical such as rivers or they can be problems in permitting.

These other systems for transmitting data to the recording truck should be considered relative to the program before selecting equipment. The newer methods also have their limitations. Telemetry signals use very high frequencies to transmit large volumes of data and require excellent transmission conditions or they require amplifiers at frequent intervals along the line to boost the signal levels. Some systems work well under dry conditions, but have serious problems or limitations under wet or icy conditions. One crew may be able to make a system work well yet another crew may not. Success or failure often depends upon maintenance of equipment and in learning how it can best be used under specific field conditions.

Multi-conductor cables are still widely used. The burden of heavy cables is partly alleviated with special cable trucks and special power equipment for laying out and picking up. Power equipment for handling heavy road cables must be used with expertise and care to avoid cable damage. In rugged country with limited surface access it may be feasible to use helicopters to move cables and geophones along the seismic line. To get across obstacles like railroad tracks or interstate highways we can use jumper cables (no takeouts) to deviate laterally to a culvert or trestle and then back to the line. An alternative is to use the regular cable segments as jumpers. In this case you need some specially-wired, very short sections to insert to avoid loss of data from the unused takeouts in the cables used as jumpers. These short sections are wired to cancel a fixed number of takeouts; therefore you need several units for various numbers of cable takeouts that may need to be canceled.

Many geophysicists may not appreciate the complexities of multi-conductor cables designed for effective roll-along recording. Cable segments are identical and yet are reversible. Signals from the takeouts from many segments connected together must reach the recorder on many different pairs of wires and in an orderly sequence. It took real ingenuity to work all of this out with hermaphroditic connectors at both ends and with the wires from a male pin at one end connecting to a female socket at the other. The male pin that mates with this female socket is not the same as the male pin that is wired to the female socket. For example, the k male pin at each end may connect to the m female socket at the other end of each cable segment. Then when two cables segments are hooked together, the k male pin of the first segment is connected through the m female to the m male pin of the second segment.

Where cables must cross roads, it is advisable to provide some protection such as a garden hose that has a spiral slit so that it can be put over the cable. For paved roads, use duct tape to fasten the hose-encased cable to the road surface.

16.5 RECORDING INSTRUMENTS

There are several manufacturers of excellent recording instruments. For records from explosive sources in rugged country the emphasis may be on portability and low power requirements. For Vibroseis operations the author strongly recommends a system having an excellent correlator with a wide dynamic range. It would be extremely useful if a manufacturer would make such a correlator that would allow you to record data with a broad band sweep and then correlate and produce separate records with restricted portions of the sweep. This would reduce the field time required to determine the effective upper-frequency limit imposed by the earth. For Vibroseis we must have an excellent system for editing noise and summing records from successive sweeps. When all of these recommendations are met, we may have most of the equipment needed to do effective field processing (at least processing of test data). Field processing can be so helpful in recording better seismic data and the added cost overall may be relatively small. The data acquisition cost may increase moderately, but the processing cost beyond the field should be reduced. The data sent to the final processing center can be already demultiplexed and correlated and the field processing information can include velocity analyses, as well as stacked data.

One of the problems with a big system is the power required in the recording truck. A big alternator driven by a big reciprocating engine produces noise that can be coupled into the earth and picked up by the geophones. A turbine engine to drive a 200 or 400 Hz generator for the instrument power supply would be a boon to seismology.

Filters are an important part of the recording system particularly for data recorded from explosive sources. For telemetry systems the filters are in circuit boxes out on the line where they are exposed to the extremes of temperature of the area. Each data channel requires its own set of filters and it is difficult to make filters which are nearly identical and which do not vary over a wide range of temperatures. In telemetry systems we must not expect filters to perform quite as well or to be as consistent either channel-to-channel or from day-to-day and from morning-to-afternoon. The specifications may be nearly as good as for filters in a central system, but keeping the units within tight specifications may not be practical. In telemetry systems the filters cannot be changed easily once the line is laid out.

With the Vibroseis technique we have our own, built-in-low-cut and high-cut filters with the selection of the sweep plus correlation. With nonlinear sweep capability this filter is even more flexible. Therefore with Vibroseis we are less dependent upon the dispersed filters of a telemetry system. If we also have equipment for canceling noise picked up from power lines (Section 16.6), then the power-line notch filters are not used and the filters for individual channels are relatively unimportant.

When we record data from explosive sources the channel filters are frequently essential for controlling noise and for field evaluation and quality control. It is a good idea to have a number of choices for the cut-off frequency of the low-cut filters. With telemetry systems we may have few choices for the filters actually in the circuit boxes, but we probably have more choices available which can be substituted for the filters in each box. But this takes time.

16.6 POWER-LINE NOISE SUPPRESSION AND CANCELLATION

Electric power companies do not like to lose power to the earth, but to some extent it is unavoidable. A trivial power loss for them is tremendous in comparison to the small electrical signal from a geophone in response to a reflection from great depth. Therefore, our geophones, strings, connectors, and cables must be insulated electrically from the earth extremely well or we are overwhelmed with electrical noise. Even under wet conditions, geophones, strings, cables, and connectors can provide a combined leakage resistance greater than a few megohms. (Some electrical noise is picked up by inductive and capacitive coupling through the air, but this is a trivial part of the problem.)

Years ago when only a dozen traces were recorded, the observer would cancel out most of the power-line pickup by adjusting a grounded center tap of a potentiometer across the secondary of the input transformer for each data channel. If he were both good and quick, this worked well. But as the channels recorded grew to 24, 30, 60, 96, 120 and more this became totally impractical. The industry changed to notch filters located at an intermediate signal level in the amplifier section of each data channel. These notch filters provide very good rejection at essentially one frequency, but they have to be adjusted from time to time to keep them precisely at the power-line frequency. (If we record high-frequency data we may need notch filters for second harmonic frequency as well.) But notch filters preclude the recording of frequencies near the notch for our reflections from the earth, the band of frequencies lost being approximately 8–10 Hz wide. Therefore it is desirable to provide some practical equipment and method to cancel most of the power-line interference so that we do not need notch filters.

The author's team once had a strong need for high-frequency data without losing data from 55–65 Hz. Efforts to get necessary equipment developed within the company were unsuccessful. Cooperative work with an instrument company named Mountain Systems Service* resulted in their Cancellation Optimization System (COS). We put the first unit they produced on an excellent crew which had little difficulty in getting it to work quite well. Other crews, with other instruments, experienced some problems that were overcome largely with improvements made by Mountain Systems Service. Since a large number of these systems have been sold, it appears that the system and equipment are being well accepted by the industry.

The concept for this system is simple. When we are ready to start a vibrator sweep we arm a gate, but do not start the sweep until it is triggered by a power line 'signal' that is purposely picked up. (This is like the starter of a race saying

*Mountain System Service, Denver, Colorado.

'ready, set' and then triggering the runners by firing his gun.) The power-line signal is amplified and then clipped to form a square wave. It is then differentiated which results in a series of spikes of alternately positive and negative polarity. The spikes are at zero crossings of the power-line signal. The negative spikes are 180° out of time phase with the positive spikes for the power-line signal. We trigger alternate sweeps with positive and negative spikes so that the power line pick-up is reversed in polarity for alternative sweeps. We use an even number of sweeps which results in effective cancellation of the power-line signal when the responses to the sweeps are summed.

You need to gain confidence in the use of power-line cancellation equipment so that you will not be tempted to revert to the use of power-line notch filters. Experience will help, but to learn you have to have enough faith to experiment. Some things to consider are:

1. The leakage resistance must be kept high to control the amplitude of the power-line pick-up. If the amplitude is too high the dynamic range will be exceeded and nothing will work (including notch filters). If in doubt, repeat a source point with the notch filters in and see if the record is satisfactory. If it is satisfactory, then the dynamic range has not been exceeded and you can proceed without notch filters. If the notch-filtered record is not satisfactory, reduce leakage;
2. The power-line interference can look bad on unsummed field records and yet everything may be under good control when summed to cancel power-line pick-up. (You may or may not have 'summed field monitors'. If you do, the field records should look pretty good.);
3. The CMP stack will also suppress power-line pick-up;
4. Processing techniques are available to further suppress the power-line pick-up if this should be required. If the dynamic range has not been exceeded, then the power-line pick-up can be handled better in processing than with notch filters in the filed. At worst, the processor can filter as sharply as the notch filters and his filter will not drift. But he can use less filter or no filter (i.e. the minimum required). Better still, he can use a program which measures the remaining power-line pick-up and subtracts it out. This program can be applied to the unstacked traces or to stacked traces. These things cost something, but they insure that we shall not lose data and the reflection signals will not be notched or will be minimally notched. The author does not recall ever having to use these programs on data acquired with the COS system of power line cancellation, but it was good to know that this insurance was there if it were needed.

With Vibroseis data we may be able to salvage data to a reasonable degree even if the power-line pick-up did exceed the dynamic range of good linearity. We should try disregarding the gain bits in the uncorrelated, and preferably unsummed data, and recorrelate. This technique amounts to a brutal noise edit of the data that also acts as an imperfect AGC and an imperfect spectrum balancing program as well. This is a last resort, but one well worth trying which helped to salvage some test data acquired with a telemetry system that was operated under very wet conditions. Telemetry systems do not always solve problems for which they are supposed to be relatively immune. They do not help if the leakage is into the geophones, strings, or other connections ahead of the telemetry system. In telemetry systems that use any conductive lines, if leakage to the lines is very bad, telemetry may not be a solution.

We must be sure that the sweeps are triggered by a good power-line pick-up signal and not by a signal from the instrument truck power supply.

The COS system of power-line cancellation can be used with shot crews. One way would be to shoot twice at each shot point and trigger the first shot on a positive-going zero crossing and the second on a negative-going zero crossing of the power line pick-up. But this would be expensive. Another approach is to record from a shot triggered on a positive-going zero crossing and record a second time with no shot but triggered on a negative-going zero crossing. Then sum the two records. This will tend to cancel the power-line pick-up, but it will increase other background noises since we would record them twice with no means to control their relationship. This technique is reasonably effective and should be used in appropriate circumstances.

16.7 COMMUNICATION EQUIPMENT

Modern seismic data acquisition is a complex operation that requires coordination of a large group of people who are doing different tasks at separate locations. Coordination is essential and this requires good communication. Communication requires enough good radios and the effective use of them. Good radios means radios that are rugged, handy to use, and have sufficient range under the field conditions in the area. Because of possible interference from other users (usually other seismic crews), several channels should be available with the crystals that are required to use these other channels.

16.8 MISCELLANEOUS EQUIPMENT

The surveyor needs good and appropriate survey equipment and this should include a small computer and software to work up his field data. We should consider outputting the survey results on magnetic tape cassettes in digital form. The data processing center can then read the survey results from the magnetic tapes, which reduces errors and speeds up the processing of the seismic data. Cans of bright-colored spray paint are quite useful for marking receiver positions and source positions on the edge of pavement. Flags are easy for vandals or thieves to move or steal. In some cases we want to have small, unobtrusive metal markers that we can implant 'permanently' at key points along lines.

The cable crew needs equipment to check cables in the field. At the very least this includes shorting plugs and ohmmeters. The cable and geophone repair man needs a quality geophone checker that measures geophone performance as well as leakage and polarity. He also needs a more sophisticated cable checker to help him identify and locate faults.

17

Practical line layout and obstacles

In this chapter we discuss the layout of lines, including: (1) starting and ending lines, (2) tying lines, (3) extending lines, (4) recording along crooked roads and trails (5) working around obstacles, (6) crossing rivers, railroad tracks and interstate highways, and (7) avoiding and minimizing data skips.

17.1 STARTING AND ENDING LINES

The top of Fig. 17.1 illustrates the starting of a line with twelve traces with a symmetrical split cable. Line direction is from left to right. (The figure also illustrates the ending of a line if the direction of the line were right to left, but the discussion will treat the starting of a line.) As illustrated there is a tail laid out behind the first source point. The tail shown has six positions but only four traces are recorded (live) for the first record because of a four-trace center gap. The second diagram illustrates the mid-point coverage. As shown the CMP fold provided by the tail primarily lengthens the line at low fold, but it does add some to the fold near the end of the coverage that would be obtained if no tail were used. In some cases the additional data provided by a tail is worth the added time and cost; in other cases it is not. If there is any doubt, make a sketch for your number of traces, gap, etc. and see just what additional data a tail would provide. Then you can better judge whether or not you should use a tail and, if so, how long it should be. Note that without a tail, full-fold coverage begins for a depth point that is half of the maximum offset from the first source point.

Suppose that the first source position is located as far to the left as possible or feasible (no permit for sources farther to the left or with vibrators or big drills having to operate in unfavorable terrain such as swamp, thick woods, etc.). Then laying out a tail beyond the solid line extends the depth point coverage farther to the left, but the fold is low at the first source point and to the left. In some cases it might be important to increase the fold to impro–

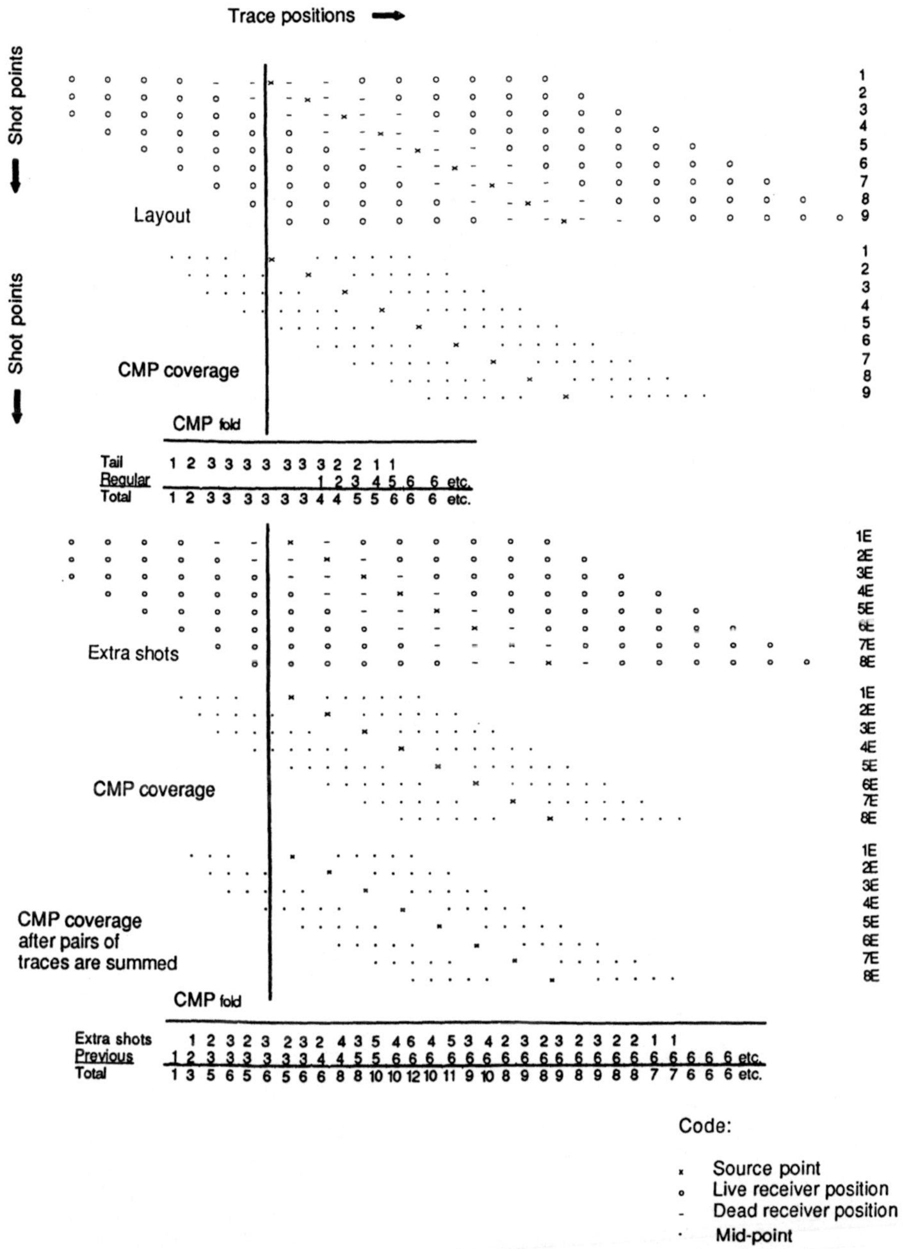

Fig. 17.1 Starting and ending seismic lines. This figure illustrates with a 12-trace recording system the starting and ending of lines both with and without 'tails' and both with and without extra source points to increase the CMP fold near the ends of the lines.

data quality at the start of a line. For example, we might have previously acquired seismic data along a north–south road that has a T-junction with an east–west road. We may want to start a line along the east–west road at the junction with the north–south road (Fig. 17.2). If we want a good tie between the new line and the old line, we need good data right at the start of the line.

Near the middle of Fig. 17.1 we added eight extra source points between the previous source points which reduced the source interval to only half the group interval. But the added CMP points do not line up with the CMP points recorded with the regular source points. Suppose that we sum adjacent traces 1 and 2, 2 and 3, 3 and 4, etc. Now our depth points do line up with the previous depth points. Figure 17.1, at the bottom, shows the added CMP fold that can be achieved with eight extra source positions. Before the adjacent traces are summed an approximate NMO correction should be applied and possibly a static correction as well.

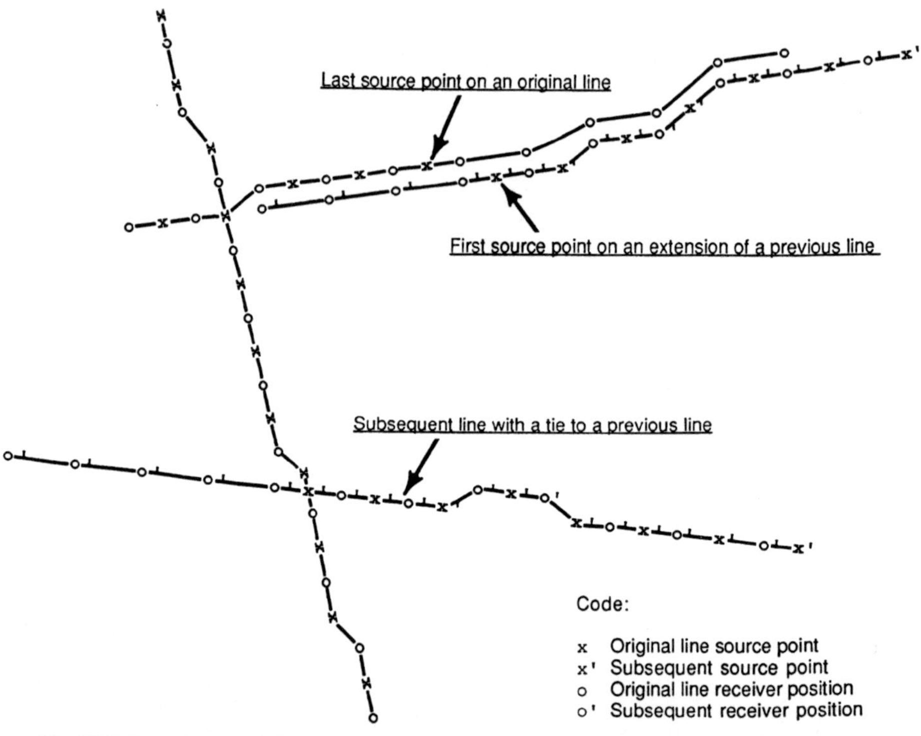

Fig 17.2 Use of cable tails to facilitate line ties or subsequent line extension. By laying out cable tails we can increase the CMP fold near the end of a line and improve the tie with a crossing line. If full tails are used in starting and in ending lines, then a line can be extended with no loss in CMP fold at the point of extension and without repeating any source points.

Would this really help improve the data quality at the tie point with the previous line along the north–south road and, if so, is it worth the cost? With this technique at the tie point we are adding reflection data with some difference in ray paths and some difference in noise. Where the CMP fold is increased substantially, the stack should be improved considerably and the tie with the previous data should be better. The value of using the technique depends upon the stakes involved, the need for a good tie, and the improvement in data quality at the tie that can be expected. Our decision in this matter will have to be based upon our evaluation of the probable improvement, the need, and the stakes. If we plan to record at less than full CMP fold, then we can shorten the source point interval to be equal to the group interval at and near the beginning of a line, and increase the CMP fold without having to sum adjacent traces.

If we are likely later to want to extend a line, then we should use a tail (Fig. 17.2). If we use a full tail we would not need to repeat any shot points (or VPs) and we would lose no fold, but if we use less than a full tail, then the fold will drop somewhat or else we would have to repeat some source points if you extended the line later.

The preceding discussion applies equally well to ending lines. Figure 17.2 illustrates the use of tails in starting or ending lines and in extending lines.

17.2 RECORDING ALONG CROOKED ROADS AND AROUND OBSTACLES

Figure 17.3 illustrates two methods for recording along crooked roads. At the top the source positions and receiver positions are laid out along the road and the intervals are measured along the path of the road. This technique is relatively easy to survey and the crew can layout using the road to take cables and geophones directly to the locations where they are to be used. This technique does result in a lot of lateral scatter in the positions of the depth points. At the bottom of Fig. 17.3 source positions are laid out along the road, but the distance along the road between source points varies. The centers of source points are surveyed half way between receiver positions and projected to the road perpendicular to the trend of the seismic line as a whole. The cable is laid out rather much along the trend of the seismic line as a whole with gentle bends to keep the lateral offset from the source from being excessive. This technique reduces the lateral scatter of depth points.

We can set up a planned layout of cable and source points and get the data processor to generate a plot of the CMP points that will result. Then we can see how much scatter we would have both laterally and longitudinally. This should help to refine the acquisition plan to record better seismic data. Lateral scatter is more serious on a strike line than on a dip line (unless the line is near a

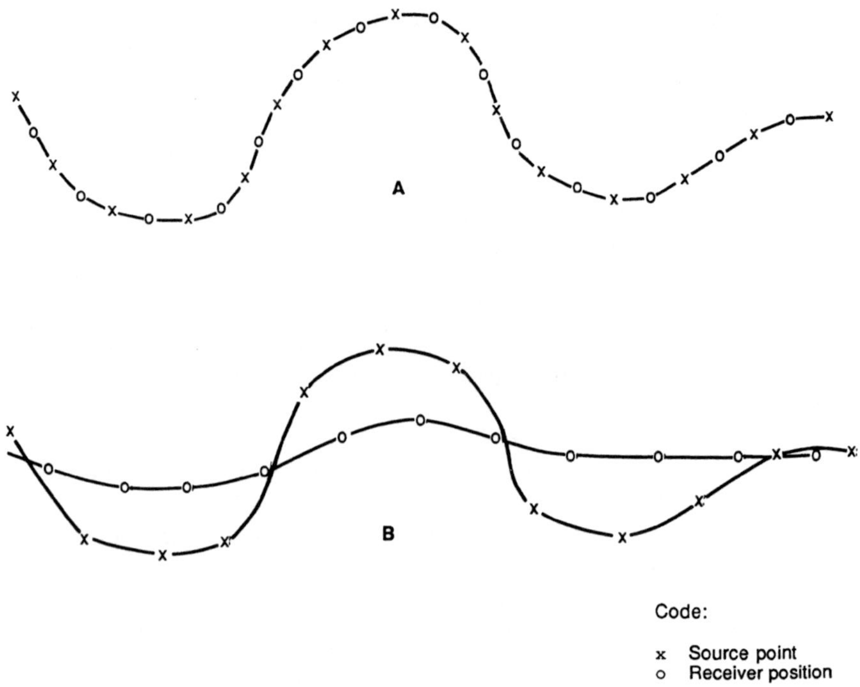

Fig. 17.3 Two methods for recording along crooked roads.
Top: The cable and source both follow the road and the position increments are measured along the bends of the road.
Bottom: The sources follow the road, but the cable follows the trend of the seismic line as a whole with gentle bends to stay closer to the sources along the road. The position increments are measured along the trend line of the seismic line as a whole.

secondary fault or other change along strike). But again, we may be able to take advantage of lateral scatter to get a measure of the cross dip of a key reflector by assuming several different cross dips and seeing what assumption results in the best stack.

Figure 17.4 illustrates a choice in layout where a gently bending road splits to avoid some obstacle and then rejoins. At the top of the figure the sources follow one fork and the cable and geophones the other. This has the advantage that it reduces the lateral scatter of the depth points. The normally used layout with the sources and cable deviated along the same fork is shown at the bottom of Fig. 17.4.

Figure 17.5 illustrates three ways to handle a sharp jog in a road. The top technique is the easiest to survey and layout. This technique also has some advantages in processing in regard to a static correction program that can be used. The middle approach is favored for jogs that are not too long. But most of the time we actually used the top technique. In the middle technique of

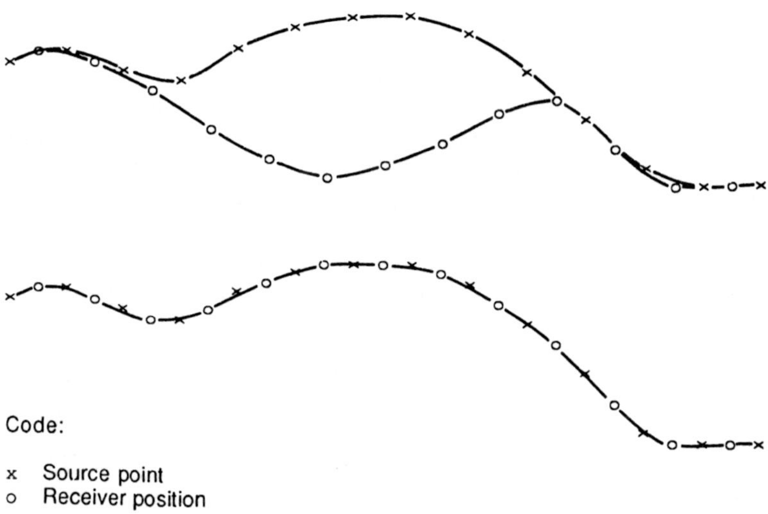

Fig. 17.4 Cable layout to keep CMP coverage along a trend line versus normal layout. Top: Deviation of cable in the opposite direction from the sources to keep CMP coverage along a trend line.
Bottom: Normal layout with cable and sources both along a winding road.

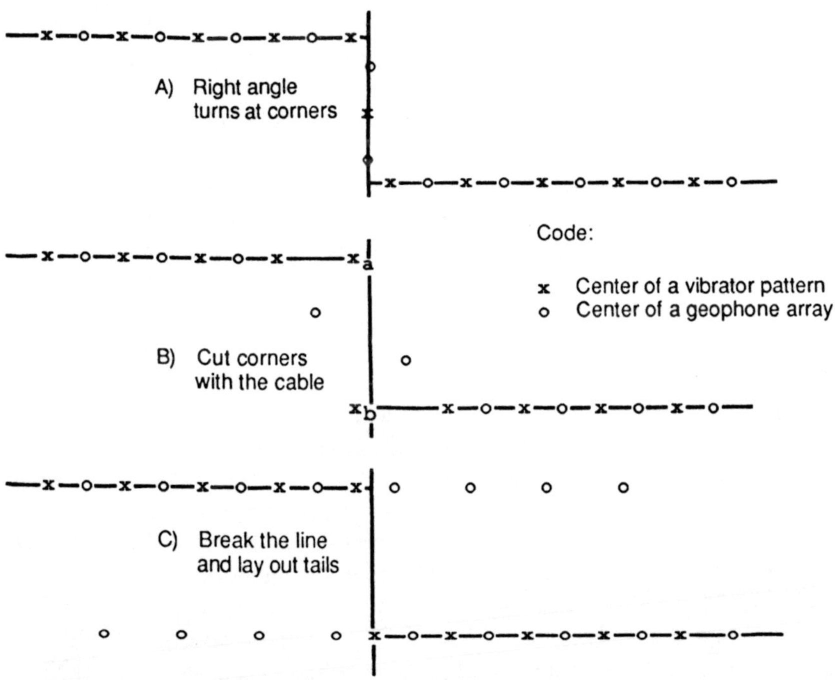

Fig. 17.5 Layout of cable and vibrator points for sharp jogs in the road.

Fig. 17.5 there is an extra source point shown at x_b which is not on the road and is actually offset from x_a. We very well may not be able to locate a source point at x_b, but if we could it would provide an opportunity to see the effects of the lateral jog of the source point with the same longitudinal offsets.

For long jogs it may be better to treat the jog as an end of a line and a beginning of another line. By laying out tails as shown, we would hope to be able to make a good correlation between the two lines where the longitudinal coverage overlaps. If the correlation is very good, then we could choose to merge the two lines in processing. If the lines are merged, this information should be prominently noted on the final section.

17.3 CROSSING RAILROAD LINES, INTERSTATE HIGHWAYS AND RIVERS

In Chapter 16 we discussed the need for jumper cables (and for special inserts if we use the regular cables for jumpers) to allow as to deviate laterally (and back) to a culvert, trestle, or other place where we can get across or around an obstacle. (If you have one of the systems with a radio link back to the recording truck, you avoid these problems. This is a key advantage of such systems.) If it is at all feasible to do so, we want to maintain a continuous cable through the obstacle zone. Otherwise, we should probably undershoot by recording a number of source points with cable and recording truck on one side and the sources on the other. Then we should interchange and repeat.

It is undesirable to have heavy traffic over a cable laid across a road. The traffic causes noise when tires roll over the cable and injury to the cable is possible even when protected by a spirally split hose and taped down with duct tape. When you cannot avoid this situation, try to shoot when vehicles are not near the exposed cable. With vibrators and multiple sweeps the noise will be suppressed considerably by editing and correlation.

Cables can cross rivers on bridges where available. Narrow rivers may be crossed by shooting an arrow across with a string attached, pulling a rope or wire across with the string and attaching the cable to the rope, and stretching the rope across the river in the air tied to trees or other objects on both banks. For rivers with modest currents and not too deep, a jumper cable can be used with any intervening cable connectors placed in styrofoam ice chests to keep them high and dry and let the cable itself sink and run along the bottom. The styrofoam chests can be attached to buoys to hold them in position and to keep them upright. Bright buoys may help to prevent problems with river traffic. An ingenious and resourceful crew can often find a way to do the job. Give them the problem to solve and praise them when they meet the challenge successfully. On a different, but related subject, one crew proposed laying cable and geophones through a railroad tunnel. We agreed after they assured us that there was a safety device that would prevent any train from entering the tunnel while they were inside. It worked out well. Where there's a will there's a way.

At times we may choose to alter the route of the line to take advantage of a favorable place to cross a river, a highway or a railroad track. Let the crew find possible crossing places then you choose the best line route.

17.4 AVOIDING AND MINIMIZING SKIPS

We should seek to prevent data skips as much as it is reasonably feasible to do so. If we are laying cable across a bridge we may be able to plant geophones on ground under the bridge for part of the way. We lay out full geophone arrays when we can, but even one string of podded geophones can help fill in data that might be better than losing a trace. The data processor can throw out the trace if it is not helpful. In towns we may be able to find a median or an unpaved strip between a sidewalk and the street where we can plant geophones. Where there are roads or driveways that are wider than the spacing between geophones, we use jumpers from one string to the next string in the same array to get across and layout as nearly as we can a good array. Where we do have to significantly shorten a string, we space the geophones out over the available distance rather than ending up with many geophones planted in one bunch.

Generally most of the data skips are with the sources. If we are shooting explosives in deep holes and some source positions are too near houses or wells, could we permit and safely use a lighter charge? If we are using mini-hole patterns, could we shoot and separately record one mini-hole charge at a time and sum the records in processing? Where we were using six-hole patterns, we could leave out some of the holes closest to a house or well. If we are using vibrators, would reducing the drive level allow us to record at the position? Could we record with fewer vibrators or sweep a portion of the source pattern? How far we push these things depends upon the importance of the data, local regulations and our good judgment. We must have responsible people carefully observing the operations when we are working through heavily populated areas.

Although we want to avoid any skips we can, it becomes more critical where there are many consecutive skip locations. If we are recording at less than full fold (i.e. using a source interval greater than the group interval), then we can add some source positions near and on both sides of the skip to help to maintain CMP fold in the area of the skip.

In some cases it is worthwhile to undershoot extended skip zones. By undershooting, what we mean is to use longer than normal offsets in the direction of the skip. Crews sometimes waste a lot of time in undershooting; therefore it is worth our time to plan anticipated undershoots and provide the crew with guidelines for unanticipated undershoots. One crew, employed for a short time only, took it upon themselves to record about twenty extra shots at the same location prior to a skip zone. For consecutive shots they simply advanced the roll-along switch one position at a time. What they did was repeat 59 traces and add one new trace per record. They could have added the 20 traces with one additional shot by advancing the roll-along switch 20 positions. They repeated 20 costly and ineffective shots on the other side of the skip (and again moved the roll-along switch one increment between shots).

In some cases we will not be able to maintain a continuous cable across the obstacle. In this case we can still undershoot by recording with the cable and truck on one side and the sources on the other side. We can record an appropriate number of source points and then switch sides and record another appropriate number of source points. Of course this only works if the undershoot interval is less than the maximum effective offset. There is no use using offsets that are way beyond what is effective for the objective. One exception would be if the target is very shallow and we have good reflections from much deeper interfaces. In this case, by using very long offsets that still record the deep data, we can maintain continuity for the deeper reflections and thereby maintain a good solution for our static correction program.

A member of the author's group wrote a computer program to help them to plan undershoots. It provided printer plots of:

1. the stacking diagram,
2. the offset distances in each CMP gather along the line, and
3. the CMP fold along the line.

This program was quite helpful in allowing them to plan more quickly a good undershoot of a river, town, no-permit area, etc.

17.5 VIBROSEIS SKIP FILL-IN WITH MINI-HOLE SHOTS

In a Vibroseis program we may be able to work along roads most of the way but have some large skips where there is no road (nor a trail that the vibrators can negotiate) or where the road loops far from the line and loops back subsequently. In these cases we should consider the advantages of filling-in the otherwise skipped source positions with mini-hole shot patterns. It takes little additional equipment to add this capability to a Vibroseis crew. We need a blaster, some caps, suitable storage for the caps, and some explosives with suitable storage. The explosives can consist of two non-explosive liquids that are mixed in the field just prior to loading to form the explosive material. We also need a few portable drills and bits. Often much of this equipment can be rented if the need is of short duration. The author used this technique extensively, with excellent success in most cases in acquiring good fill-in data.

Fill-in with mini-hole data presents a challenge to the data processor to make the fill-in data from explosive sources compatible with the main vibrator data and to merge the two sets of data in common CMP gathers. It would be easier if we used a minimum-phase correlation for the Vibroseis data, but this may or may not be acceptable to the interpreter (Chapter 5 and Appendices G and H). In addition to phase considerations, the processor should make the two amplitude spectra as similar as possible (Chapter 19). It takes a team effort to acquire data that the processor can handle well to produce displays that your interpreter will find most useful in solving his exploration problems. Communication, cooperation, mutual understanding, and mutual respect are required for maximum success.

18
Putting it all together

This chapter covers miscellaneous problems including crew management, public relations and philosophy and it summarizes much of previous discussion.

18.1 THE JOB OF THE FIELD GEOPHYSICIST

The job of the field geophysicist is larger than just acquiring seismic data. The purpose of his work is to help the interpreting explorationist so he can find and sell drillable prospects which, over a period of time, will be successful and will much more than repay the company for the cost of the exploration program. Both we and the data processor are in the service business and our mutual client and customer is the interpreting explorationist. Both of us should seek ways to help the interpreter. This starts with the prospect form. We need to be sure we know what the explorationist needs and what would be most helpful for him. We can make sure that the data processor also knows the needs of the interpreter and what would be most helpful to him. We certainly want the processor and the interpreter to communicate directly, but there are many reasons why this communication might be incomplete. Since we will work directly with the processor, we can easily take opportunities to relay information, suggestions, and ideas between the interpreter and the processor that will be helpful to both. This is discussed more fully in Chapter 20.

Whereas the interpreter is our client and customer, this does not mean that he tells us what to do or how to do it. We are responsible for providing him with the data that will best meet his needs at the minimum cost. If he insists on data at 2 ms sample rate even though the sweep only went to 85 Hz, try to convince him that this is a waste of money unless there is some valid reason. For example, the plan may be to do some beam steering (Chapters 11 and 19) in processing and a 2 ms sample interval would facilitate the beam steering. If he persists in wanting a 2 ms sample rate without any valid reason, you will have to take the issue to higher management. Hopefully it will not go that far, particularly if the interpreter is higher management. We all win some battles and lose some battles, but we must have some room to operate in to be effective in our job. If that room is too restricted, our opportunity to contribute is small. But we may be able to gradually increase our operating room by patient discussion and growing success when we are given opportunities to improve and innovate. It is unlikely that you will be given the perfect job; rather you will make your job more perfect as you perform well in serving the needs of interpreters.

18.2 PLANNING ONE OR MORE ACQUISITION PROGRAMS

It is rare that the start-up of a new crew on a new project is smooth and without delays; and there are expenses in mobilizing a new crew. Most important, it takes time to develop good communication and good working relationships with a crew. For these reasons there are advantages to planning a series of acquisition programs, within an area of reasonable size that can be effectively carried out with a crew with essentially the same equipment and personnel. Flexibility in being able to move from one project to the next is advantageous if circumstances warrant or require a change in plans. From the outset we may plan to test and perhaps acquire some stackable data in an area, do another project, and return to record the bulk of the program after the initial data have been processed and reviewed by ourselves and by the interpreter. Flexibility is important if weather or permit problems arise in one area; we may have permitted ahead, but permits may be rescinded and we may need time to recover. When crews are kept for a period of time they improve; if they do not improve, we do not keep them long. In good times, when the company has plenty of work to do and an adequate budget, it is recommended that we keep the best crews indefinitely, working somewhere for the company. This requires effort, planning and coordination, but the rewards can be significant in terms of data quality and cost.

In planning a program or a big group of programs we must take into account the time of year, the weather, the crops and other land activities. A farmer is much more likely to give a permit when his land is fallow than just before a crop is ready to harvest. Even if he is paid generously for the crop loss, he hates to see the fruits of his hard work destroyed. The hunting season, lambing season, and bird nesting season must not be forgotten in our planning.

At the outset it is prudent to check if we have good maps and if not, arrange to get what we need. Topographic maps may be out of date; old roads and trails may no longer be passable but there may be new roads that would be helpful. Where we plan to run a line through apparently undeveloped country, there may now be a subdivision. We must get the latest maps, but also check things out on the ground or have a crew member check them. We need to pay particular attention to lands controlled by multiple governmental agencies. Permits may be quick and easy or they may be difficult; and, if given, they may require many months to secure. Just because it was quick and easy last year does not mean it will be this year.

18.3 AFTER THE TESTS

In Chapter 7 we conducted some tests. The next step is to make our final decisions on technique and select recording parameters. These decisions must be given to the party manager of the crew. Be certain that he understands. It is best to give written instructions, including diagrams, and to go over the written instructions orally either in person or by telephone. Written instructions and data can be transmitted by telephone with relatively inexpensive equipment. If instructions must be given orally (by telephone), the party manager should repeat back what he understood was said then written instructions and diagrams should follow.

Particularly in a new area we should monitor the data closely. We may find it advisable to have the crew send us two representative field records every day by mail or by telephone facsimile. This gives us a chance to note problems quickly and to react quickly. If we see a way to improve data quality, the change can be tested by recording two or more records both with and without the change. If the change proves to be helpful it should probably be adopted immediately rather than waiting to do so at the start of the next line. Of course, the change should be well documented and the information should be noted on the final sections. Where our play is stratigraphic and we are looking for a subtle change in character, we might need to wait until the start of the next line to make a change.

From past experience in an area we may anticipate that data quality will probably degrade when we come to a previously troublesome outcrop. Consider the last example in Chapter 7 where previous data on the dolomite outcrop were poor. The test was planned to evaluate the best parameters for both on and off the dolomite. We had already decided that we may well increase the CMP fold on that portion of the line that will be on or near the dolomite. We may decide from the test results to make other changes, when the crew gets on the dolomite, in the geophone arrays, source patterns or both. In this hypothetical case the objective was to map structure. We have had success with programs like this where major changes were planned and made along the line based upon test results and previous work in the area.

If data quality should deteriorate unexpectedly along a line, we should consider the possible advantages of stopping work on that line. The crew could start another line or even another program if one is available not too far away.

The purpose would be to have some time to investigate what the problems might be and develop one or more possible solutions to try.

Consider the second test example in Chapter 7 where we had an important 80-mile program in a bad record area. The test results might be inconclusive and yet we would have had to make a decision and get the drillers started right away. Suppose we decided to use charges in holes about 100 feet deep. With more time to work with the test data we might later conclude that six hole patterns of charges in holes ten feet deep would work better. The main purpose of this initial program was to see if we could find a practical method to explore the area. We should definitely try the patterns of shallow holes. In this case we would make a major change in the program. The only question is when we should make the change. Certainly we should shoot and record the deep holes already drilled. Also we should get enough stackable data with the deep holes to evaluate that technique. (It is still possible that it may prove to be the best approach when the data are fully processed and displayed.) We would need to get enough mini-hole drills and drill crews to make an efficient operation and the drill crews would have to get enough patterns drilled to keep well ahead of the recording crew. We may well decide to start the mini-hole crews on the second line and go ahead and finish the first line with deep holes if the first line is not too long.

Should we repeat some portion of the deep hole line with mini-hole patterns for a direct comparison? That would be nice, but is that the best use of the time? Probably we should defer this question until near the end of this phase of the program, as by that time we may have a better idea of the results we are getting. If we are really uncertain about which is the best source, we may choose to record more program with deep holes where we have already recorded with the patterns, rather than repeating part of the first line with patterns. This would be particularly appropriate if the chosen line segment showed some interesting data. But we may be quite pleased with the pattern results, in which case the interpreter should be involved in deciding whether it is more important to repeat part of the first line with patterns (for a direct comparison) or to record some other line with the mini-hole patterns.

18.4 CREW MANAGEMENT

The party manager is the field manager and we must always work with and through him (or a crew member designated by the party manager). If he is not available we may well need to discuss matters with other crew members and proceed to take appropriate actions, but we need to be sure that the party manager is informed of what has taken place and why. At times the surveyor may call us directly with a surveying problem and he may be miles from the rest of the crew. Just close the conversation with an instruction to the surveyor (or permit man, etc.) to inform the party manager of your conversation and your decisions. Then the next time you talk to the party manager refer to the conversation and be sure that he knows what was decided and why.

It is important to keep communications open. The party manager should call us if he has any questions or if a significant problem arises. We should have him give us a formal report by telephone once a week. It helps to have a printed crew report form (Appendix A) that he fills out before he calls and you can fill in the blanks on a copy of the same form as he gives you the information. Obviously communications should not be limited to the information on the form.

It takes a great deal of communication and coordination to get things running smoothly and to keep them that way. This is particularly true in a portable crew operation where we record mostly short lines (because of the high costs). Also the program may be scattered if we are recording final lines on a number of previously found prospects. We may have a permit man in one place, a surveyor somewhere else, the drill crews at a third location, and the recording crew at a fourth—and the poor party manager has to try to keep up with everything and keep us informed as well. Fortunately it is seldom quite that bad.

The crew needs to know that safety is extremely important to us and to the company. They should also understand that it is important that we establish and maintain good relationships with the public and with officials.

We must stress repeatedly the need for all crew members to be careful in what they disclose about crew operations. If someone wants to know about the recording parameters or technique, the crew must refer him to the industry scout group or tell him that they cannot divulge such information. Certainly any information or hearsay about results is strictly confidential. This problem may be more critical on a crew that does data processing in the field. We should tell and remind the data processor to shred discarded results and we need to set up strict controls on who is and who is not allowed to see processed data and for what purposes. During wartime a drunk sailor's loose lip can sink a ship. In our operations, a careless crew member can ruin an otherwise good crew and this must be understood by every crew member and by crew management at all levels.

We should not expect crew members to be angels—angels are scarce anywhere here on earth. But we can require them to be responsible on the job and reasonable in their conduct and actions off the job.

One of the things we should always be looking for and asking others to note is what are the bottlenecks that hold up production? Once a bottleneck is identified we need a solution, but a solution that does not require more people, if possible. If more people is the only reasonable answer, try to figure out when and how the crew can be cut back at a later date. Crews, like bureaucracies, grow big and fat unless carefully watched and occasionally pruned.

Surveying is one of those essential things that we may not pay much attention to until there is a problem. Of course we need accurate surveying, accurate and complete notes, and quick final survey results. But the surveyor is also the key to laying out the field program in the most effective manner from both an operational standpoint and from the standpoint of the best seismic data to meet the exploration objectives. We can help the surveyor by seeing that he has the appropriate equipment and by sending him the best maps possible as early as is reasonable. We need to work with him to help him understand our preferences in line layout and when he should call for our input in specific situations. We want him to study the terrain and cultural obstacles and to suggest alternatives, but we do not want him, on his own, to make decisions which may affect the suitability of the seismic data to meet the needs of the interpreter.

18.5 PUBLIC RELATIONS

It is imperative for the company and the industry that we all do whatever we reasonably can to develop and keep good relations with the public and with governmental agencies. We do not want to get the reputation of being suckers who will pay dearly for fraudulent, frivolous, or minor claims. We want to develop a positive reputation of being good citizens engaged in essential work who are reasonable and are concerned about people, property, safety, and the environment. If someone's car is stuck in a ditch, we can take a little time to help to get them out. If we rut up a unpaved road, we can see that it is promptly smoothed better than when we arrived at our cost in cooperation with the local road agency. We can avoid littering and we must clean up thoroughly along our lines when we are through. If feasible, we can use local people and local merchants for services and supplies to make ourselves more welcome. Most of all we can be alert to any adverse reaction or damage we may cause and seek to solve the problem quickly before it gets worse. We can show respect for everyone and for their property. We can take opportunities to explain our operations and the importance of them without giving away information that is confidential.

18.6 SUMMARY OF RECOMMENDATIONS FOR LAND DATA ACQUISITION

This summary is brief to minimize redundancy; therefore general cases will be emphasized at the expense of exceptions.

1. Use a symmetrical split cable, shoot full-fold with source centers half way between receiver positions. If we have to shoot only half maximum fold, then shot points may be centered at receiver positions if double group interval arrays are to be used depending upon our preference.
2. Determine or estimate the maximum effective offset for reflections from the maximum depth of keen interest. Frequently this offset is approximately equal to the maximum depth of interest. If offsets are too long, reflections may become refractions or the reflected signal may be weak and may compete with strong noise from first breaks or reverberating first breaks. But the greater the maximum offset the greater the group interval we can use with a given number of channels. The longer the group interval the fewer records required per mile, the greater the production rate, and the lower the costs per mile. If we are working along very crooked roads we may choose to limit the maximum offset to reduce the lateral displacement between reflection depth points. Also we may reduce the maximum offset to limit loss of higher-frequency components particularly if long arrays are required to suppress noise.
3. Determine or estimate the minimum effective offset considering the objective, but also consider the value of mapping shallow events for geologic information and also for static corrections. The minimum effective offset is frequently one that results in a few hundred feet (150–250 m) of separation between the nearest live geophone and the closest source position. For a dynamite shot in a single, deep hole the minimum effective offset may be shorter.
4. Determine or estimate the maximum frequency that can effectively be recorded from depths of significant interest. Where vertical resolution is quite important in meeting the exploration objective of the work, try for somewhat higher frequencies (but be reasonable). Try to achieve an effective frequency bandwidth of at least two octaves. If we are pushing for superior vertical resolution data with the Vibroseis technique, we try to expand the effective bandwidth to 2.5–3 octaves. We should use an appropriate, nonlinear sweep to achieve an adequate S/N ratio over the sweep band in the minimum amount of recording time.
5. Select a group interval that together with the number of channels available will yield a maximum offset that is approximately equal to the maximum effective offset. Check to see if this group interval is short enough so that it will not exceed half of the apparent wavelength of the highest frequency component of reflections from your objective. If this group interval is too long, consider reducing it by adding more channels, shortening the cable, or lengthening the minimum offset.
6. Use linear, equally spaced and equally weighted geophone arrays with effective lengths of twice the group interval. If such arrays would severely limit the response at needed higher frequencies, use arrays with effective

lengths equal to the group interval. In very good record areas where source-generated noise of moderate wavelengths in a minor problem, group interval arrays may be optimum. If long-wavelength noise is an overwhelming problem, consider triple or quadruple group interval arrays. (If a quadruple group interval array is selected and we are shooting half-fold, we put shot points on trace centers.)

7. If rather steep dips are to be mapped then either the geophone array or the source pattern must be short. If dips to be mapped are steep only in one direction we may want to consider the use of a single-ended cable or a non-symmetrical split cable. In this case we could use a long geophone array in the updip direction of the expected steep dips to be mapped (or a long source pattern updip from short geophone arrays).
8. Determine or estimate the wavelengths of coherent, source-generated noises that may be most troublesome at objective times and formulate an overall plan for suppressing these noises.
   (a) Extremely long wavelength noises may be attacked with the CMP stack, but other ways of suppressing such noise should also be considered. Frequently ground roll and other long wavelength noises are the main noise problem and these may be reduced substantially by shooting deeper with explosive sources. Use of nonlinear sweeps with the Vibroseis technique is often quite helpful since this tends to suppress the lower-frequency components which have longer wavelengths. Many times this will allow reflections with higher-frequency components to stand out.
   (b) Noises with moderate-to-long wavelengths of 50 ft to perhaps 800 ft (15–240 m) may be attacked with the geophone arrays and source patterns.
   (c) Noises with short wavelengths must be attacked by close spacing between geophones or between source positions. Alternatively, we may be able to use a minimum offset long enough to record data from the objective before the short-wavelength noise arrives.
9. Adjust the geophone array and the source pattern to suppress wavelengths of the stronger noises near objective times. Once the geophone array is set it is often desirable to adjust the source pattern relative to the geophone array so as to provide notches of good noise suppression at wavelengths not well suppressed by the geophone array. This is most critical for wavelengths of strong noises. Consider the overall suppression of noise by geophone arrays, source patterns and CMP stack.

18.7 PHILOSOPHY

A data acquisition program to get better seismic data may be compared with being sent to a mountainous area with the mission to take some mountain-top pictures. We have a limited amount of time and money, but we want to get some great shots. First we have to decide which mountain to climb. From the roads winding through the valleys it is hard to tell which mountain is the tallest, and of the taller mountains it is difficult to say which we can climb with our gear in the time available and within our budget. We decide quickly that there may be no right mountain. Rather there are several mountains that will afford good shots. So we quickly drive the available roads and pick out a couple of likely prospects. We study a topographic map and pick out a mountain, a path and a location to reach. Near our location we may find a better spot or perhaps we cannot actually get precisely to the planned location. We find a good spot and take our pictures. From our lofty perch we can survey the nearby mountains more clearly, and we take time to visualize the shots we could get from them. Also we can better see the problems faced in climbing other mountains. If we ever return to this area, we shall get even better pictures.

Similarly in seismic data acquisition there may be no right answer to the question of how to get the best seismic data. But hopefully there are some answers that will give you better seismic data. Your job is to find a near optimum answer, so look at the possible approaches. You quickly discard some of the foothills and some of the lofty peaks as being economically impractical. But you find a few mountains that hold promise and survey the difficulties and opportunities for each. You may test more than one idea in the field by climbing part way up to see the view and to see the obstacles; then you make your choice and begin. After you start you may shift your location, but you are now too committed to change mountains. But having stood on one mountain you can better see what you should do to get better seismic data if you should return.

19

Seismic data processing

In this chapter we briefly discuss seismic data processing from the standpoint of the field geophysicist and what he needs to know about processing to do his job well.

19.1 OPTIMUM INTERRELATIONS BETWEEN ACQUISITION AND PROCESSING

Computers and processing programs are so versatile and powerful that data processors sometimes brag that all they need from the field are records with adequate sampling and CMP fold and with reflections not mutilated in the field. There is some truth in their point of view. But give them field data with reflections completely buried by many different noises and see how often they spend the time and effort to find the keys required to successfully process the data with all their wonderful tools. They are overwhelmed with options, and the combinations of options and processing parameters are staggering. If it takes the right combination of several procesing options and sets of parameters to reveal reflections, then the processor may never be so lucky or so patient as to find the complex key to success.

Where noise is very strong, better field techniques such as more effective geophone arrays and source patterns may be required to record both the noise and weak reflection signals from depth within the dynamic range of the recording system. Where this is the case, the strong noise must be suppressed in the field, or the project is doomed, no matter how skillful, lucky, and persistent the data processor may be.

The author's view is that in data acquisition we should do what is feasible to suppress noises and bring out the reflections so long as we do not significantly damage our reflection signals from the zones of exploration interest. Some may think he goes much too far in Chapter 4 when he said that, with reluctance, he would accept a 30 per cent loss in amplitude for the highest frequency components for the CMP gather as a whole if it appeared to be necessary to handle strong noise. But he feels much more comfortable when at least glimpses of reflections near objective times can definitely be seen in the field data. But even when this is not the case, processed results may still be quite satisfactory.

Our field techniques and parameters must be selected so that the processing that is likely to be needed will be most successful. An example here is the recommended use (with full-fold recording and split spreads with even gaps) of double group interval arrays or arrays with effective lengths that are an integral multiple of the group interval. This recommendation is to give the CMP stack the optimum opportunity to suppress noise. It is the field geophysicist's job, in consultation with the explorationist and the processor, to select the techniques and parameters that will result in the best seismic data after it is fully processed. To do this he must know the exploration objective, the needs of the interpreter, what processing is likely to be required for best results, and what field techniques and parameters would optimize this processing. That is a tall order, and that is the reason for this chapter and part of the reason for the next chapter on communication.

Seismic data processing of digitized data by huge computers may seem mysterious to those of us who have not specialized in data processing, but mostly it involves the same computations that have always been required to utilize seismic data. With computers, with their speed, accuracy and memory, it is practical to make routinely detailed calculations and analyses that an interpreter or group of interpreters would have done only very selectively before the computer age. Without computers it would not be practical to record so many channels of data. The common mid-point technique with high fold would not be practical without computers. Our data is recorded in multiplexed format that would not be feasible without computers to unscramble (demultiplex) the data.

With computers, data processors make elevation static corrections, NMO corrections, and residual static corrections. They migrate the data when reflections show significant dip and to collapse diffractions so as to resolve faults and internal structure better within stratigraphic sequences. Most of these were done rather routinely before we had computers. To make elevation static corrections we had to estimate a correctional velocity and specify a datum. For NMO corrections we had to extract moveout velocities from the data. There is little essential difference in these steps today. Formerly we noted strong noise with linear moveout on our records and figured out trace mixes that would suppress these noises. This is analogous to multichannel filtering with the computer.

Automatic gain control (AGC) was essential for analog records. With digital recording and computers there are many ways to control gain. We avoid the distortion of AGC in the field and in our processing until, and unless, we want to apply some AGC.

The fact that these types of computations and corrections were made prior to the use of computers hopefully takes away some of the mystery of computer processing of seismic data. But it should not obscure the fact that computer processing has allowed great improvements in seismic data and many of these improvements are routinely accomplished with relative ease. The residual static correction program and velocity analyses are good cases in point. With the computer, massive amounts of statistical data are compiled and utilized to extract static corrections from reflections, refractions or both. These corrections are constrained to prevent unwarranted undulations or drift in the static corrections. In velocity analyses, common mid-point gathers (or sums of a number of gathers) are NMO corrected for a near continuum of velocities and the coherence is calculated and plotted. This is done over finely sampled time intervals. (Measurement of coherence is a component of many processing programs. The method used is similar to correlation which is described in Appendix C and the result is called semblance.)

Certainly computers have added tremendously to our frequency filtering and wavelet shaping capabilities as compared with the analog filters that formerly were more crucial. Formerly, additional filtering was accomplished by the geophones and by the galvanometers in the recording camera. Special geophones with various resonance frequencies and damping were keys to more satisfactory records. Even the tension on the galvanometer strings was adjusted to change the filtering effect of the galvanometer to suit the user. Today field filters are much less critical. In fact, their main use is to provide a field record that is more useful in terms of quality control.

19.2 DECONVOLUTION AND SPECTRAL WHITENING

One of the major advances available with computers involves deconvolution and other techniques for wavelet shaping (which includes spectral shaping and phase adjustment). Perhaps this is the area of greatest mystery to those not intimately involved in seismic data processing.

To make a simple synthetic seismogram we calculate the reflection coefficients for the earth as a function of two-way travel time and we require a reflection wavelet. At the time of each reflection coefficient we generate a reflection wavelet with the amplitude proportional to the reflection coefficient and with the polarity the same as that of the reflection coefficient. Normally the reflection coefficients will be closely spaced in time as compared with the length of the reflection wavelet, so that there will be massive overlaps of the many wavelets. All of this can be accomplished with the computer using a process called convolution. Deconvolution is the inverse of convolution. For deconvolution of field data the processor selects a wavelet shape that he wants the reflections to have. A digital operator (which acts like a filter) is generated by the computer which is convolved with the field record traces to produce traces with reflection wavelets of the desired shape.

Sounds great, but there are limits on what can and cannot be done. What deconvolution does is to alter the amplitude and phase spectra of the data to that of the desired wavelet. If the desired wavelet has frequency components that are not present in the reflection wavelets (or are below the noise level), then they cannot be manufactured. If the processor asks for a wavelet that has strong frequency components that are above the effective frequency limit for reflections from your objective, then the deconvolution operator can either bring up noise for those components or it can simply leave these components at a low level. (The weak or missing components may also be those in notches resulting from ghosts, Chapters 8 and 9.) The processor chooses the extent to which the operator will attempt to bring weak components up to the level of the chosen wavelet. If the processor pushes too hard, the operator brings up noise for those frequencies. If he does not push hard enough, the resultant spectra will be unnecessarily deficient for some frequency components.

The deconvolution program must be based upon some assumption regarding the phase of the input data. It is relatively easy to adjust the amplitude spectra of the reflections wavelets to match the amplitude spectra of the desired wavelet. But phase is more difficult to measure and more difficult to handle well. For data recorded with impulsive sources, the usual assumption is that the data are minimum-phase (Chapter 5 and Appendix G). This is a reasonably good assumption if all processing is minimum-phase and linear. (Nonlinearity at the source may be a problem, but at least the wavelet near the source will be causal.) Originally the digital filters used in seismic data processing were designed to be minimum-phase. However in recent years there has been a tendency in some processing centers to use zero-phase filters on all seismic data. This is unfortunate for the handling of data from impulsive sources. If only one phase were to be used for all filters, it should be minimum-phase because filtering is less necessary with Vibroseis data. This is because of the frequency control exerted by the limited bandwidth of the sweep and because of the filtering that can be done by correlation in the Vibroseis technique.

For Vibroseis data the assumption usually made is that the input data are zero-phase. But the data came from a geophone that was minimum-phase and rarely, to my knowledge, is a correction made for the phase effects of the geophone. (This correction would be small for frequency components that are well above the resonant frequency of the geophone, but would be substantial for frequency components near the resonant frequency of the geophone.) Also the earth filter is minimum-phase; so the Vibroseis data are actually mixed-phase (Chapter 5, and Appendices G and H).

Fig. 19.1 Example of data recorded with mini-hole charges as originally processed with deconvolution. Note the low-frequency noise with steep dips and note the generally poor S/N ratio. These data were supplied by Grant-Norpac, Inc.

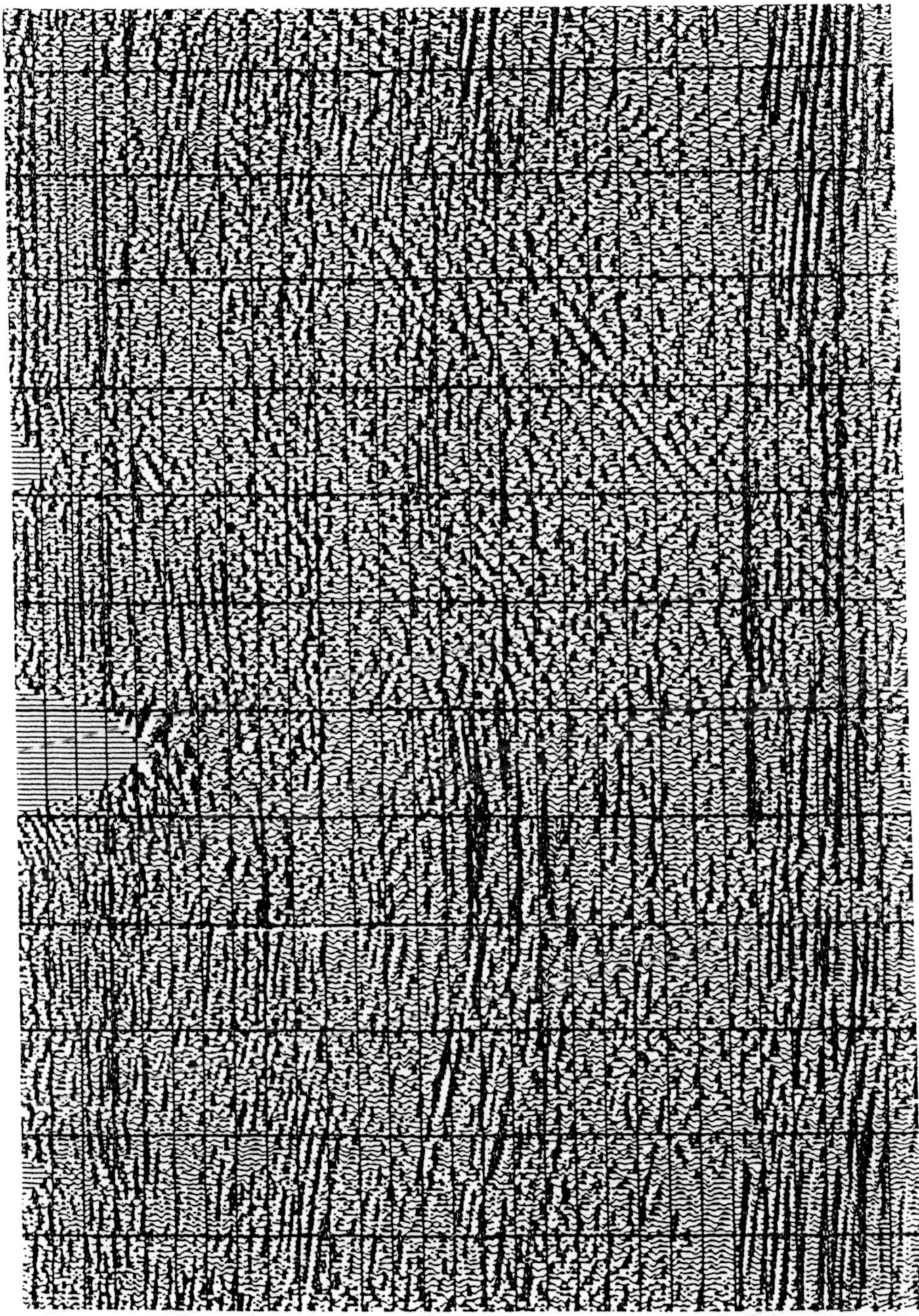

Often in marine seismology the energy trapped in the water layer results in strong reverberation that obscures the reflection data (Chapter 8). A minimum-phase ('spiking') deconvolution is the effective means for de-reverberating data. Ringy data is more prevalent in land data than is generally recognized. With Vibroseis data, if a zero-phase correlation is used, there is a severe problem because the reverberation is minimum-phase. This is another point in favor of using a minimum-phase correlation with Vibroseis data so that a minimum-phase deconvolution can be used both to de-reverberate the data and to flatten the spectra of the reflection wavelets. But if a non-linear sweep is used and we want a minimum-phase wavelet the processor should correct the phase for the zero-phase filter effect of the nonlinear sweep.

For marine work it is helpful to measure the source signature accurately and to use this source signature in what is called 'signature processing'. It is relatively simple to measure the far field source signature in deep water or to measure this signature in the near field and make corrections to estimate the far field signature. Signature processing would also be helpful with land data but it is much more difficult for land data to measure a signature. Also, on land the signature varies markedly with variations in the surface of the earth and in the coupling of the source to the earth.

The design of deconvolution operators is beyond the scope of this book. More information is provided by Robinson and Treitel (1980). Jurkevics and Wiggins (1984) and Galbraith and Wiggins (1968).

The spectral whitening program discussed in Chapter 7 (and elsewhere) that worked so well for us was developed by Western Geophysical Company. Other programs may differ in some details but may yield comparable results. The program splits the spectra into a number of relatively narrow frequency bands with zero-phase filters which overlap at the half amplitude levels. Then each seismic trace (usually a stacked trace) is applied to the set of filters and a separate AGC is applied to each of the separate outputs. This serves to equalize the amplitudes for the spectra. The separate signals are then recombined and scaled to the amplitude of the input signal. Typically about one per cent white noise is added to provide some limit to the restoration of very weak signals. Western Geophysical Company of American can give more specific information about their program.

Initially we used this spectral whitening primarily on high-resolution Vibroseis data acquired with nonlinear sweeps. It proved to be so helpful that we later tried it on a variety of data from many different areas. In many cases it

---

Fig. 19.2 Same example as in Fig. 19.1 but reprocessed with spectral whitening. Note the suppression of low-frequency noise yielding a better S/N ratio. Also note the improved resolution.

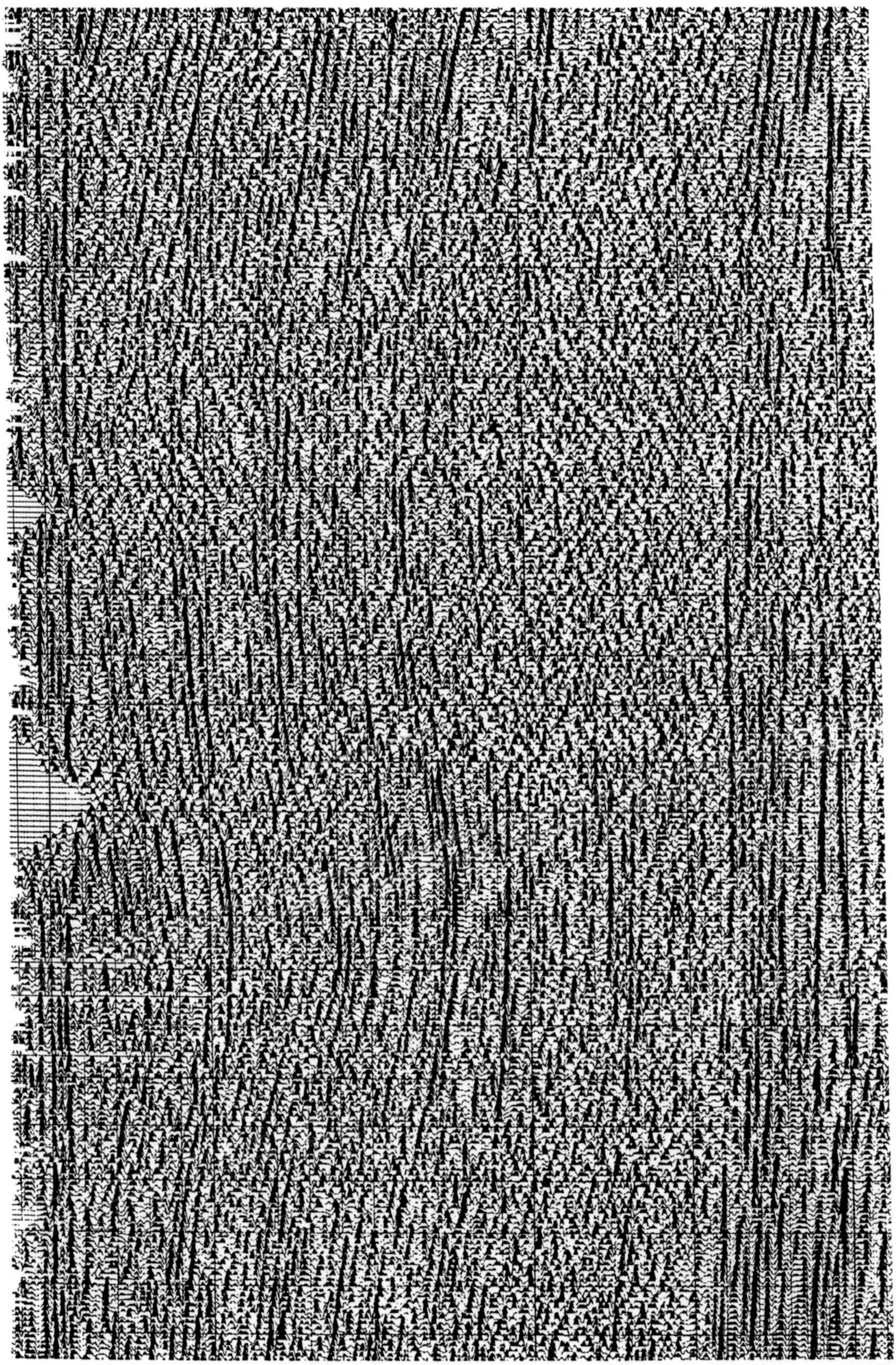

improved the resolution significantly and revealed evidence of geologic details that were not visible or were less visible before. In a number of cases it improved the S/N by emphasizing reflections which had higher-frequency components than most of the noises. It was used successfully on data recorded with both vibrators and impulsive sources. Figures 19.1 and 19.2 provide an example of its use with data recorded from mini-hole patterns of charges. Figure 19.1 shows the data as it was originally processed with deconvolution; note the low-frequency noise with high dip that obscures reflection data. In Fig. 19.2 note the suppression of this low-frequency noise and the improvement in the section that was achieved by reprocessing the same data with spectral whitening. On the author's computer the cost and time for running this program was no more than the cost and time for deconvolution.

The AGC used in the program does alter the phase of the wavelets and this alteration cannot be defined easily but it is not significant if the AGC used is moderate. In addition, it is possible to minimize this unknown phase alteration by preceding this program with a deconvolution that should be minimum-phase for minimum-phase data. For Vibroseis data that is correlated with a zero-phase correlation, a zero-phase deconvolution is used, unless reverberation is a serious problem in which case a minimum-phase deconvolution may be the best choice despite the inconsistency. The idea here is to accomplish as much of the spectral balancing as we can with a program that handles phase in a known manner that is consistent with the way we want the phase to be handled. Inverse Q-filters are another option prior to or instead of spectral whitening. In a number of cases such as the example in Figs. 19.1 and 19.2, data that had been previously recorded and previously processed were reprocessed with the spectral whitening program. Frequently the resolution and general quality of the data were improved. Yet the previous processing had included a deconvolution that was supposed to balance (or whiten) the spectra. In some cases after the spectral whitening program produced improved results, the processors were able to achieve essentially the same results with a deconvolution program. But without the spectral whitening program showing what could be done with the data, we did not achieve equivalent results with normal deconvolution. In the author's experience there were no cases where deconvolution yielded significantly better results than the results obtained with the spectral whitening program. At the very least this program, or a similar one, should be used to give the processors a goal to shoot for with their deconvolution programs. Since this program utilizes a nonlinear AGC, it should not be followed by deconvolution.

It is probable that the spectral balancing program and techniques for its use have been improved. It is also possible that even better programs are now available. Even if this is true, the program served the author's team quite well and for them it was an important step along the way to better seismic data.

19.3 MULTIPLE SUPPRESSION

Multiples posed many problems for which we had no good answers before the CMP method and computer processing were available. Typically multiples have lower velocities than primary reflections at the same record times (because velocity typically increases with depth). If there is a significant difference in the moveout over the range of offsets in a CMP stack between a multiple and a primary, then the stack will suppress the multiple relative to the primary. We need relatively long offsets and an adequate range of offsets to allow the processor to suppress multiples well.

Sometimes multiples are so strong or their velocity is so close to that of the primaries that the CMP stack does not do an adequate job of suppressing the multiples. One of the problems is that the difference in moveout is relatively minor for the shorter offsets and may not be very significant for even the longest offsets. The stack would then be dominated by traces where the difference in movement is trivial, thus the suppression of the multiple would be minor. One way to improve the result is to weight the stack with lower weightings for the shorter offset traces and higher weightings for the longer offset traces. This, however, reduces the effectiveness of the CMP stack in suppressing other noises.

Another approach is to pick the multiple moveout velocities from normal velocity analyses and to make NMO corrections with the multiple velocities. Then in each CMP gather the multiples should line up at constant record times. Static corrections may also be applied to improve the line up of the multiples. The apparent velocity of the multiple is then infinite. A velocity filter is applied that filters-out events with very high velocities and consequently suppresses the multiples. The NMO correction is backed out and the data are then processed in a normal manner using primary velocities for the NMO corrections. But if there is too little difference in the moveout of a multiple and a primary, then the primary is also suppressed by the velocity filter.

The bottom line for the field geophysicist is that if multiples are thought to be a serious problem, then we would like to have offsets as long as the maximum effective offset for your objective. Ideally we would like a wide range of offsets which means we also want a short minimum offset. But if the group interval is constant, a short minimum offset means there will be a lot of traces with little difference in moveout between multiple and primary (if the velocities are not much different) and this is bad. So if the multiple velocity is close to the velocity of the primary, we may want to use a long minimum offset. To choose the minimum offset, we first estimate the primary and multiple velocities and calculate the differences in the moveout as a function of offset at the record time of our objective. This should help us to select an offset range that will be effective in suppressing both multiples and other noise.

19.4 MULTICHANNEL FILTERS

Have you ever seen strong, ringy noise on records and noticed in a particular case that the peaks and troughs of traces rather close together interlaced in a very regular manner? Traces that are adjacent or a few traces apart may be almost precisely out of phase for the noise. If we sum pairs of such traces we will suppress the noise markedly. This is a simple two-channel filter made possible by the fact that an integral multiple of the trace spacing resulted in a

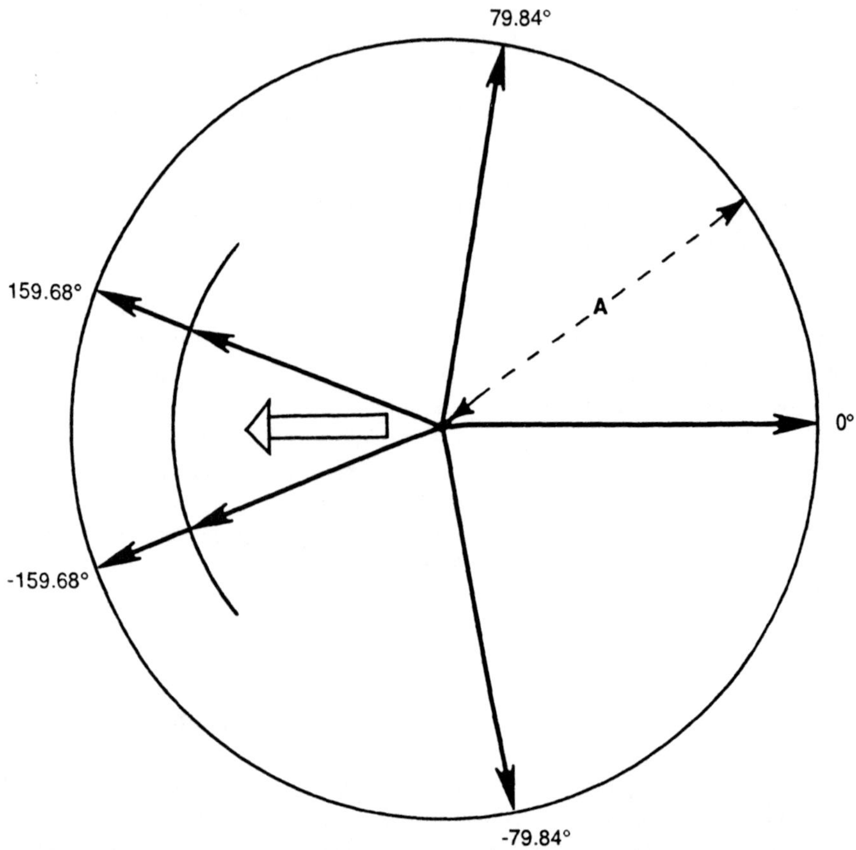

Fig. 19.3 Suppression of a refraction by a five-trace, multichannel filter. The vectors represent refraction noise on five adjacent traces. Without special weighting the refraction noise on five traces would sum up to the broad open arrow. The suppression of the refraction noise can be improved greatly by reducing the weighting of the extreme traces (at angles of 159.68° and – 159.68°) as indicated by the shorter arrows. The weighted sum of traces in groups of five should cancel the refraction noise.

half-wavelength (or half-period) shift of the noise. With a computer the processors do not have to be lucky enough to have the right trace spacing. They can find some number of traces and a weighting for these traces that will suppress any ringy event that persists for many traces.

Perhaps an example would be helpful. Remember the refraction in Chapter 13 from the unconformity below the low velocity surface? The apparent velocity of this refraction was about 16 000 ft/s (4877 m/s). Suppose that we filtered our data with a 16 Hz low-cut filter, after which the period of the strong refraction was 0.062 seconds. If our trace interval were 220 ft (67 m), then the refraction arrival time difference between adjacent traces in a shot gather would be 220/16 000 = 0.01375 s. The time difference in periods of the refractions would be 0.01375/0.062 = 0.2218 periods. The phase difference of the refraction between adjacent traces would be (0.2218)(360°) = 79.84°. Figure 19.3 shows five vectors with amplitudes A with an angle of 79.84° between adjacent vectors. These five vectors represent the refraction noise on five adjacent traces. For the five vectors the relative phase angles are 159.7, 79.8, 0, −79.8, and −159.7 degrees. A simple summation of these five vectors is the short, wide vector with an amplitude of (0.52274)A. We want to find weightings we can use with these five vectors such that the sum will be zero as follows:

$$w_1 \cos(159.7°) + w_2 \cos(79.9°) + w_3 + w_2 \cos(-79.9°) + w_1 \cos(-159.7°) = 0$$

$$2w_1 \cos(159.7°) + 2w_2 \cos(79.9°) + w_3 = 0$$

$$2w_1(-0.93777) + 2w_2(0.17640) + w_3 = 0$$

In an actual case we would probably decrease the weights somewhat on all but the center vector and have $w_1 < w_2 < w_3 = 1$. But to make things simple let w_2 and w_3 be 1 and solve for w_1. We get:

$$w_1 = \frac{1.3528}{1.8755} = 0.72129$$

The shorter arrows in Fig. 19.3 at angles of +159.7° and −159.7° represent the outer traces after being weighted. The sum of the five vectors is zero with these weightings. Therefore running five-trace sums of our data with weightings of 0.7213, 1, 1, 1, and 0.7213 would suppress this refraction. (For example, if we started with consecutive traces numbered 1–5 then the 0.7213 weighting would be applied to traces 1 and 5 for the first sum, traces 2 and 6 for the second sum and traces 3 and 7 for the third sum, etc.) But we would be likely then to find that higher-frequency components of the refraction with shorter wavelengths and shorter periods would still be a problem (unless our arrays and source patterns had taken care of these). The technique can be applied to any type of gather where coherent noise is consistent and displays a linear moveout.

When a technique like this does a good job of suppressing noise, it enables the processor to improve the selection of velocities for NMO corrections which, in turn, allows a better static solution and correction. In some cases the final stack is not much better with or without the multichannel filter, yet the filter may have been required to allow the processor to make good corrections needed to produce a good stack. In other cases the velocity filter is essential to getting the best stack and must be used in the final processing for best results.

Did we try this technique on the data with the low-velocity surface and the bad refraction from the unconformity? It was suggested but was probably not tried. However, velocity filters (discussed next) were tried with limited success on this extremely strong refraction.

Instead of a simple, multichannel weighting, which is a wavelength filter, the processor may choose to use a multichannel velocity filter which would attack all frequency components of the refraction and many other noises as well. For stacked data the processor sets upper limits on the dip that he assumes reflections can have. Dip of an event on a time section can be thought of as a change in time for a change in distance which is the reciprocal of velocity; hence the name velocity filter. The allowed dip may be different in one direction along the seismic line than it is in the other direction. Obviously the processor should be sure his dip limits are acceptable to the interpreter. (At this stage of the processing the processor may be able to look at his stacked section and see what maximum dips are apparent in the data.) This filter discriminates against all events with dips that exceed the selected dip limits. It is a powerful tool and can be effective when properly used, but like most powerful tools, when it is misused it can do more harm then good.

In the previously described multichannel filter, different weights were applied to each trace before summing. For the velocity filter each trace is filtered with a different filter before the traces are summed to provide one output trace. Figure 3.2 (Chapter 3) illustrates an f-K plot and shows where noises of various types appear on such plots (where k is the reciprocal of wavelength). Reflections with vertical or near-vertical arrivals would plot along lines through the origin and along or near the f axis (the ordinate axis). High-velocity noise plots along lines through the origin at only slightly larger angles to the f axis and cannot be cleanly separated from the reflection data using either arrays or frequency filters. A velocity filter is designed to provide a passband for a pie-shape slice of velocities within the reflected signal range and to reject velocities or dips outside this passband. This pie-shape slice is illustrated by the oblique, dashed lines in Fig. 3.2. The mathematics and further information about velocity filters are given in Embree (1963), Kanasewich (1973), Sengbush (1968), and Galbraith (1968).

19.5 CONSTANT-VELOCITY STACKS

When the quality of unstacked data is very poor, velocity analyses may not be satisfactory. In these cases computers make it feasible to stack significant amounts of data with a number of different velocities for the NMO correction. The processor may also display many NMO corrected CMP gathers sampled along the line so that he can see how any coherent events that may be present line up with different NMO velocities. He then has to decide which coherent events are most likely to be primary reflections and pick the best velocities for a number of different record times to form a velocity function. An alternative procedure is to stack the data with a sequence of velocity functions. For example, we may have a sequence as follows:

Velocity function	*Time* (s)	*Velocity* (m/s)	*Time* (s)	*Velocity* (m/s)	*Time* (s)	*Velocity* (m/s)
1	0	2000	0.8	2500	1.3	3000
2	0	2200	0.8	2700	1.3	3200
3	0	2400	0.8	2900	1.3	3400
4	0	2600	0.8	3100	1.3	3600
etc.						

From this point the procedure is identical. The advantage is that if we have estimated correctly the general shape of the correct velocity function, then we shall generate some pretty good stacks over a wider range of times, which may facilitate the final selection of a good velocity function.

As is often the case in seismic data processing, the processor has to make key decisions that involve his interpretation of the data with which he is working. If the interpreter were present when velocities were being picked, he might have a different opinion as to which events were primary reflections and therefore he might choose a different velocity function which would then emphasize different events. It is therefore imperative that the data processor know the exploration objectives and that he confer with the interpreter frequently. That is the main subject of the next chapter.

19.6 BEAM STEERING

In Chapters 8 and 11 we discussed beam steering both in the field and in processing. Beam steering in processing is flexible and is an important tool that has not often been used to full advantage. Beam steering can be used to prevent aliasing without having to record traces so closely spaced. Beam steering requires judgment and knowledge and at present cannot be done routinely. Perhaps this is a fertile field for the use of artificial intelligence in processing?

One problem is how to display the data for different beam angles for best utilization by the interpreter. It is hoped that much more progress is now being made in this area.

19.7 LARGE THREE-DIMENSIONAL SURVEYS

Computers and computer software had to grow enormously before large three-dimensional seismic projects became feasible, and this has been a development of staggering proportions. Because of the massive amounts of data, even after processing, the effective display and utilization of the data have required extensive developments as well. It has become essential for the interpreter to utilize a computer work station in his interpretation of the data (Brown 1986).

19.8 MIGRATION

Migration has always been an important step in the utilization of seismic data particularly where reflecting interfaces are steep and where the geology is structurally complex. Even though migration is an expensive process it is being used more frequently in cases where dips are quite moderate to collapse diffractions. In complex areas, three-dimensional data are often needed and acquired and three-dimensional migration is performed. There are many different methods for migrating data with a computer. Some are designed to work well even for very steep dips of 60° and more. Others will not handle such steep dips adequately but are less costly to run. Migration is a extra process and display that is usually made after the data are fully processed, stacked and initially displayed. Ideally, data should be migrated before stack for maximum benefit but this means migrating such more data requiring much more time and expense. As computers and computer software continue to improve, pre-stack migration will become more practical.

The main concern about spatial aliasing is in regard to migration. We need traces along the surface of the earth to be spaced closely enough so that we get at least two samples per wavelength of the highest frequency components of our data. The steeper the dip, the closer the required spacing of our traces along the surface. One other concern for the field geophysicist is that the lines along the surface be long enough at both ends to allow migration of the steep events over the length of subsurface coverage that is needed by the explorationists. Not only do we need some extra line length at the ends, but some extra record length as well.

REFERENCES

Brown, Alistair R. (1986). 'Interpretation of three-dimensional seismic data.' *AAPG Memoir 42.*

Embree, P., Burg, J. P. and Backus, M. M. (1963). 'Wide-band velocity filtering—the pie-slice process.' *Geophysics* **50**, 948–74.

Galbraith, J. N. Jr. and Wiggins, R. A. (1968). 'Characteristics of optimum multichannel stacking filters.' *Geophysics* **33**, 36–48.

Jurkevics, A. and Wiggins, R. (1984). 'A critique of seismic deconvolution methods.' *Geophysics* **49**, 2109–16.

Kanasewich, E. R. (1973). *Time Sequence Analysis in Geophysics*, University of Alberta Press.

Sengbush, R. L. and Foster, M. R. (1968). 'Optimum multichannel velocity filter.' *Geophysics* **33**, 11–35.

20 Communication

The subject of this chapter is full, three-way communication between the seismic acquisition group, the interpreters, and the data processors. In some cases they may be separated by hundreds of miles or more. Yet no single group can do its job well without adequate and timely communication with the other two. We can help immensely by doing our part to fully inform the others about all matters related to their work and to let them know what information we need from them. We can also help by passing along appropriate information we get from one to the other. In the event that part or all of the processing is done by another company, an immediate problem arises in regard to what information should be withheld from the processors. At best, communication is hampered by this concern; at worst communication is limited to a few grunts and nods. If the data are processed inside the company, communication should be much better and this is one reason why having most data processed within the company is preferable. But there are good reasons to use outside processing for at least part of the company's data. A contractor may have developed a better program or technique for handling a particular problem. Also, the company processors will tend to be more careful in processing our data if they know that we sometimes compare their processing with that of competent contractors.

For inside company processing we start by sending copies of the prospect forms and the recording parameter forms to both other groups. We should also send copies of these forms to any field support group within the company. Explorationists naturally are concerned about security and everyone must clearly understand that some of the information is for internal use only. No copies of these forms should ever be exposed to anyone outside the company unless their company is a partner in a project and they are thereby entitled to the information. The interpreter needs to know that this is our normal practice so that if we need some very tight information, he can give it to us without including it on the prospect form.

Our client and customer is the interpreter and our contacts with him should be so frequent that it is highly desirable that we be located in offices within the same building. If not, we will need to make extra efforts to keep in close touch so as to know fully the needs of the interpreter and what improvements would be most valuable to him. We really need a meeting of the minds. If we are fortunate enough to be located in close proximity to our customer, the main communication problems are likely to be for the two of us with the data processor. Unfortunately the data processor will probably be located at a central processing center with huge computer facilities and hence probably not close to his client and customer. The data processor needs to know the needs of the interpreter and our needs as well. Both of us must tell him our needs and discuss with him what we can do to help him. Somehow we must bridge the gap and communicate.

20.1 WHAT THE FIELD GEOPHYSICIST NEEDS FROM THE PROCESSOR

Basically what the field geophysicist needs from the processor is a partnership relationship dedicated to the acquisition, processing, and display of data that will best meet the needs of the interpreter. Specific examples are:

1. Are we using the best range of offsets in acquiring data?
2. What is our effective upper-frequency limit for our objectives and for other reflectors as well?
3. Is there some strong noise that is particularly troublesome that might be better suppressed in acquisition?
4. Is there a multiple problem, particularly near objective time?
5. Are there problems in the recording of the seismic data that need to be handled?
6. Are there surveying problems that need to be handled?
7. Does data quality appear to be normal, unusually good, or unusually poor? Does it change along the line or from line to line?
8. Does the data indicate unusual or unexpected geology? Do we need to lengthen our records, or should they be shortened?
9. Does the processor have any suggestions as to how we could improve the data so as to yield better final sections?
10. Does the processor have any suggestions as to how we could save money or time in acquisition or processing?

The processor has a number of clues with regard to the relative effectiveness of the longer and shorter offsets and he may develop other clues or follow up on clues he has, if he knows that this would be helpful to us. Even if the shorter offsets are not effective for deep objectives, they may still be useful for static correction using shallow reflectors or shallow refractors. The longer offsets may not be too effective for low dip reflections, but may be helpful for reflections from steep interfaces (Chapter 11). This question of effective offsets may not be easy to answer, but the processor has the best opportunity to extract the necessary information from the data to answer the question.

The question of the effective upper-frequency limit may not be answered accurately until the final processing is complete, but narrow-band filters run on unstacked data (source gathers or CMP gathers and with or without NMO) can often yield good clues. Certainly static corrections and deconvolution or spectral whitening may reveal higher frequencies than are apparent at an early stage (particularly where data quality is not good). Even after the final stack, the question may not be well answered unless the processor runs some narrow-band frequency filters on one or more selected portions of the final stack. In these narrow-band filter displays the data will be very ringy, but all we need to see is coherent energy that is consistent with reflections near objective time.

We may or may not be aware of certain strong noises, or the noises may have changed and may no longer be well suppressed. In either case, if it is a significant problem for the processor, we need to know it and re-evaluate our recording parameters. We shall need to know the period (or frequency), the apparent velocity, and the apparent wavelength. If there is a multiple problem, we may or may not be able to do anything more than we did originally to help. But we do need to know.

Problems in recording data may include parity errors, incorrect geometry, identical traces, open traces, polarity reversals, cross talk, bad power-line pick-up, etc. Incorrect geometry will be indicated by a consistent lack of symmetry in the first breaks despite the use of a symmetrical cable. Identical traces indicate traces shorted together (usually because the geophone strings for two traces are connected). Cross-talk is most evident on far offsets before the first breaks from the first breaks on near offsets. Bad power-line pick-up usually means high leakage current or there may be a problem with the observer's sheets being incomplete or inaccurate. All of these problems require discussion with the party manager and the recording crew, followed by the development of ways to reduce or eliminate them.

Surveying problems can include tardy receipt of the survey data, errors, inconsistencies in the data, or incomplete data. They require discussion with the surveyor and the party manager and the development of solutions.

Particularly at the outset of a program in a new area, we want a quick appraisal of data quality and of any changes in data quality along the line. If we are not getting satisfactory data, we must do something about it as soon as possible. We can make some changes, or stop recording or record something else while we assess the problem and come up with possible solutions.

Unexpected geology may include very late reflections, evidence of steep dips when we did not expect them or just interesting events. These could heighten interest in the area and might call for addition to or modification of the program. Late reflections may mean we should record longer records. Steep dips could call for a reconsideration of your recording technique.

Saving time or money may involve such things as demultiplexing the data in the field or buying equipment for the surveyor to provide processed survey data on magnetic tape cassettes that could be read directly into the computer.

20.2 WHAT THE PROCESSOR NEEDS FROM THE FIELD GEOPHYSICIST

Not only should the field geophysicist ask the processor what he needs from him, he should also suggest things that may prove helpful. The following items are examples:

1. Prospect forms and recording parameter forms;
2. Any changes in recording parameters or in the prospect forms;
3. Test plans and need for quick displays and processing;
4. Plans, particularly as they relate to future work and the resultant load on the processors;
5. Crew problems that could affect the processors or problems that the processor should be alert to spot and report back;
6. Any items beyond the ordinary that should be put on the final section to define the recording technique and parameters accurately (or for any other purpose).

The processor needs the prospect forms and the recording parameter forms. The prospect forms may prompt questions that he will take up directly with the interpreter. He needs the recording parameter forms to fill out the label for the seismic sections.

Test plans that require quick turnaround by the processor need discussion to see what can be done to accomplish the goals. The author hates to get into a rigid recording plan just so that the processor can set up his computer run ahead of time. But on rare occasions that may be acceptable to expedite the processing of the data. The biggest problem with a rigid test recording plan is that early test results often can be used to alter the test plan to expedite the test or to acquire better test information. The processors, like everyone else, need to be able to project the demand for their services for many reasons. Our best guess as to the future program and the impact on the processors should be better than his guess. We must tell him what we expect and update this information frequently.

If you have a new observer on the crew (or a temporary replacement) you might expect more problems in communication. Be sure the processor has appropriate telephone numbers so that communication problems are solved directly and quickly. If the head linesman quits the crew, we might expect more problems with line layout and should warn the processor. If the crew has been having equipment trouble this may show up in the data and the processor should be warned.

20.3 WHAT THE INTERPRETER NEEDS FROM THE PROCESSOR

Examples of the needs of the interpreter from the processor are:

1. Status of the processing and an estimate of when final sections will be available;
2. Displays (marked 'preliminary') of intermediate data including samples of velocity analyses, a few constant-velocity stacks, decon tests, etc.
3. Processor's tentative and final decisions on velocity functions, decon type, decon parameters, filters, etc;
4. Early and updated estimates of final data quality particularly at objective time;
5. Any shallow discontinuities in the data that might be due to bad static problems not correctly handled, but could also be due to shallow faults;
6. Possible special processing or special displays that might be most helpful;
7. Any unusual or unexpected geology indicated by the data.

The amount of intermediate results the interpreter wants will vary among different interpreters and will vary with the projects. He may very well have insights that could be helpful in making good choices regarding processing parameters and techniques. Whatever methods are used to determine the velocity function, it is a key item and the interpreter should participate in one way or another. If the processor is using a velocity inversion (a velocity decrease with increasing depth), the interpreter needs to know this and express his opinion on the probability that an inversion is correct. Likewise the interpreter needs to assess the potential geologic significance of such a velocity inversion. Naturally the interpreter wants to keep up with progress and with estimates as to the probable final data quality.

The interpreter needs to know about possible static correction problems so that he can assess the chances that the problem is not with the data or the processing but represents a shallow earth fault or other condition. If it could be a static problem the interpreter may be able to offer a clue as to the possible cause and hence insight as to how it should be corrected.

If the processor tries some special processing technique and it appears to be useful, he needs to talk to the interpreter and get his ideas before proceeding. The interpreter may have valid reasons for not wanting the special processing or he might have information or insight that might affect the choice of parameters for the special processing.

Unusual or unexpected geology may represent an opportunity that should be addressed immediately. Line changes, line extensions, additional lines, and land plays are all possible actions that could result.

20.4 WHAT THE PROCESSOR NEEDS FROM THE INTERPRETER

Examples of the needs of the processor from the interpreter are:

1. A processing request form with basic information including the exploration objectives, the desired datum, and the desired correctional velocity to this datum;
2. Any changes or additions to the exploration objectives of the project or any changes in the needs of the interpreter;
3. Any changes in the information about the play or the geology, etc;
4. Locations along the line where geologic changes may affect the processing of the data and the appearance of the data;
5. Any geologic condition that may affect the appearance of the data;
6. Any seismic markers that should be consistent (with only regional dip).

Locations along the line where geologic changes may affect the data are very important to the processor. Shallow faults may cause the appearance of error in the static corrections. The processor can waste time and effort trying to correct a static bust that is actually valid information about the shallow earth; or worse, he may be successful in hiding the fault and in degrading the section in the process.

In one instance there was a bad static bust on a line with little other static problem. The bust was precisely where the line crossed a small valley which turned out to have some glacial fill. This sort of information is sometimes helpful in modeling the static problem, which does not necessarily mean a formal model, but rather a geologically reasonable concept of the static problem leading to a reasonable static solution that works.

Other geologic changes that affect the processor include thrust faulting that causes abrupt changes in the velocity function and may lead to velocity inversions. Knowledge of these possible changes allows the processor to relax restrictions that he might otherwise place on his velocity function. Such relaxation of normal restrictions may be required to reach a correct solution for the processing of that particular line. Changes in surface outcrops can change the seismic wavelets as well as affect the statics.

Geologic conditions that the processor needs to know about include evaporite sections with possible slumping and possible reef development, both of which can cause undulations in reflections that might cause concern for the processor. Also angular unconformities should be pointed out. Seismic markers that should be essentially planar can be a big help to data processors for static corrections and for noting changes in the velocity of overlying beds that affect moveout velocities.

20.5 WHAT THE INTERPRETER NEEDS FROM THE FIELD GEOPHYSICIST

Much has already been covered about the communications needed between the interpreter and the field geophysicist. The needs of the interpreter include:

1. Permit, terrain, and cultural problems along planned lines;
2. Possible line changes either because of problems or to take advantage of previously unknown favorable situations;
3. Appraisal of data quality and changes in data quality;
4. Costs and projected costs;
5. Field processing results (where there is field processing);
6. Updated recording parameter forms to show any changes;
7. Landowner problems and opportunities including possible leasing.

The initial line locations may have looked good on available maps, but a check on the ground may reveal major problems. On the positive side, a check on the ground may reveal a better line route. This is particularly true for regional lines where precise location is relatively unimportant as compared with good continuity of high CMP fold and relatively straight lines.

The field geophysicist can often estimate final data quality from the early field records and can update his estimate as more data are collected. This may encourage the interpreter to expand his program or discourage him so that he decides to reduce his program. It may be wrong to make decisions like this too hastily, but we do not want to waste a lot of time and money collecting data that is not satisfactory. If costs are running either high or low, this may have an impact on what the interpreter wants to record. If costs are low he may want to add to the program. If costs are high he may face a hard choice as to what he should record now versus what can wait. In either event he needs frequent updating on costs and on projected total costs for the project.

Where we have field processing, we certainly want to get field displays to the interpreter immediately upon receipt. In fact we will probably have the field processor send two copies of many items so we can keep one and give the other to the interpreter.

Neither the field geophysicist nor the seismic crew are in the business of scouting the activities, successes, or failures of competitive companies. But if interesting information is freely received, it should be passed along to whoever in the company may need the information. But we should strongly discourage the trading of information; this is a proper function of the scouts in our company and in other companies. If a land owner indicates an interest in leasing his land this information should be passed along to the interpreter immediately.

20.6 WHAT THE FIELD GEOPHYSICIST NEEDS FROM THE INTERPRETER

As was said earlier, the field geophysicist really needs to know the mind of the interpreter. He needs to know the exploration objective and the needs of the interpreter to best solve the exploration problems. He needs to know of any changes in the objectives or in the information that is covered generally by the prospect form. Additionally he needs to know about:

1. Possible additional work on all projects including projects in just the initial stage of consideration;
2. The nature of possible additional work including objectives, also the type of crew and equipment that might be most suitable;
3. Possible timing of such future work.

It takes time to line up good seismic crews with the right equipment and people to do a job. It also takes planning and consideration of many factors including time of year, weather, etc. The above information is essential to good planning and to success.

21
Miscellaneous theory

This chapter discusses the fundamental theories and concepts of superposition, reciprocity, wavefront healing, and the suppression of random noise.

21.1 THEOREM OF SUPERPOSITION

The theorem of superposition states that in a linear system the response to a number of signal inputs applied simultaneously is the same as the sum of the responses to the signals applied separately (one at a time).

Application of the theorem of superposition is most meaningful for signals resulting from sources we control in actuality or in theory. To a limited degree it can be applied statistically to random, background noise from wind, atmospherics, cars, etc. which noises are variable and relatively unpredictable. Noise picked up from power lines and repetitive noises from well pumps may or may not be essentially random relative to our measurements. If we trigger our measurement time from repetitive noises (like we do with the COS system for power-line noise suppression), then these noises are controlled to a limited degree and the theorem of superposition can be applied more effectively.

As an example, suppose we have a geophone at point A and two vibrators that sweep at separate points B and C. The measured response to the vibrator at point B and the vibrator at point C sweeping simultaneously will be equal to the sum of the responses to separate recordings from the two vibrators Superposition is a fundamental theorem that we use several times in this book. Examples of its use include the development of a source pattern response in Appendix D and the simultaneous recording of pairs of VPs in Chapter 14. This theorem applies to linear systems in general including linear electrical circuits and electromagnetic fields and waves as well as seismic waves.

Frequently when we apply the theorem of superposition and the theorem of reciprocity (Section 21.3), we ignore the effects of noises from uncontrolled sources. Often, however, we can consider the effects of these noises separately by using statistical concepts.

21.2 SIGNALS PLUS RANDOM NOISE

In the previous example if we want to include the effect of random, background noise for two separate recordings, we must include two measurements of the background noise versus one measurement of that noise for simultaneous recording. Since background noise is a variable over which we exercise no control, we have to treat it statistically. If we measured background noise over several 10 s periods, no two measurements would be identical. But the amplitude spectra may be similar and the average amplitude spectra may be usable as an estimate of the amplitude spectra of the background noise for measurements at about the same time at the location. On this basis, ten sweeps measured separately from a single vibrator will include ten measurements of the background noise. But the phase of this noise, from sweep to sweep, cannot be predicted—unless we trigger our sweep by the noise as we do to suppress power-line pick-up (Chapter 16). We can treat any frequency component of the ten samples of background noise as a vector (Appendix D) with a random phase angle. Statistically, the sum of ten vectors of equal amplitude and random phase is a vector with an amplitude equal to the product of the amplitude of one vector times the square root of 10 (with a random phase angle). Let the average background noise (all frequency components together) for one record time be represented by a vector $N_b(t)$. Then an estimate of the sum of ten recordings of this background noise would be $(10^{0.5})N_b(t)$. If the earth response to a single sweep is a signal $s(t)$ then the sum of the signals from ten separate sweeps would be $10s(t)$. An estimate of the total response of signal plus background noise would be $10s(t) + (10^{0.5})\,N_b(t)$. An estimate of the total response to a single sweep by ten identical vibrators would be $10s(t) + N_b(t)$.

Random events such as background noise add up statistically as the square root of the sum of the squares.

21.3 THEOREM OF RECIPROCITY

The theorem of reciprocity is also used in electrical circuit theory and in electromagnetic field and wave theory. It is a rather simple theorem, but it is sometimes misused. In electrical circuitry this theorem says that in a linear system if we apply a voltage V to a circuit at point A and measure a current I at point B then if we apply the same voltage V at point B, we will have the same current I at point A. Note that the theorem involves applying a voltage and measuring a current. It says nothing directly about applying a voltage and measuring a voltage nor about applying a current and measuring a current. If we want to deal only with voltages or only with currents, we must take the ratio of the electric impedances at A and B into account.

The analogy of electric circuits to seismic compressional waves in the earth relates voltage with pressure, current with displacement velocity, and electric impedance with acoustic impedance. (Acoustic impedance of a compressional wave is equal to the ratio of the pressure to the displacement velocity. In an infinite, homogeneous, isotropic medium the acoustic impedance is equal to the product of the density of the medium and the velocity of sound in the medium. Near the boundary between two semi-infinite homogeneous media the acoustic impedance is intermediate between the density and velocity products of the two media.)

For seismology the theorem of reciprocity says that in a linear system if we apply a pressure signal at point A and measure a resulting velocity at point B, then if we apply the same pressure signal at B, we will measure the same velocity at point A. If we apply a velocity signal at point A and measure a velocity at point B and then apply the same velocity signal at B and measure a velocity signal at point A, the two responses will not be the same unless the acoustic impedances at A and at B are identical. To relate the two responses we must take the ratio of the acoustic impedances into account.

Reciprocity allows us to interchange sources and receivers without changing the result except for any nonlinear effects. For example, a star source pattern at point A exciting a linear geophone array at point B will have the same response as a linear source pattern at point B exciting a star geophone array at point A except for any nonlinear effects near the source. But it is important that in a heterogeneous earth or a laterally varying earth that the source and receiver positions have to be interchanged. The elements of the star source pattern must be replaced by receiver elements at the same locations, and the elements of the linear receiver array must be replaced by source elements at the same locations. These requirements are sometimes overlooked.

In seismic applications of the theorem of reciprocity we must realize that the coupling of the source must be consistent i.e. the pressure signal coupled into the earth at point A must be the same as the pressure signal coupled into the earth at point B. Similarly, the coupling of the receiver must be consistent such that the coupling of the earth velocity signal to our sensor must be the same at point B as at point A. For example, point A might be on a surface with a thick soil layer whereas point B might be on an anhydrite outcrop. If you look at an uncorrelated sweep record from a vibrator at a point such as B on a hard rock outcrop (a similarity record or a response from a nearby geophone), you may be shocked by the distortion and the amplitude variations particularly at the lower frequencies. What is involved here is a difficulty in coupling the vibrator to the earth in combination with shallow earth resonances. It is true that the response of a geophone at point B on the hard rock outcrop will also be influenced by the same shallow earth resonances, but the effects of imperfect coupling will be quite different. For the geophone the coupling may be quite good, or it may be poor particularly if the geophone has a long spike that barely penetrates into thin soil or into a crevice. In this case the geophone is on an inverted pendulum with a resonance determined by the length of the pendulum. The coupling effects on the anhydrite outcrop will not be the same for the sources and for the receivers, and therefore, the coupling effects will not be offsetting.

21.4 WAVEFRONT HEALING

There is theory called 'wavefront healing'. The concept of this theory is that wavefronts perturbed by inhomogeneities near the source will tend to heal and straighten out by diffraction as the wavefront propagates farther from the source. Consequently, scattering near the source is less disturbing to a measurement than the same type of scattering approximately mid-way between source and receiver.

For the theorem of reciprocity to hold there must be some reason why scattering near the receiver is also less serious than the same type of scattering approximately mid-way between source and receiver. Scattering is dominantly forward, i.e. at small angles of deviation. Scattering at short distances from the receiver dominantly results in relatively little lateral displacement as compared with scattering approximately mid-way between source and receiver. There is no essential conflict between the concept of wavefront healing and the theorem of reciprocity, since scattering is less significant near both source and receiver than it is near some midpoint.

In seismology the shallow earth section is frequently where scattering is most severe due to weathering (including leaching and precipitation of solids from aqueous solutions), variable consolidation, variable pore fluids (water or air), and variable topography. It is fortunate that scattering in the shallow earth near our sources and receivers is less serious than it would be in a deep earth section. Even so, scattering in the shallow earth section is frequently a serious problem.

22
Wave propagation, attenuation and dispersion

22.1 INTRODUCTION

This chapter discusses seismic wave propagation and the alteration of seismic wavelets during propagation through sedimentary rocks and hence the effects on our seismic data.

Here we will deal only with body-wave dispersion as opposed to dispersion associated with waves guided by interfaces. Exploration seismologists have recognized and dealt with dispersive surface waves, but we have tended to ignore the dispersion of body waves. Yet body-wave dispersion is finite for any earth with a finite Q or any earth with inhomogeneities that scatter seismic waves. The effects of body-wave dispersion are often subtle, but as seismology becomes more sophisticated we should no longer ignore dispersion.

With dynamite records seismologists measure times from the shot time (the onset time of the input wavelet) to the onset times of reflected wavelets. But we do not really use the onset times of received wavelets (except in wells shot for velocity), because onset times are difficult to pick even for reflections from interfaces between very thick and homogeneous beds. Instead we use a large crest or trough of the reflection wavelet and refer to it as the crest or trough for a reflection interface of interest such as the reflection from the top of the Devonian. Therefore the reflection travel times we use are really the reflection onset times plus the time from the onset (beginning of the wavelet) to the chosen wavelet crest or trough.

But wavelets change progressively as they propagate through the earth due to changes in the amplitude and phase spectra. Near an explosive source the time from the onset to a high-pressure crest is a fraction of a millisecond. As a wavelet propagates significant distances through the earth, the time from the onset to a major high-pressure crest increases progressively to several milliseconds and more. At seismic frequencies dynamite shots should produce an input wavelet that is very nearly minimum-phase. Since the earth filter is minimum-phase (as are all physically realizable filters) and if the processing of the data is minimum-phase, the reflection wavelets should be nearly minimum-phase (Chapter 5 and Appendix G).

With Vibroseis records with a zero-phase correlation there is no onset time and we measure times from the center of the autocorrelation of the sweep to the center of the correlated reflection wavelets. If we are using zero-phase and zero-degree wavelets, we measure a crest-to-crest time directly (or a crest-to-trough time for reflections from interfaces with negative reflection coefficients). For the same reflection interfaces, is there any difference between the times measured with the Vibroseis technique and those using dynamite? Yes there are differences. For the zero-phase, zero-degree Vibroseis measurement we do not have the added time from the onset of the received wavelet to the large crest or trough that is identified with the reflection event. Well then, must Vibroseis reflection events always be earlier in time than corresponding events on dynamite records, and must this time difference increase progressively with depth? No, in fact, Vibroseis reflection events may be significantly later than corresponding events on dynamite records. In many field cases the times may be nearly the same. But along some lines recorded by both techniques one or more corresponding events may drift with respect to each other. Where dip is slight, this drift may reverse the apparent dip of beds between the two sections.

Many seismologists will disagree with much of the above, and the author might also were it not for a well velocity survey conducted several years ago that provided some very puzzling results. In an effort to understand these strange results the author started with basic wave theory and developed a rational explanation that has been substantiated by computations using earth models based upon measured earth properties. Dispersion plays a key role. It causes the group velocity and the phase velocity to be different; this causes some surprising differences between measurements with zero-phase wavelets and measurements with minimum-phase wavelets.

In the well survey with the puzzling results we locked a geophone at many different depths in a borehole. At each locked depth we recorded data from both a dynamite charge in a shallow hole and from vibrators. The vibrator data were correlated to yield zero-phase wavelets of zero degree. With the geophone locked at very shallow depths the onset time of the dynamite data was very nearly equal to the measured time of the central crest of the correlated Vibroseis records (after minor corrections for the uphole time for the dynamite records). But as we went progressively deeper into the well the onset time of the dynamite wavelet came in progressively earlier relative to the time of the central Vibroseis crest. At the greatest depth the difference was about 45 ms. Since we had two different recording systems, we were 'sure' that there was a discrepancy between the clocks of the two recording systems. But at one depth near the bottom of the well (where the discrepancy was near the maximum 45 ms) we recorded both sets of data on different channels of our seismic recording truck instruments. Subsequent analysis of these data proved that the time discrepancy was real and that both clocks were accurate. If the maximum time discrepancy had been only 5 or 10 ms we might have been able to dismiss it as being due to slight differences in the frequency spectra and or differences in the ray paths. But the near-surface layer was a high-velocity limestone and corrections for path differences were a few milliseconds at most.

There was relatively little difference between the amplitude spectra of the received wavelets at any given depth. We used digital filters (minimum phase for the dynamite wavelets and zero phase for the Vibroseis wavelets) to make the wavelets have the same amplitude spectra but the resulting changes were minor. Even at the greatest measurement depth the delay time from onset to the main crest of the dynamite wavelet was only about 15 ms so that the crest for the dynamite wavelet preceded the main crest measured in response to the vibrator by as much as 30 ms!

22.2 CAUSES OF BODY-WAVE DISPERSION

To many seismologists dispersion is associated only with surface waves. Fortunately dispersion for seismic body waves is often small and for some purposes can frequently be ignored (which greatly simplifies seismic theory). But dispersion is finite (non-zero) in any earth with a finite Q (not lossless). Also, dispersion is finite for any earth where there are inhomogeneities that scatter seismic waves. As seismology becomes more sophisticated the effects of dispersion become more significant. Sasche and Pao (1978) discuss a number of causes of dispersion. For seismic body waves in sedimentary earths the dominant causes of dispersion are attenuation and scattering.

22.3 WAVE PROPAGATION THEORY

If dispersion is zero, then the phase and group velocities are equal, but where dispersion is finite they are not equal. The group velocity is the velocity with which energy is transferred between two points. Phase velocity is the velocity with which wave crests appear to move in the direction of propagation. Lighthill (1965) provides an excellent development of both the mathematical theory and a physical understanding of wave propagation of all types and in all kinds of media. (Lighthill's lecture applies equally to seismic waves, radio waves, light waves, etc.) The following quotations are pertinent and may be surprising to some geophysicists:

1. 'For the vast bulk of wave motions occurring in nature, however, the phase velocity, with which the crests and troughs are propagated, and the group velocity, with which the energy is propagated, have quite different magnitudes...'
2. 'The magnitudes of the phase and group velocities are, however, not equal

for any waves whose phase velocity takes different values for waves of different length.'

3. In this process of dispersion, the energy associated with waves of a given length is propagated at the group velocity, say U, of these waves. Hence, after a time t has elapsed, waves of that length will be found a distance Ut farther on. Anyone who imagined that, because the crests travel with the phase velocity v, those waves should be observed at the quite different distance vt, would fail to find them.'
4. 'Thus, although it is variations in the phase velocity which cause the phase and group velocities to be different, it is variations in the group velocity which produce the dispersion.'

In seismology, since dispersion is slight, the effects will be less dramatic; but that does not mean that these effects will not be significant. We will not fail to find waves where we expect to find them, but the precise arrival time and precise shape may be altered.

Sachse and Pao (1978) give the following expressions involving group and phase velocities:

$$\omega = kv, \text{ and} \tag{1}$$

$$U = \mathrm{d}\omega/\mathrm{d}k = v + k(\mathrm{d}v/\mathrm{d}k) \tag{2}$$

where

ω is the angular frequency which is 2π times the frequency f,
k is the wave number (defined here, like many physicists, as $2\pi/\lambda$)

v is the phase velocity, and

U is the group velocity.

Since $\mathrm{d}k = \mathrm{d}\omega/U$, equation (2) can be written:

$$U = v + kU(dv/\mathrm{d}\omega). \tag{3}$$

$$U = v/[1 - k(\mathrm{d}v/\mathrm{d}\omega)]. \tag{4}$$

Equation 4 shows that the group velocity is faster than the phase velocity any time the phase velocity increases with an increase in frequency (Sheriff 1982). Conversely, the group velocity would be slower than the phase velocity if the phase velocity were to decrease with an increase in frequency. The two velocities are equal only if there is no dispersion. At seismic frequencies dispersion will be finite for conditions that are typical of the earth we explore (finite Q and some scattering). It follows that normally there will be some difference between the phase and group velocities. The differences may be negligible for some purposes and for some earth sections. But for other earth sections the differences may be significant, and for some applications of seismic data, the effects of such velocity differences are critical. The effects of velocity differences may be annoying, but these effects may also provide opportunity if we can relate measured velocity differences to important geological information.

Lighthill (1965) illustrated group and phase velocity with a very simple, two-frequency, plane-wave example that he attributed to Stokes (1876). Consider two input sinusoids, the first being:

$$s_1(t, x) = \cos(2\pi f_1 t - k_1 x) \tag{5}$$

The phase velocity of this wave is:

$$v_1 = \mathrm{d}x/\mathrm{d}t = 2\pi f_1/k_1$$

Let the second sinusoid be:

$$s_2(t, x) = \cos(2\pi f_2 t - k_2 x) \tag{6}$$

with a phase velocity of:

$$v_2 = 2\pi f_2/k_2.$$

Let the two waves propagate simultaneously and we get:

$$s_1 + s_2 = \cos(2\pi f_1 t - k_1 x) + \cos(2\pi f_2 t - k_2 x).$$

It can be shown that the value of $(s_1 + s_2)$ in eq. 7 is equal to the product of two terms as follows:

$$s_1 + s_2 = \mathrm{G}_1 \mathrm{G}_2$$

Where:

$$\mathrm{G}_1 = 2\cos[2\pi(f_1 - f_2)t/2 - (k_1 - k_2)x/2] \text{ and}$$

$$\mathrm{G}_2 = \cos[2\pi(f_1 + f_2)t/2 - (k_1 + k_2)x/2]$$

$$s_1 + s_2 = \cos(2\pi f_1 t - k_1 x) + \cos(2\pi f_2 t - k_2 x). \tag{7}$$

first term second term

$$s_1 + s_2 = 2\cos[2\pi(f_1 - f_2)t/2 - (k_1 - k_2)x/2]\cos[2\pi(f_1 + f_2)t/2 - (k_1 + k_2)x/2] \tag{8}$$

The first term has a relatively low frequency which is $(f_1 - f_2)/2$ whereas the second term has a higher frequency of $(f_1 + f_2)/2$ which is the average of the two input frequencies. The velocity of the low frequency, first term is:

$$U = 2\pi(f_1 - f_2)/(k_1 - k_2). \tag{9}$$

The velocity of the second term is intermediate to v_1 and v_2 and is:

$$v_{12} = 2\pi(f_1 + f_2)/(k_1 + k_2). \tag{10}$$

The velocity of this second term is the apparent phase velocity of the apparent frequency $(f_1 + f_2)/2$.

Quoting from Lighthill (1965):

> 'The first term on the right is a slowly varying amplitude for the rapidly varying second term, so that the formula can be interpreted... as a series of packets traveling with the velocity U and incapable of exchanging energy with one another through the nodal points where the amplitude is zero.'

If we applied two sinusoidal waves to a linear earth the resultant wave would have an apparent intermediate frequency with an apparent intermediate phase velocity which would be modulated by a term which would separate the higher frequency waves into packets. Furthermore these wave packets would travel at a different velocity, which is called the group velocity. If we were to take this resultant and extract an amplitude spectrum we would find that only the two original frequency components f_1 and f_2 were present. We could apply a filter to remove one of the two components and we would then recover the other component as a continuous sinusoid of constant amplitude.

For theoretical non-dispersive earths, phase shift increases linearly with increasing frequency, which means that $k_1/f_1 = k_2/f_2$. For non-dispersive earths it can be shown that v_1, v_2, v_{12}, and U are all equal.

22.4 BODY WAVE DISPERSION IN THE PIERRE SHALE

McDonal *et al.* (1958) measured seismic waves at closely spaced depths in the Pierre Shale. Wuenschel (1965) computed phase velocities from these data for frequencies ranging from 25 to 400 Hz. He used a theory of wave propagation developed by Futterman (1962) to show that dispersion was significant for compressional waves propagating in the Pierre Shale. Since then better theory and models are available from the work of Einar Kjartansson (1979). But Wuenschel's results based upon the earlier theory are used here with the expectation that they will be satisfactory for defining a reasonable model of a dispersive earth.

After correction for divergence, Wuenschel (1965) showed that the propagated compressional waves could be described by the expression:

$$s(t, x) = s(t_0, x_0)e^{-\alpha x} \sin(\omega\tau - kx), \quad (11)$$

where for $f > 1$:

$$k = (\omega/c) - 2(b_1/\pi)(\omega)[\ln(\omega/2\pi)]\ \text{ft}^{-1}\ (\text{or m}^{-1}) \quad (12)$$

where:

$$\alpha = 1.38(10^{-5})(f) = (0.2196)(10^{-5})(\omega)\ \text{ft}^{-1}$$

$$\text{or } [\alpha = (0.72047)(10^{-5})(\omega)\ \text{m}^{-1}]$$

$$b_1 = 0.2196(10^{-5})\ \text{ft}^{-1}$$

$$\text{or } [b_1 = 0.72047(10^{-5})\ \text{m}^{-1}]$$

$$c = 6522\ \text{ft/s} (1988\ \text{m/s})$$

f = frequency in hertz,

$\omega = 2\pi f$ is the angular frequency,

t_0 = time zero in seconds,

x_0 = wave position in ft (or m) at time t_0

k = wave number in ft^{-1} (or m^{-1})

The phase velocity for the Pierre Shale is:

$$v = \omega/k = 6522/\{1 - 0.009118[\ln(f)]\}\ \text{ft/s}\ (f > 1) \quad (13)$$

$$(v = 1987.9/\{1 - 0.009118[\ln(f)]\}\ \text{m/s})\ (f > 1)$$

Wuenschel's results are readily extended to yield the group velocity as follows:

$$U = d\omega/dk = 6522/\{1 - 0.009118[1 + \ln(f)]\}\ \text{ft/s}\ (f > 1) \quad (14)$$

$$(U = 1987.9/\{1 - 0.009118[1 + \ln(f)]\}\ \text{m/s})\ (f > 1)$$

Table 22.1 shows calculated values of wave number and of phase and group velocities as a function of frequency for the Pierre Shale. Note that the phase velocity increases with frequency and the the group velocity is about one per cent faster than the phase velocity at any given frequency. A difference in reflection times of one per cent between seismic sections could cause difficulty in tying events on seismic records. For example, if the two-way reflection time were 3 s, a one per cent change would be 30 ms. Figure 22.1 plots phase and group velocities as a function of frequency for the Pierre Shale.

We can illustrate group and phase velocities for the simple, two-frequency case governed by equation (9) using values of wave number from the Pierre Shale work. Let $f_1 = 140$ Hz and $f_2 = 100$ Hz, then $k_1 = 0.128797\ \text{ft}^{-1}$

Table 22.1 Propagation of compressional waves in Pierre Shale

Frequency f (Hz)	*Wave number* k (ft^{-1})	*Phase velocity* v (ft/s)	*Group velocity* U (ft/s)	$100\frac{(U-v)}{v}$ (%)
10	0.009432	6661.9	6724.5	0.940
20	0.018741	6705.2	6768.6	0.946
25	0.023378	6719.2	6782.9	0.948
30	0.028005	6730.7	6794.7	0.951
40	0.037239	6749.0	6813.3	0.953
50	0.046451	6763.2	6827.8	0.955
60	0.055645	6774.9	6839.7	0.956
70	0.064825	6784.8	6849.8	0.958
80	0.073992	6793.4	6858.6	0.960
90	0.083148	6801.0	6866.3	0.960
100	0.092293	6807.9	6873.3	0.961
110	0.101431	6814.0	6879.6	0.963
120	0.11056	6819.7	6885.3	0.963
130	0.11968	6824.9	6890.7	0.964
140	0.12880	6829.7	6895.6	0.965
150	0.13791	6834.2	6900.2	0.966

Frequency f (Hz)	*Wave number* k (m^{-1})	*Phase velocity* v (m/s)	*Group velocity* U (m/s)	$100\frac{(U-c)}{v}$ (%)
10	0.030945	2030.6	2049.6	0.940
20	0.061486	2043.7	2063.1	0.946
25	0.076699	2048.0	2067.4	0.948
30	0.091880	2051.5	2071.0	0.951
40	0.122175	2057.1	2076.7	0.953
50	0.152398	2061.4	2081.1	0.955
60	0.182562	2065.0	2084.7	0.956
70	0.212680	2068.0	2087.8	0.958
80	0.242755	2070.6	2090.5	0.960
90	0.272795	2072.9	2092.9	0.960
100	0.302798	2075.1	2095.0	0.961
110	0.332778	2076.9	2096.9	0.963
120	0.362729	2078.6	2098.6	0.963
130	0.392650	2080.2	2100.3	0.964
140	0.422571	2081.7	2101.8	0.965
150	0.452460	2083.1	2103.2	0.966

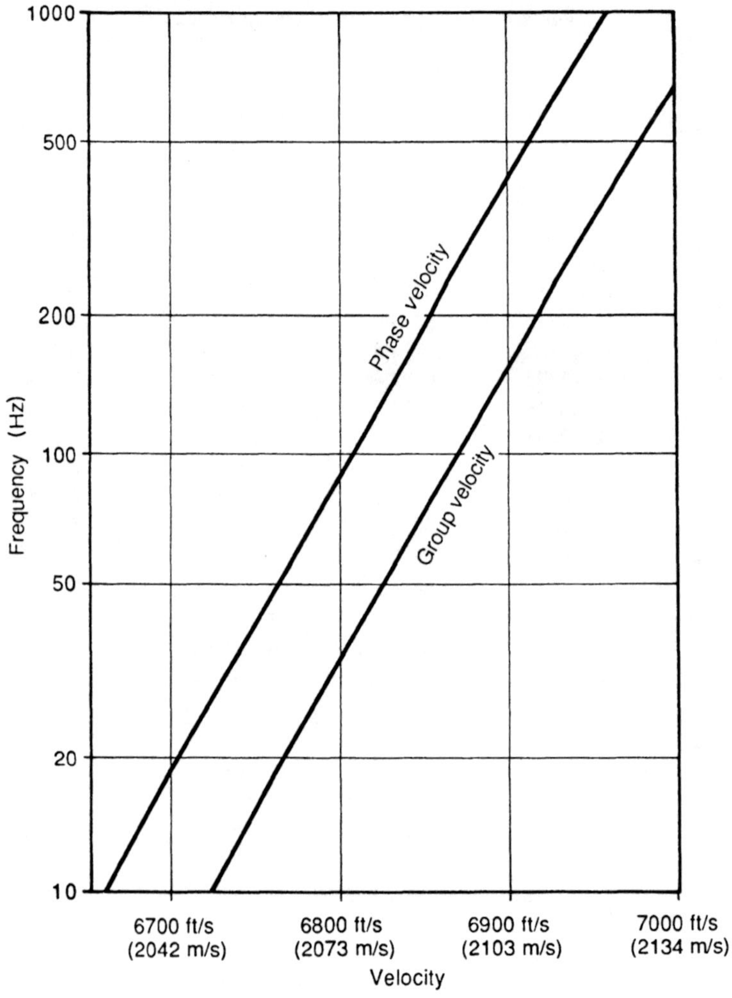

Fig. 22.1 Phase velocity and group velocity of the Pierre Shale. The phase velocity of the Pierre Shale increases with frequency which means that the group velocity is faster than the phase velocity as shown.

(0.422560 m^{-1}) and $k_2 = 0.0922931$ ft^{-1} (0.302798 m^{-1}). The group velocity for the wave packets is:

$$U = [2\pi(140 - 100)]/(0.1287966 - 0.0922931) = 6885.0\ \text{ft/s}\ (2098.5\ \text{m/s})$$

(which is very nearly equal to the average of the group velocities of the two frequency components).

For this simple case of two plane-wave sinusoids propagating through Pierre Shale, the phase velocities are 6829.7 and 6807.9 ft/s respectively for the 140 Hz and the 100 Hz components (2081.7 and 2075.1 m/s). But the velocity of

Table 22.2 Propagation of a wave with two sinusoidal frequency components through 10 000 ft (3048 m) of Pierre Shale

Note:
$f_1 = 140$ Hz, $v_1 = 6827.7$ ft/s (2081.1 m/s), $k_1 = 0.128797$ ft^{-1} (0.422562 m^{-1})
$f_2 = 100$ Hz, $v_2 = 6807.9$ ft/s (2075.0 m/s), $k_2 = 0.092293$ ft^{-1} (0.302798 m^{-1})

Item	*Apparent travel time for a 10 000 ft (3048 m) path length (s)*	*Velocity (ft/s)*	*(m/s)*
140 Hz frequency component	1.4642	6829.7*	2081.7*
100 Hz frequency component	1.4689	6807.9*	2075.0*
Combined resultant with an apparent frequency of 120 Hz	1.4661	6820.6*	2078.9*
Wave packets (energy)	1.4524	6885.0**	2098.6**

*Phase velocity
**Group velocity

the crests of the apparent 120 Hz sinusoid is:

$$v = [2\pi(140 + 100)]/(0.1287966 + 0.0922931) = 6820.6 \text{ ft/s} (2078.9 \text{ m/s})$$

The travel times through a 10 000 ft (3048 m) path length are given in Table 22.2. The propagation time for the energy (and for the wave packets) is 14 ms less than the apparent travel time of the crests and troughs of the apparent 120 Hz sinusoid.

Figure 22.2 plots the resultant wavelets for this case of two plane-wave sinusoids of frequencies 140 Hz and 100 Hz propagating through Pierre Shale as given by Equation (11) except that attenuation is ignored. (Or we could consider that attenuation has been offset by a zero-phase correction.) Figure 22.2 plots the wavelets as functions of time for depths of 0, 688.5, and 1337 ft (0, 209.86, and 407.5 m). Any given wavelet packet propagates both downward and forward in time at the group velocity of 6885.0 ft/s (2098.6 m/s). The diagonal lines represent traveling through space and time at this velocity. Since the phase velocity increases with frequency in this dispersive earth, the phase velocity is less than the group velocity and the wavelet within the envelope of the packet lags behind in phase. The wavelet started out as a zero-phase wavelet of zero degree (cosine symmetry with one main central crest). But the lag resulting from the dispersion causes the wavelet to become mixed-phase as the main crest moves to a position later in time within the moving packet. At some distance the wavelet will display a crest and a trough of equal amplitude and will then look much like a wavelet of ninety degree with sinewave symmetry. Indeed this is approximately true in Fig. 22.2 at a depth of 1377 ft (419.7 m). But the wavelet is now mixed-phase and is no longer a zero-phase wavelet.

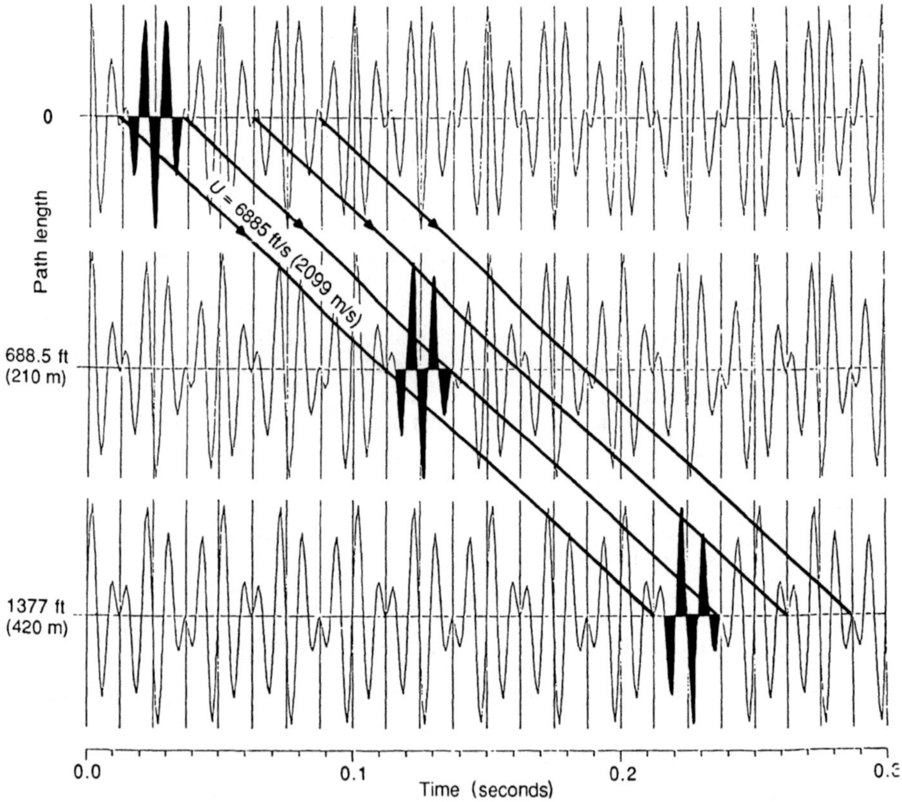

Fig. 22.2 Propagation of continuous, sinusoidal waves of two different frequencies through Pierre Shale. The wave packets travel at a group velocity of 6885 ft/s (2099 m/s) as indicated by the diagonal lines. The initial wavelet has 'cosine symmetry', but the wave shape is altered as the wavelet propagates through the shale because the phase velocity is less than the group velocity, causing individual peaks and troughs to lag behind within each wave packet.

Perhaps the effects of dispersion in the Pierre Shale can be better appreciated by comparing with a theoretical, non-dispersive model. Table 22.3 defines a theoretical earth where the wave numbers k_1 and k_2 have been changed to make this theoretical earth model non-dispersive with a group velocity equal to the group velocity of the packets for the previous Pierre Shale example. For comparison we consider the same two-frequency, plane-wave example and again ignore attenuation. Figure 22.3 shows the wavelets as functions of time at the same depths as in the previous figure. As a wavelet propagates along diagonal lines, there is no lag of the higher frequency wavelet within the envelope, and the initial zero-phase wavelet of zero degree remains a zero-phase wavelet of zero degree. This is because there is no dispersion in this

Table 22.3 Propagation of a wave with two sinusoidal frequency components through 10 000 ft (3048 m) of a non-dispersive earth model

Note:

The earth model used was patterned after the Pierre Shale with the same group velocity, but with the wavenumbers adjusted such that the model is non-dispersive.

$f_1 = 140$ Hz, $v_1 = 6885$ ft/s (2098.6 m/s), $k_1 = 0.127762$ ft^{-1} (0.419166 m^{-1})
$f_2 = 100$ Hz, $v_2 = 6885$ ft/s (2098.6 m/s), $k_2 = 0.091259$ ft^{-1} (0.299406 m^{-1})

Item	*Apparent travel time for a 10 000 ft (3048 m) path length (s)*	*Velocity (ft/s)*	*(m/s)*
140 Hz frequency component	1.4524	6885.0*	2098.6*
100 Hz frequency component	1.4524	6885.0*	2098.6*
Combined resultant with an apparent frequency of 120 Hz	1.4524	6885.0*	2098.6*
Wave packets (energy)	1.4524	6885.0**	2098.6**

*Phase velocity
**Group velocity

theoretical model and therefore no difference between phase and group velocities.

22.5 TRAVEL TIME TO A WELL GEOPHONE

Suppose that we had a geophone in a well at a considerable depth, and we applied signals at the surface near the well. The measured travel time with an impulsive source would be the depth divided by the group velocity, which would be the time required for the energy to propagate from the surface to the geophone. This conclusion is supported by Lighthill (1965) and in particular by quotations (1) and (3) cited previously (Section 22.3). Practical problems can arise, particularly if dissipation is great such that the signal received at depth is weak relative to the background-noise level. In such cases we may be tempted to try to pick a crest or trough of the received wavelet rather than the onset time. If we compare the signal received at different depths, the appearance may be so changed that the selection of corresponding points on the wavelet may be quite difficult and, consequently, differential travel times measured between corresponding crests or troughs may be either inaccurate or ambiguous. Velocities calculated from differential travel times will be equal to the group velocity only if the corresponding points are the onset times. Use

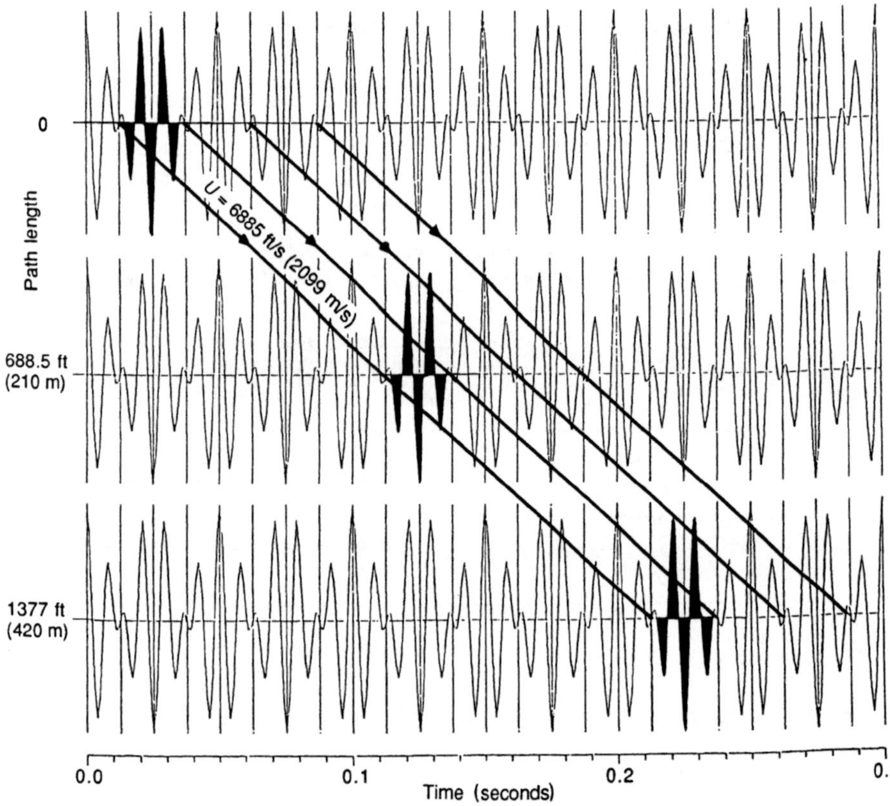

Fig. 22.3 Propagation of continuous, sinusoidal waves of two different frequencies through a non-dispersive model of the Pierre Shale. The wave packets travel at a group velocity of 6885 ft/s (2099 m/s) as indicated by the diagonal lines. The initial wavelet has 'cosine symmetry', and this symmetry is preserved as the wavelet propagates through the shale because in a non-dispersive medium the phase velocity is equal to the group velocity. Individual peaks and troughs do not lag behind within each wave packet.

of prominent crests or troughs will give velocities that are slightly slower than the group velocities.

However, usually in well surveys the energy or signal propagation time can be measured with adequate precision and without ambiguity when an explosive source is used (Section 22.8). Suppose that we also sought to measure the travel time using continuous, sinusoidal waves of some fixed frequency. We could measure the time delay between input crests and the corresponding crests of the waves received at geophones at very shallow depths. This would give us a travel time corresponding to the depth divided by the phase velocity. As we made measurements at greater depths, we would be concerned about which crest of the received wavelet corresponded to a given crest of the input wavelet. We could eliminate the ambiguity, either by making measurements at closely spaced depths and by following each crest very carefully, or by making measurements at many different sinusoidal frequencies. By such methods, we could measure an apparent travel time corresponding to the depth divided by the phase velocity.

Instead of using continuous sinusoidal waves of fixed frequencies, let us slowly change the frequency with time, as is done in the Vibroseis technique. If we used a very long sweep with a very slow rate of change of frequency then the Vibroseis measurement would approach the same result as measurements with one or more fixed sinusoidal frequencies. In the Vibroseis technique the correlation of the received wave with the input wave is a means for measuring the time delay of the received crests relative to the input crests. Correlation is an unambiguous method for accurately determining this time difference. Actually with a zero-phase correlation the rate of change of frequency is immaterial and the velocities measured with the Vibroseis technique are measurements of phase velocities, and travel times are equal to path lengths divided by the phase velocity. But if we were to use a minimum-phase correlation with the Vibroseis technique and if we measured onset times, we would then measure group velocities and travel times would be equal to the path lengths divided by the group velocity.

22.6 LABORATORY MEASUREMENTS

Measurements of wave propagation through the earth are complicated by the hetrogeneity of the earth and by limited access for measurement and observation. Can measurements be made in the laboratory on physical materials that illustrate the effects of attenuation and dispersion on phase and group velocity and on changes in wavelet shape?

Sachse and Pao (1978) show laboratory measurements of phase and group velocities as well as input and output wavelets. Measurements of compressional waves in aluminum showed no measurable dispersion and phase and group velocities were equal. But measurements of shear waves propagating in a laminated medium of parallel, boron fibers bonded with epoxy showed severe dispersion and unequal phase and group velocities at the same frequencies. Their Figure 4 is reproduced here as Fig. 22.4. The shear wave was propagated in the direction of the fibers and the particle motion was in a direction perpendicular to the fibers and parallel to the laminations. The frequency range over which data were recorded was about 0.5–6 MHz. Wave number is not precisely proportional to frequency and phase velocity increases with frequency. Attenuation is significant particularly for the higher-frequency components. The group velocity is significantly faster than the phase velocity at any given frequency.

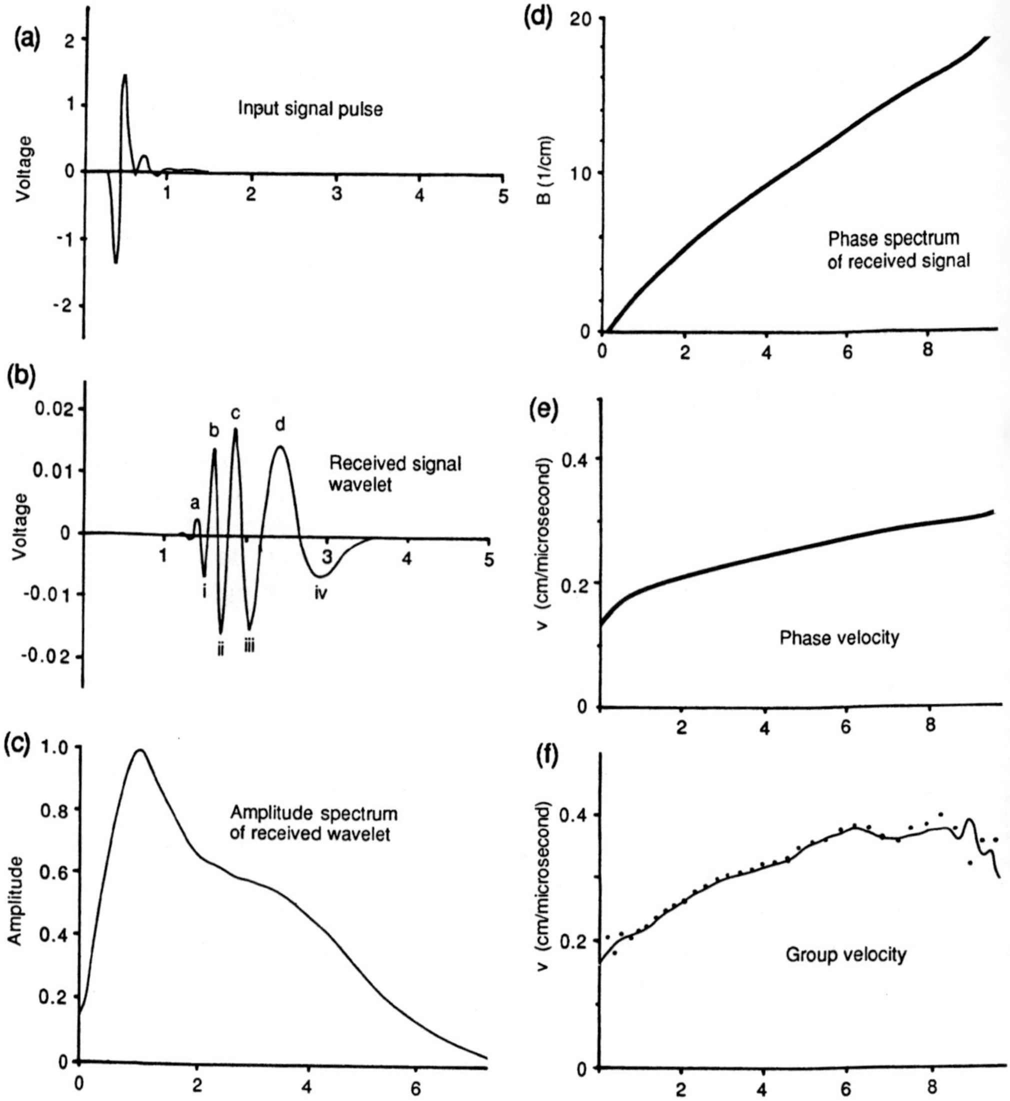

Fig. 22.4 Propagation of a shear wave in a composite material as measured by Sachse and Pao (1978). These laboratory measurements were made on a very dispersive material and show effects that are exaggerated as compared with the effects of dispersion of seismic waves in the earth. Note that the phase velocity shown in (e) increases with frequency and the group velocity shown in (f) is faster than the phase velocity. Note that the received wavelet show in (b) is 'front weighted' with the higher frequencies since the higher frequencies have higher velocities.

Table 22.4 Laboratory measurements* of wave propagation in a highly dispersive medium

Frequency (MHz)	*Travel times through 0.546 cm of boron & epoxy laminate*			*Velocities from travel times*	
	From Fig. 4e at phase vel. (μs)	*From Fig. 4f at group vel.* (μs)	*From Fig. 4b* (μs)	*Fig. 4e Phase* (cm/μs)	*Fig. 4f Group* (cm/μs)
1.18**	2.91	2.48	2.42**	0.188	0.220
1.875**	2.70	2.14	2.06**	0.202	0.255
3.00**	2.41	1.83	1.88**	0.227	0.299
3.75**	2.29	1.74	1.72**	0.238	0.313
4.76**	2.15	1.60	1.61**	0.254	0.341
5.45**	2.08	1.51	1.51**	0.263	0.362

*From Sachse and Pao (1978), Figure 4.
**Calculated from measured periods and times of Figure 4b to show that the received wavelet is 'front-loaded' with the higher frequencies.

The received wavelet is 'front-weighted' with the higher-frequency components. The author labeled the crests and troughs of the received wavelet of Fig. 22.4 and measured the time intervals between successive crests and between successive troughs. The reciprocals of these single-cycle periods were used as estimates of the dominant frequency components for that portion of the received wavelet. The arrival times of these dominant frequency components were taken as the time of the intervening crest or trough. For example, the differential time between crests *a* and *b* was found to be about 0.183 μs, yielding an estimated dominant frequency of 5.45 MHz with an estimated arrival time of the intervening troughs labeled *i* at a time of 1.51 μs. Table 22.4 shows the estimated frequency components in the first column and the corresponding estimated arrival times in the fourth column. For each of these frequency components the phase and group velocities were obtained from the appropriate plots in Fig. 22.4, and these velocities are shown in the fifth and sixth columns of Table 22.4. Travel times for the 0.546 cm length of the material were calculated for both the group and phase velocities (second and third columns of Table 22.4).

Figure 22.5 is a plot of these calculated times along with the measured travel times. For this impulsive measurement, the various frequency components are found to travel at the group velocities and not at the phase velocities. This experimental result is consistent with the previously cited third quotation from Lighthill (1965), (Section 22.3). Suppose that a measurement had been made on this material using two sinusoidal frequencies of 2.5 and 3.5 MHz. From Fig. 22.4 the phase velocities would be 0.219 and 0.237 cm/μs respectively. From these values, the calculated phase velocity of the apparent 3 MHz center frequency would have been 0.229 cm/μs, which would mean that the apparent

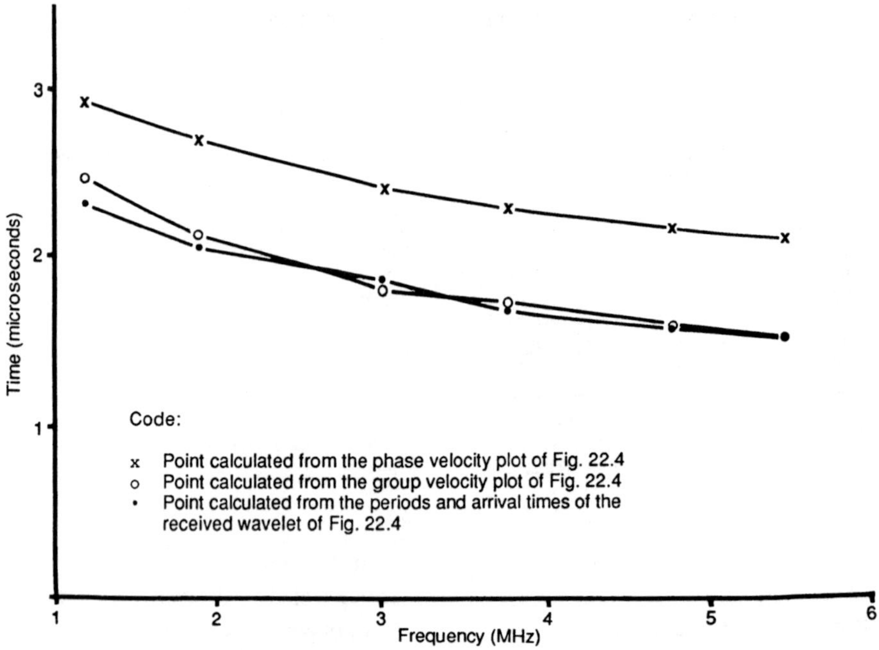

Fig. 22.5 Laboratory measurements of wave propagation in a highly dispersive medium. This figure shows that the arrival times estimated from the received wavelet of Figure 4(b) are consistent with arrival times calculated from the group velocities from Figure 4(f) and are not consistent with the arrival times calculated from the phase velocities from Figure 4(e).

travel time would have been 2.38 μs. If a zero-phase Vibroseis measurement had been made the measured travel time would also have been 2.38 μs. This compares to an impulsive arrival time for this frequency of 1.85 μs.

These laboratory measurements on this physical material illustrate effects of dispersion which are exaggerated as compared with the effects of dispersion of seismic waves propagating through sedimentary earth. But this exaggeration may serve to illustrate and substantiate more subtle effects in seismology that we seismologists have tried to ignore for too long. As we become more sophisticated in our analysis, interpretation, and practical use of seismic data, we need to be more aware of the effects of dispersion where it is significant. Past misconceptions need to be corrected.

22.7 CALCULATED EXAMPLES OF PROPAGATED BROAD-BAND WAVELETS

Many have argued that in a linear earth Vibroseis reflection times must be identical to reflection time measured with impulsive sources. They correctly state that correlation is a linear mathematical process. Therefore the response of a linear earth to a correlated wavelet is identical to the response to an uncorrelated wavelet that is correlated after it is received (i.e. the order of the operations has no effect). This is all true. They continue that the frequency components of the correlated Vibroseis wavelet and the same frequency components of the impulsive wavelet will be subject to precisely the same phase shifts during propagation. This is also true and is consistent with general wave theory. This does not prove, however, that Vibroseis reflection times must be identical to reflection times measured with impulsive sources. With calculated propagation times for an earth patterned after the Pierre Shale we will show that propagation times measured with zero-phase, zero-degree wavelets are longer than propagation times measured with minimum-phase wavelets with an identical amplitude spectrum. Since zero-phase wavelets have no onset time, measured propagation times must be made from the central crest of the input wavelet to the corresponding crest of the received wavelet. For measurements made with minimum-phase wavelets the propagation times are measured from the onset time of the input wavelet to the onset time of the received wavelet.

Our Pierre Shale model, patterned after the extended results of Wuenschel, is linear, attenuative, and dispersive. The calculated examples show that travel times measured with zero-phase wavelets are equal to the path length divided by the phase velocity, whereas travel times measured with minimum-phase wavelets are equal to path length divided by the group velocity. Propagation involves simultaneous phase shift and attenuation. We chose to apply these separately and observe the effects separately before combining the effects. We did this to try to gain insight into the propagation process. The only physically meaningful results are those with both effects applied since these effects are not physically separable.

We used a correlated sweep for the input wavelet to represent Vibroseis measurements. The correlated sweep was zero-phase and zero-degree. To represent impulsive sources we converted the zero-phase wavelet to minimum-phase. This insured that the amplitude spectra were identical for both wavelets. For additional insight we also used a unit impulse as an input wavelet. The unit impulse is a zero-phase wavelet and in this sense is like a zero-phase Vibroseis wavelet. It is equal to the normalized auto-correlation of a sweep of infinite duration over the frequency range from zero to infinity. The unit impulse response corresponds to the limiting case of a zero-phase Vibroseis measurement with a zero rate of change of frequency and therefore equivalent to measurements with continuous sinusoidal waves where apparent travel time is path length divided by phase velocity. To represent a Vibroseis measurement with a finite bandwidth and a zero rate of change of frequency we could simply filter the unit impulse response with a zero-phase, band-pass filter. Then measured travel times would still be the path length divided by the phase velocity.

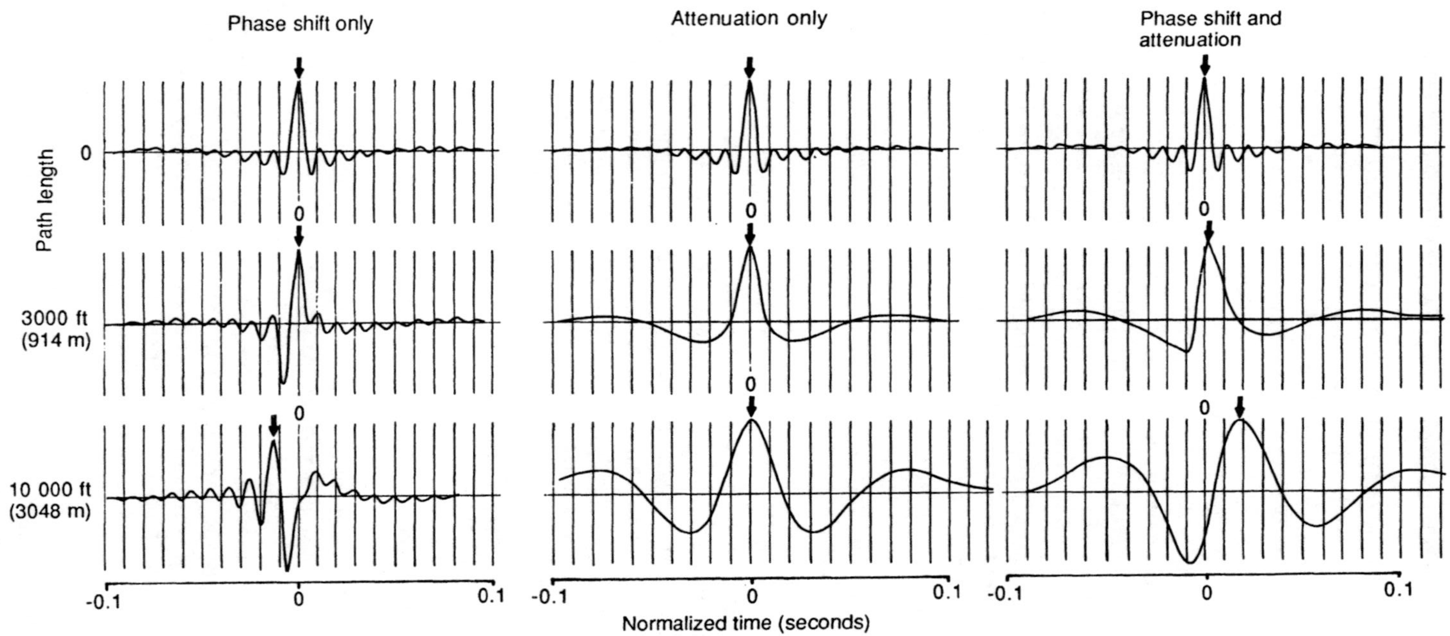

Fig. 22.6 Propagation of a broad-band, zero-phase wavelet in a Pierre Shale model. The effects of attenuation and phase shift are applied separately for this study, but physically the two effects are inseparable; therefore only the combined effects shown in the last column are physically realizable. The arrival time is normalized by a time shift equal to the propagation distance divided by the group velocity of the 25 Hz frequency component.

But the unit impulse is also like a minimum-phase wavelet in the sense that it does have a discrete onset time. Since energy propagates at the group velocity the onset time must propagate at the group velocity for both the unit impulse and for a minimum-phase wavelet. The unit impulse must simultaneously display an onset time traveling at the group velocity and a major crest traveling at the phase velocity. As this wavelet propagates in a dispersive medium, the phase of the wavelet becomes mixed and the crest become progressively more and more displaced from the onset time. These effects are illustrated by the calculated examples.

In Figs 22.6, 22.7, and 22.8 time zero is shifted by times equal to the path length divided by the group velocity of a 25 Hz frequency component. We will refer to times with these shifts as 'normalized' times. Also the amplitudes were normalized to offset the effects of attenuation so that the wavelet amplitudes would not be too small to be seen clearly. Problems were encountered in generating these figures. One problem resulted from the fact that the unit impulse has a finite component at zero frequency. The results obtained are adequate to illustrate the effects of dispersion including unequal phase and group velocities.

Figure 22.6 shows the propagation of a 10–100 Hz, zero-phase wavelet which is the autocorrelation of a 100s sweep. (Essentially identical results were obtained using the autocorrelation of a 1 s sweep.) The input wavelet at a distance of zero is shown along with wavelets at distances of 3000 ft (914.4 m) and 10 000 ft (3048 m). When only phase shift is applied the wavelet crest actually moves ahead in normalized time because it contains frequency components of up to 100 Hz for which the velocities are faster than the 25 Hz group velocity. When only attenuation is applied the wavelet broadens but there is no shift of the crest in normalized time. But when both phase shift and attenuation are applied the main crest of the wavelet moves back in normalized time corresponding to propagation at the slower phase velocity rather than at the group velocity.

Figure 22.7 shows the combined effects of both phase shift and attenuation for input wavelets of a unit impulse, a 10–100 Hz zero-phase, zero-degree wavelet, and a 10–100 Hz minimum-phase wavelet with the same amplitude spectrum as that of the zero-phase wavelet. We were a bit surprised by the onset times for the minimum-phase wavelets at 3000 ft (914.4 m) and 10 000 ft (3048 m) since they seem definitely to precede the normalized time zero. This would be expected for frequency components higher than 25 Hz, but the shape of the wavelet just after onset time does not give the appearance of being the result of high-frequency components. But appearances can be deceiving and the summation of higher frequencies with various attenuations and phase shifts must account for the early wavelet shapes as shown. The onset times at a distance of 30 000 ft (9144 m) for the unit impulse and for the minimum-phase wavelets are hard to pick and the picks shown by arrows are not precise. But

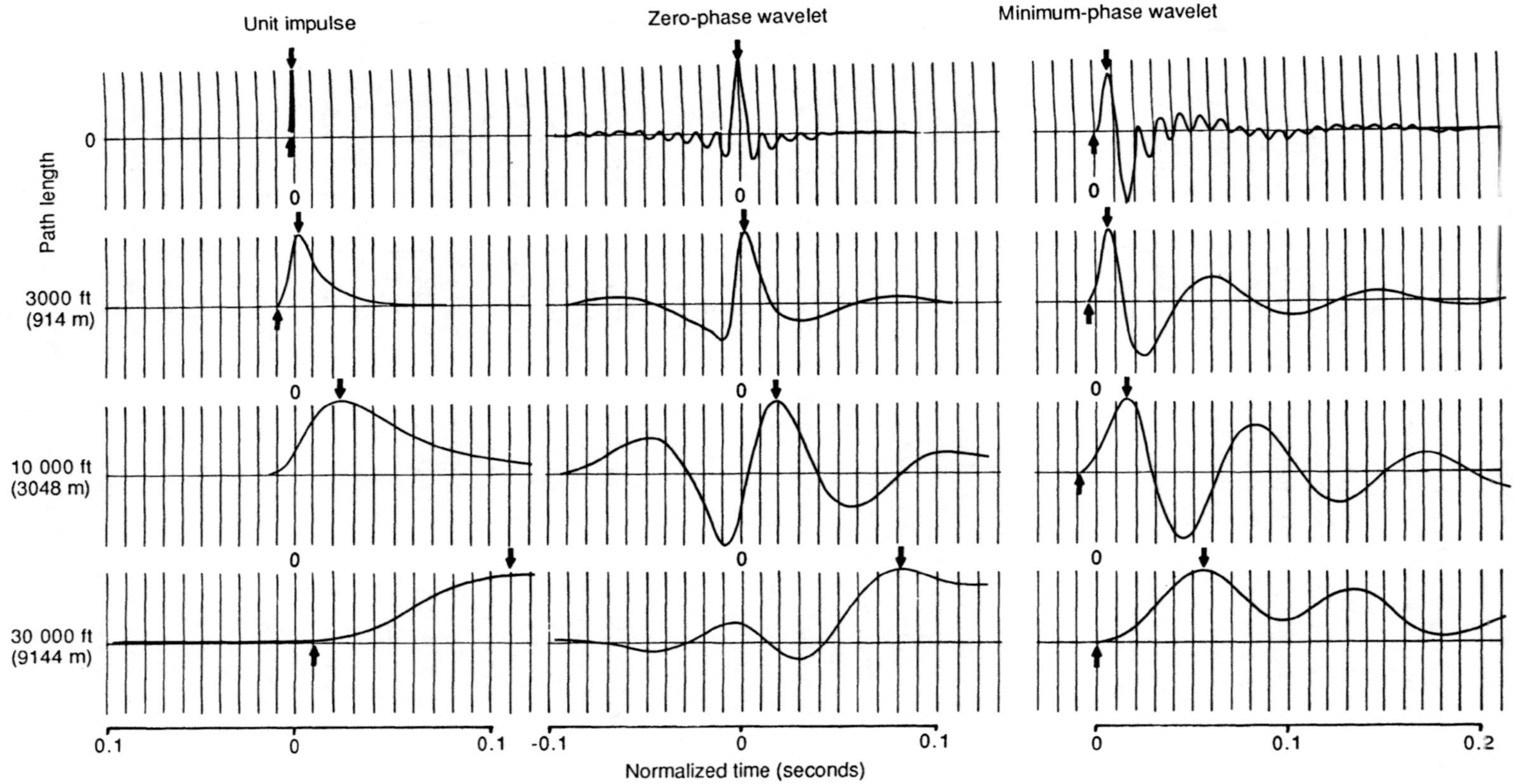
Unit impulse
Zero-phase wavelet
Minimum-phase wavelet
Path length
0
3000 ft
(914 m)
10 000 ft
(3048 m)
30 000 ft
(9144 m)
0.1
0
0.1
-0.1
0
0.1
0
0.1
0.2
Normalized time (seconds)

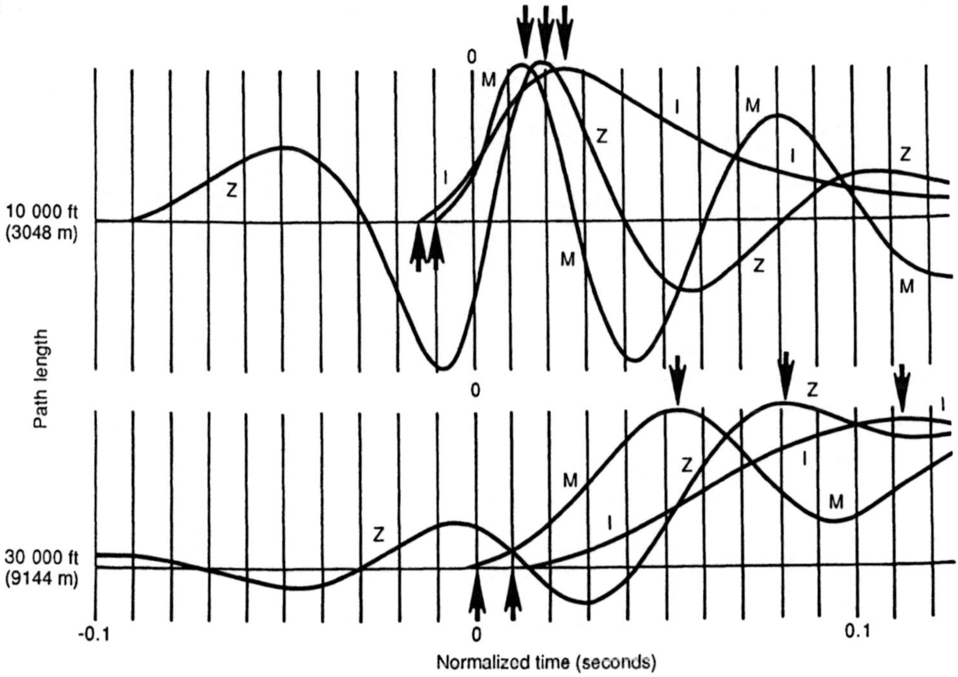

Z = Zero-phase, zero degree input wavelet
M = Minimum-phase input wavelet
I = Unit impulse input wavelet

Fig. 22.8 Wavelets after propagation through a dispersive Pierre Shale model. This enlarged display of the superimposed wavelets illustrates the approximate onset times for the unit impulse and for the minimum-phase wavelet and the arrival times of the comparative peaks for the unit-impulse and for the zero-phase wavelets. At a propagation distance of 10 000 ft (3048 m) the onset times precede the normalized time zero because both wavelets contain frequency components higher than 25 Hz (used for normalization) and these higher-frequency components have faster group velocities. The arrival times of the comparative peaks of the unit-impulse and of the zero-phase wavelet lag progressively behind because the phase velocities are slower than the group velocities. The lag is greater for the unit impulse because it is dominated by frequency components that are below 10 Hz which have slower phase velocities. This domination of the unit-impulse response by very low frequency components is most pronounced at the longer propagation distance where the effects of earth attenuation have greatly reduced the higher-frequency components.

Fig. 22.7 Propagation of three broad-band wavelets in a Pierre Shale model. The combined effects of phase shift and attenuation are illustrated for a unit impulse and for broad-band, zero-phase and minimum-phase wavelets propagated through a Pierre Shale model that is dispersive. The onset times of the unit impulse and of the minimum-phase wavelet are equal to the propagation distance divided by the group velocity of the highest significant frequency component in the received wavelets. The arrival time of the comparative peak of the zero phase wavelet and of the unit impulse is the propagation distance divided by the phase velocity. The comparative peak lags progressively behind because the phase velocity is slower than the group velocity. (Fig. 22.8 has an enlarged display).

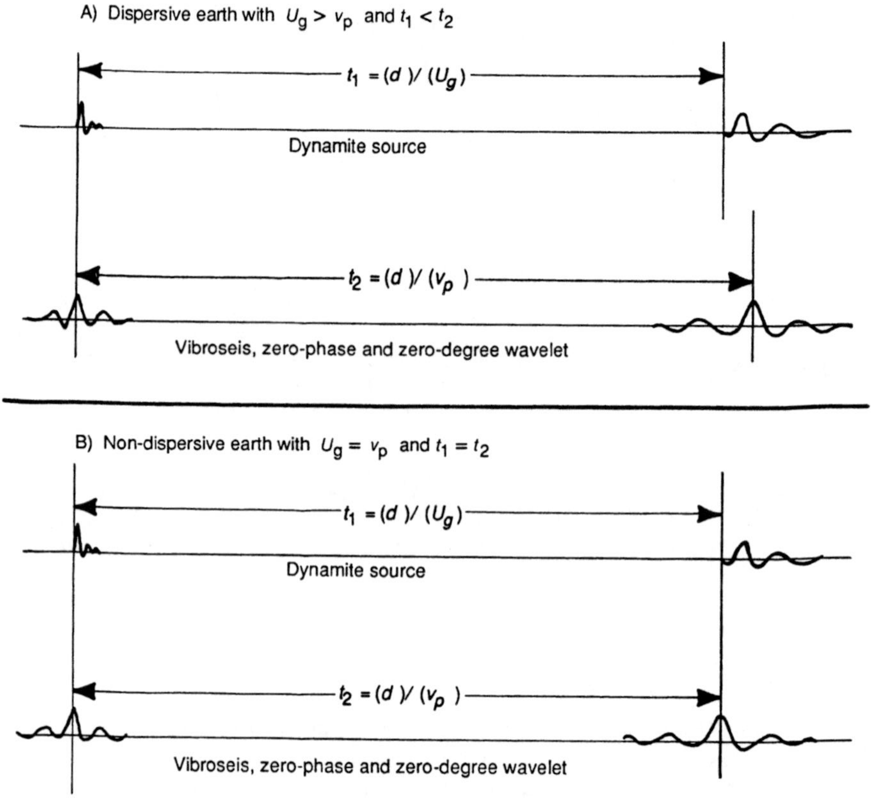

Fig. 22.9 Comparative arrival times for reflection events. In dispersive earths (top) a reflection arrival time for a dynamite record is shorter than for a Vibroseis record because the group velocity is faster than the phase velocity. Prominent crests or troughs on the dynamite records may also arrive before the main crests or troughs of the Vibroseis records despite the delays between onset times and prominent crests or troughs on dynamite records.

In theoretical, non-dispersive earths (bottom) the arrival times would be the same for dynamite and Vibroseis records, but corresponding crests and troughs on a dynamite record would lag behind those on a Vibroseis record because of the delays between onset times and prominent crests or troughs on dynamite records. Actual earths are not non-dispersive, but dispersion may be minor for some earths.

the approximate onset times are reasonably consistent with wave theory that predicts propagation at group velocities.

For the unit impulse and for the zero-phase wavelet the crest arrival times are consistent with wave theory that says they should propagate at phase velocities. Figure 22.8 shows an enlargement of the arrivals at distances of 10 000 ft (3048 m) and 30 000 ft (9144 m) for all three input wavelets superimposed.

Reflection 'event' times on minimum-phase dynamite seismic sections are equal to the path lengths divided by the group velocities plus the delay times from the onset to the picked crest or trough. Reflection times on zero-phase Vibroseis sections are equal to the path lengths divided by the phase velocities (Fig. 22.9). The corresponding event times may be nearly equal if the delay times on the dynamite sections are approximately offset by the group velocities being faster than the phase velocities; in many cases this may be approximately true. If there were no dispersion, then corresponding events would always be earlier on zero-phase Vibroseis sections than on minimum-phase dynamite sections. But where dispersion is relatively severe and velocity increases with an increase in frequency, events on minimum-phase dynamite sections will precede corresponding events on zero-phase Vibroseis sections because the phase velocities will be slower than the group velocities.

For dispersive earths there will be similar differences in event times between Vibroseis sections with zero-phase and minimum-phase correlations and processing. The minimum-phase correlation used may or may not result in some time shift which would have to be taken into account.

22.8 WAVEFRONT VELOCITY

In the seismic literature, the term wavefront velocity has been used to denote the velocity calculated by detecting the onset of the arrival of a propagating wavelet from an impulsive source. But as was pointed out previously, the propagation velocity is the group velocity. Therefore the wavefront velocity is the group velocity of the frequency component which has the fastest group velocity. A good seismic model of a typical earth is one where the phase velocity and attenuation increase with increasing frequency and hence the group velocity is faster than the phase velocity. In such a model, the highest frequency component measurably present in the received wavelet has the highest group velocity and this velocity is the wavefront velocity. In a well velocity survey where the S/N ratio is good, we can turn up the amplifier gain and be likely to pick a slightly earlier onset time and hence calculate a slightly faster wavefront velocity. What we are doing in such cases is to offset the effects of earth attenuation so as to make somewhat higher frequencies detectable. (For dispersive earths, the time–depth relationship from such a velocity survey would not be precisely correct for interpreting seismic data with a lower upper-frequency limit.) In summary the proposed earth model is linear, and has finite dispersion such that phase velocity increases slightly with frequency and group velocities are faster than phase velocities.

22.9 BACK TO THE PUZZLING WELL VELOCITY SURVEY

We were not the first to be puzzled by arrival time discrepancies in well velocity surveys. A similar variation between dynamite and Vibroseis arrival times was reported by Ward and Hewitt (1977) in an AMOCO test well. At the maximum well depth of only 2500 ft (760 m) they observed a discrepancy of 10 ms which was about what we observed at similar depths. If they had made measurements at depths of more than 10 000 ft (3048 m), then perhaps they would have noted discrepancies as large as we did. We might have accepted a 10 ms discrepancy as being the result of a combination of various little things like errors in picking, errors in measurements, differences in the spectra, differences in paths, etc. But a 45 ms discrepancy simply could not be brushed aside.

It is probable that the earth section at our well is unusually dispersive. It is an area where compressional-wave reflection data quality is poor and the section is underlain by a large gas field. It is possible that there could be a small percentage of free gas in the pores of low permeability rocks in much of the section. This might have increased attenuation, increased dispersion, and reduced the quality of seismic reflections. But the time discrepancy was approximately linear with depth, which means that much of the earth section was dispersive rather than just a few, particular intervals.

We know that earths with finite Q must be dispersive. The earth at our well site and the Pierre Shale are dispersive. It appears that the earth at the AMOCO well site is dispersive. Where reflection events on zero-phase correlated Vibroseis data wander with respect to corresponding events of data recorded with impulsive sources, it is probable that the earth is dispersive and the dispersion varies. If we paid more attention to dispersion we might find that it is commonly measurable with seismic data. With Vibroseis data we could use both a zero-phase correlation and a minimum-phase correlation and use appropriate processing. We might find that measurements of dispersion could be related to useful geologic information.

22.10 CONCLUSIONS

Dispersion is a normal and common occurrence in the propagation of seismic body waves through sedimentary earth sections. Dispersion causes the group velocity to be different from the phase velocity. Time–depth relationships will not be identical for Vibroseis data with a zero-phase correlation and data acquired in response to an impulsive source where dispersion is significant. Where phase velocity increases with frequency, reflection onset times will be shorter for impulsive source data than corresponding reflection event times for zero-phase data. But the added time from onset to a picked crest or trough for the dynamite reflections may frequently result in picked event times that are nearly equal to corresponding event times on zero-phase Vibroseis sections. In other cases the corresponding event times may be significantly different and the dynamite reflection events may precede the corresponding Vibroseis event times.

In a typical earth where dispersion is significant, phase velocity increases with frequency, group velocity is faster than phase velocity, and the wavefront velocity is the group velocity of the highest frequency component that is significant in the received wavelet. Under these circumstances the received wavelets from an impulsive source will be somewhat 'front-weighted' with the higher-frequency components. We assume that the earth is linear at low energy levels.

REFERENCES

Futterman, W. I. (1962). 'Dispersive body waves'. *J. Geophys. Research* **67**, 5279–91.

Kjartansson, E. (1979). 'Constant *Q*-wave propagation and attenuation'. *J. Geophys. Research* **84**, 4737–48.

Lighthill, M. J. (1965) 'Group velocity'. *J. Inst. Maths. Applics.* **1**, March, 1–28.

McDonal, F. J. *et al.* (1958). 'Attenuation of shear and compressional waves in Pierre Shale'. *Geophysics* **23**, 421–39.

Sachse, W. and Pao, Y.-H. (1978). 'On the determination of phase and group velocities of dispersive waves in solids'. *J. Appl. Phys.* **49** (8), August, 4320–27.

Sheriff, R. E. and Geldart, L. P. *Exploration Seismology, Volume 1: History, theory, and data acquisition.* Cambridge University Press (1982), pp. 55–7.

Stokes, G. G. (1876). 'Smith's Prize examination question no. 11', reprinted in *Mathematics and Physics Papers* (1905). **5**, p. 362. Cambridge University Press.

Ward, and Hewitt, (1977). 'Monofrequency borehole traveltime survey'. *Geophysics*, **42**, October, 1137–45.

Wuenschel, P. C. (1965). 'Dispersive body waves – an experimental study'. *Geophysics* **30**, 539–51.

23
Economics, future trends, and conclusions

In this final chapter we discuss economics, seismic crew efficiency, future trends in exploration seismology, and some final conclusions.

23.1 COST OF SEISMIC DATA AND SEISMIC CREW EFFICIENCY

Costs for acquiring and processing seismic data vary widely for many reasons. Frequently cost is expressed in dollars per mile. Managers need these numbers to plan big exploration programs. Cost numbers such as these change rapidly with time, conditions, locations, etc. But some approximate cost numbers are given in Table 23.1 because they may have relative value.

Marine costs per mile or per kilometer are relatively low because of high production rates that can be achieved under favorable conditions. Yet a boat in harbor waiting for suitable weather is a big overhead item which adds to the average cost. Since marine acquisition costs on average are low, the processing costs are a larger percentage of the total costs.

On land, Vibroseis work along straight, rural roads can allow rather high production rates particularly if the source and group intervals are long and the background noise level is low. Again, if production rates are high, processing costs become a larger percentage of the total. By contrast, portable crew work in rugged and remote terrain can result in very high acquisition costs.

Dollars per mile (or per kilometer) is a poor measure of crew and operational efficiency, particularly for land work. We judged crew efficiency based upon the number of records per day and per month. We took into consideration the number of traces, the geophone array, the terrain, the accessibility, the average length of line, etc. For Vibroseis crews we calculated the percentage of time the vibrators were sweeping and data were being recorded. We included in the recording time the move-ups between sweeps but not the move-up from one VP to the next. Conditions are never the same, but with some effort you can get a good feel for the efficiency of crews and a sense of

Table 23.1 Illustrative costs for seismic data

Crew type/conditions	*CMP fold*	*Production (mi/month)*	*Acquisition (dollars/mi)*	*Processing (dollars/mi)*	*Total cost (dollars/mi)*
Vibroseis/good	high	50	3500	800	4300
Mini-hole portable/good	high	35	5000	800	5800
Deep hole/good	medium	25	8000	800	8800
Portable/fair	medium	20	10 000	800	10 800
Portable/tough	medium	10	20 000	800	20 800
Marine/good	high	1000	800	800	1 600
Marine/average	high	500	1500	800	2300
Crew type/conditions	*CMP fold*	*Production (km/month)*	*Acquisition (dollars/km)*	*Processing (dollars/km)*	*Total cost (dollars/km)*
Vibroseis/good	high	80	2175	500	2675
Mini-hole portable/good	high	56	3105	500	3605
Deep hole/good	medium	40	4970	500	5470
Portable/fair	medium	32	6215	500	6715
Portable/tough	medium	16	12 430	500	12 930
Marine/good	high	1600	500	500	1000
Marine/average	high	800	930	500	1430

whether the efficiency of a given crew is increasing or decreasing. We always emphasized data quality ahead of production, but we sought to improve production too.

Where our objectives are relatively shallow, costs tends to rise because the number of records per mile increases. Yet in one area we maintained a high production rate in miles per day despite the use of VP intervals as short as 110 ft (33.5 m) which required 48 VPs per mile (30 VPs per km). In this case our acquisition costs per mile were moderate, but since we were getting so much data our processing costs rose to 40 per cent of the total cost. This was before we were challenged by a program with even shallower objectives, which we attacked by the simultaneous recording of pairs of VPs. Many of our Vibroseis programs with shallow-to-moderately deep objectives could have been recorded very successfully with pairs of VPs recorded simultaneously, with significantly lower acquisition costs. Particularly for rather shallow objectives you should try simultaneous recording of pairs of VPs if you can use a Vibroseis crew (Chapter 14).

23.2 FUTURE TRENDS

In seismic data acquisition the future trend will be toward telemetry systems particularly using radio linkage. Beam steering in data acquisition should be used widely as more equipment is developed to make this feasible. More computer control should be incorporated into the instruments in the recording truck, out on the line, and in vibrators and other sources. This computer control equipment will facilitate field beam steering as well as other variations in recording techniques. Source beam steering is relatively simple but receiver beam steering will also be practical by use of delay and sum equipment out on the line. Geophone array and vibrator pattern changes will be made during the sweep, with longer arrays and patterns being used when lower frequencies are being swept and shorter arrays when the higher frequencies are being swept. Again this can be relatively simple for the vibrators pattern as was suggested in Chapter 13 using different sweep lengths with a common start frequency and a common rate of change of frequency. Circuitry should be developed to make it feasible to utilize codes such as those discussed in Chapters 10 and 14 to allow the simultaneous recording of multiple data sets. Means will be developed to allow better separation of decoded signals from multiple data sets that are recorded simultaneously, so that larger differences in signal levels can be handled successfully.

If the past is any guide, the number of data channels will continue to expand. But for what purpose? If we are simply recording one line of data at a time, we have a limited range of effective offsets and more channels require a shorter group interval which greatly increases costs. Recording compressional-wave and shear-wave data simultaneously will triple the number of useful channels. Using multiple source positions and simultaneously recording multiple lines of data enables us to use more channels. If we are using receiver patches and in particular if we use a roll-along patch we can effectively use more channels (Chapter 12). But for single-line, compressional-wave data it is likely that there will still be a strong need for recording systems with less than 100 data channels.

Perhaps accelerometers (or geophones with feedback coils with characteristics similar to accelerometers) will come to be used extensively to help increase dynamic range and to improve the phase response of the detectors. Surely field processing with adequate display will become the norm, at least for quality control and for tests. Much more use will be made of earth-orbiting satellites for navigation and location at sea and on land.

In marine work there will be expanded use of digital cables utilizing fiber optics with many more channels, which will provide opportunity for flexible summation under computer program control to form various geophone arrays. Time shifts before summation should provide receiver beam steering capability which should vary with distance from the guns and with record time. This would augment less sophisticated source beam steering with time shifts between gun groups which are operated either at the same depths or at different depths.

'Black-box' exploration techniques will continue to pop up and promoters will make false claims of super effectiveness and low cost. These black-box techniques, based upon either wishful thinking or pure hoax, will extract money from the naïve before each fades into oblivion. But seismology will remain a major tool of exploration so long as men and women search for oil and gas.

23.3 ACQUISITION, PROCESSING AND INTERPRETATION

A seismic exploration program is like a stool with three legs. The three legs are data acquisition, processing, and interpretation. A seismic program will not support much weight unless all three legs are sturdy. All three jobs are important and require good people to perform them well to achieve an effective exploration team. Unfortunately, it sometimes happens that one leg comes to think it is the important leg and that it carries most of the load. Debates about which leg is most important make as much sense as which came first, the chicken or the egg? In an effective exploration organization each leg appreciates the importance and value of the others and each reinforces its own position by its own superior performance and results. Similarly geophysicists, geologists, landmen and engineers must work together not only with cooperation, but with real appreciation of the contributions and importance of each. It is worth spending a modest amount of time learning what the others do and the problems they face.

23.4 HIGH-TECHNOLOGY 'PSEUDO EXPERTS'

During the Second World War the military did a superb job of taking farm boys with little background in science and mathematics and in a matter of weeks or months training them to calibrate, maintain, repair and operate sophisticated equipment and systems including radars. (The successful trainees were smart.) The military succeeded by getting bright and well-trained people to prepare excellent classes and profusely illustrated manuals that focused sharply on the jobs to be performed. They did not study Einstein's theory of relativity, but the successful trainees did learn everything they needed to do their jobs well.

Today the geophysics profession is overflowing with extremely bright men and women who have excellent backgrounds in basic science and mathematics. Even those without advanced degrees are well prepared to learn whatever they need to do their jobs quite well. Yet in some organizations there are a few pseudo experts who try to seize some area of high technology and claim it for their own. These are high-technology pseudo experts (Feyerabend 1978). In effect they claim that only they are capable of properly using the seized technology. Hogwash! If they were genuine experts that really had the company's interest at heart, they would analyze the technology and how it could be effectively applied. They would then teach and show others so that the technology would be quickly and widely utilized by the whole organization. Then for a brief time their company would be ahead of its competitors in the use of the technology. If you let a pseudo expert lock up technology and, like a medical doctor, dispense it by prescription only your company will lag behind.

High-technology pseudo experts may be sincere but they are blinded by their overestimation of their own capabilities and their underestimation of the capabilities of others. How do they succeed personally at the expense of their companies? They make the technology and its application as complex as possible instead of as simple and straight-forward as possible, even when simple solutions are better. The more complicated they can make things appear the easier it is for them to rationalize their contention that only the elite (themselves) could possibly use the technology successfully.

23.5 BUMBLEBEES

We can thwart the pseudo experts by successfully using some technology and teaching others to do likewise. First, we must understand the technology and where it may be used effectively in our work. We need to consider the possible problems and dangers, but look for simple and practical ways to avoid the problems or at least ameliorate the dangers. Avoid excessive concern about the problems and dangers. Years ago, sophisticated experts proved, by accepted theories of aerodynamics, that the bumblebee could not fly. But being ignorant of these theories, the bumblebee flies quite well. We may not know how he does it, but he does not do it by focussing on all the possible problems and the possible dire results if he should try but fail.

It is our job to adapt and utilize available technology and equipment to acquire better seismic data. In considering the use of some technology define precisely what you want it to do for you. If the technology seems to make sense for better meeting your needs, then look for ways and opportunities to try it with minimum cost and minimum risk. If early results are disappointing, determine why it was not successful and then look for ways to make it work, discarding what cannot be made to work successfully, but aggressively pursuing encouraging results. When successful, carefully document why the technology was helpful defining when and how it can best be used and teaching others so that your company gains maximum benefit as rapidly as possible. Be receptive to others when they share their successes with you. If some 'expert' says you cannot fly, be a bumblebee and fly despite his theory—you may even teach him to fly.

Of course, neither the bumblebee nor we can violate valid laws of physics. But working within these laws, motivated, creative geophysicists who are familiar with the problems from first-hand experience can often find practical solutions that 'experts' further removed from the operation are not likely to discover. The subject of this book is really geophysical engineering and the book is primarily addressed to geophysical engineers.

REFERENCES

Feyerabend, P. *Science in a Free Society: a book of essays* (1983). Schocken Books.

Appendices

APPENDIX A—RECOMMENDED FORMS

Examples of recommended forms on the following pages include:

1. Separate onshore and marine prospect forms;
2. Separate Vibrator and dynamite recording parameter forms;
3. Separate weekly crew report forms for vibrator and dynamite crews;
4. Separate English units and metric units forms for use in designing geophone arrays and source patterns. Actually either form can be used with either English or metric units depending upon the range of noise wavelengths of concern. The forms shown used K & E semi-logarithmic graph paper #46 4770 which has six small divisions per major division of the linear scale. For the form normally used for metric numbers this graph paper was split and rearranged.

These recommended forms should be adapted to fit requirements and should be adjusted as needs change.

Forms should be designed to cover the equipment available on each crew along with the currently used recording parameters. The latter will overlap information on the recording parameter forms, but will serve to document any changes made in the recording parameters during the recording of a project. These forms will also serve to verify that the recording parameters in use by the crew are those currently specified.

All of these forms except the prospect forms should be distributed to the field crew and they should keep them until a project is completed and work in the area is completed. Then the field crew should shred their copies.

We should keep permanent files of all of these forms, preferably in loose-leaf notebooks. Copies of all of the forms should be distributed to anyone in the company that may need the information, including the interpreter, the data processor, and all field-crew contact and quality control people.

FM1 Onshore Seismic Prospect Form

Prospect ______________ Line prefix _________ Auth. no. ______________________
Miles of program ____________ Date _____–_____ 19 ____ By ________________
Country ____________ State of Prov. ____________ Counties ________________
Quads ____
General objectives ____
____
Topography _____________ Vegetation ______________ Culture ______________
Seismic datum _______________ Correctional velocity ______________________

Specific objectives

Priority/ Horizon	Record time	Depth	Stacking velocity	Lithology	Dip	Resolution Vertical	Resolution Horiz.
1 ______	______	______	______	______	______	______	______
2 ______	______	______	______	______	______	______	______
3 ______	______	______	______	______	______	______	______
Basement ______	______	______	______	______	______	______	______

Geology

Surface ____
____
Near surface ____
____
At depth ____
____
Key wells ____

Previous seismic data

Line	Prospect name	SP range	Tie points	Quality	Problems	Improvement required
______	______	______	______	______	______	______
______	______	______	______	______	______	______
______	______	______	______	______	______	______

Additional comments or suggestions ____
____
____

Preferred order for recording lines? Yes/No. If 'yes' give details including reasons, on a separate sheet.

Please fill out this form as completely as is feasible. Attach additional sheets if necessary.
Distribution: Prov. Mgr. Interpreter Processor WPC KMK JPB SRS

FM2 Marine Seismic Prospect Form

Prospect ______________ Line prefix _________ Auth. no. ____________________
Miles of program _____________ Date ______–______ 19 ____ By ________________
Country _____________ Nearest state and county or area name __________________
Quads ____
General objectives ____
____
Water depth range ________________ Water bottom — Soft — Medium — Hard

Specific objectives

Priority/ Horizon	Record time	Depth	Stacking velocity	Lithology	Dip	Resolution Vertical	Horiz.
1 _______	_______	_____	_________	_________	____	_______	_______
2 _______	_______	_____	_________	_________	____	_______	_______
3 _______	_______	_____	_________	_________	____	_______	_______
Basement _______	_______	_____	_________	_________	____	_______	_______

Geology

Shallow ____
____
At depth ____
____
Key wells ____

Seismic problems ____ Multiples (multiple velocity _____________) ____ Irregular bottom
____ Narrow bandwidth (ringy) ____ Penetration (or random noise)
____ Vert. resolution ____ Source-generated noise ____ Shallow gas pockets

Operational problems ____ Surface obstacles (what? ______________________________)
____ Sea floor obstacles or reefs (what? _______________) ____ Shallow water
____ Currents ____ Rough seas ____ Political (what? ________________________)

Previous seismic data

Line	Prospect name	SP range	Tie points	Quality	Problems	Improvement required
_____	_________	_________	______	________	_________	_________
_____	_________	_________	______	________	_________	_________
_____	_________	_________	______	________	_________	_________

Additional comments or suggestions ____
____
____

Preferred order for recording lines? Yes/No. If 'yes' give details including reasons, on a separate sheet.

Please fill out this form as completely as is feasible. Attach additional sheets if necessary.
Distribution: Prov. Mgr. Interpreter Processor WPC KMK JPB SRS

FM3 AJAX Exploration Company: Vibrator Recording Parameters

Prospect ____________ Line prefix ____________ HO- ______-______

Crew ____________ State ____________ Date ______-______-198____ By ______

Trace interval ____________ ft VP interval ____________ ft Number of traces ______

(VP's to be located at midpoint between trace locations.)

Spread definition ________________________________

Spread description ________________________________

DISCONNECT the dead traces in the gap. No. of traces in gap ____________

GEOPHONE ARRAY: No./Trace ____________ Type ____________ Spacing ________ft.

Energy source:

1. No. of vibrators ________ Configuration ________ Pad-to-pad spacing ________ ft.
   (No. of vibs) (spacing) = ________ ft + Move-up of ________ ft = ________ ft.
2. No. of vibrators ________ Configuration ________ Pad-to-pad spacing ________ ft.
   (No. of vibs) (spacing) = ________ ft + Move-up of ________ ft = ________ ft.
3. No. of vibrators ________ Configuration ________ Pad-to-pad spacing ________ ft.
   (No. of vibs) (spacing) = ________ ft + Move-up of ________ ft = ________ ft.

Number of sweeps/VP ______ with ______ vibrators Move-up __________ ft/sweep

______ with ______ vibrators Move-up __________ ft/sweep

______ with ______ vibrators Move-up __________ ft/sweep

DO NOT WORK with less than ______ vibrators

Vibrator drive level __________% ________________________________

Reduce drive level according to cultural conditions.

Sweep frequency ________-________ Hz ________________________________

Sweep length ____________ sec. Start Taper ________ sec. End Taper ________ sec.

Listen time ____________ sec.

Total time ____________ sec.

Sample rate ____________ sec.

Low-cut filter ______ Hz ______ dB/octave High-cut filter ______ Hz ______ dB/octave

Notch filter ☐ in ☐ out Static filter ☐ in ☐ out COS ☐ in ☐ out

Correlation ☐ Minimum-phase ☐ Zero-Phase

Summation: Make ______ summation(s) per VP. Sum in groups of ______ sweeps with ______ sweeps/VP or in groups of ______ sweeps with ______ sweeps/VP.

Noise edit: ________________________________

Record FILTERED and UNFILTERED sweeps for each file, if possible, on auxiliary channels. If this is not possible due to instrument limitations, record FILTERED and UNFILTERED sweeps on data channels at the beginning of each workday.

Additional instructions ________________________________

____

____

____

Crew Province Manager Interpreter WCP MKK

JBP RSS WCM

FM4 AJAX Exploration Company: Dynamite Recording Parameters

Prospect ______________________ Line prefix ______________

Crew ______________ State ______________ HO– ____________ – ____________

Trace interval ____________ ft. SP interval ____________ ft. Number of traces ________

☐ (SPs to be located at midpoint between trace locations.)

☐ (SPs to be located at trace locations.)

Date ________ - ________ -198 __ By ______________

Spread definition ______________________________

Spread description ______________________________

DISCONNECT the dead traces in the gap. No. of traces in gap ____________

GEOPHONE ARRAY: No./trace ______________ Spacing ______________ ft.

Type ______________________________

Geophone Array

SHOT ARRAY: Hole Depth ____________ ft. Charge size ____________ lb.

Array type ______________________________

Number of holes ____________ Hole spacing ____________ ft.

Shot Array

Record Length ______________ sec. Sample rate ______________ sec.

Low-cut filter ____________ Hz ____________ dB/octave

High-cut filter ____________ Hz ____________ dB/octave

Notch filter ☐ in ☐ out Static filter ☐ in ☐ out

Additional instructions ______________________________

____

____

____

Crew	Province	Manager	Interpreter	WCP	MKK
JBP	RSS	WCM		BSC	

FM6 AJAX Exploration Company: Dynamite Crew Weekly Status Report

Crew		From			To			Date		
Date	Line	Stations	No. vp.	Miles	Skips	Miles less skips	Hours	Hours to date	Vibs working	Comments
		to								
		to								
		to								
		to								
		to								
		to								
		to								
		to								

PROGRAM

Tapes shipped and dates

Survev Notes? Y N

Visitors' names and dates

Comments

This week's plans

Break? M T W R F S S

Gp. int. ft	No. traces	Geop/gp.	Gp. length ft	Sweep length sec	Listen time sec	No. sweeps

Sweep frequencies Hz- Hz	Sweep type	VIBRATOR PATTERN	Pad-to-pad spacing ft	Total moveup ft

 AJAX Exploration Company: Dynamite Crew Weekly Status Report

Crew		From			To			Date	
Date	Line	Stations	No. sp.	Miles	Skips	Miles less skips	Hours	Hours to date	Comments
		to							
		to							
		to							
		to							
		to							
		to							
		to							

PROGRAM		DRILLS					
Line	Miles	Date	Line	No. drills	No. patterns	Hours	Comments

Tapes shipped and dates	
	Survey notes? Y N
Visitors' names and dates	
Comments	
This week's plans Break? M T W R F S S	

Gp. int. ft	Sp. int. ft	No. traces	Record length sec	Geop/gp.	Gp. length ft		
PATTERN	Type	Number	Spacing ft	Length ft	CHARGE	Size lb	Depth ft

FM8

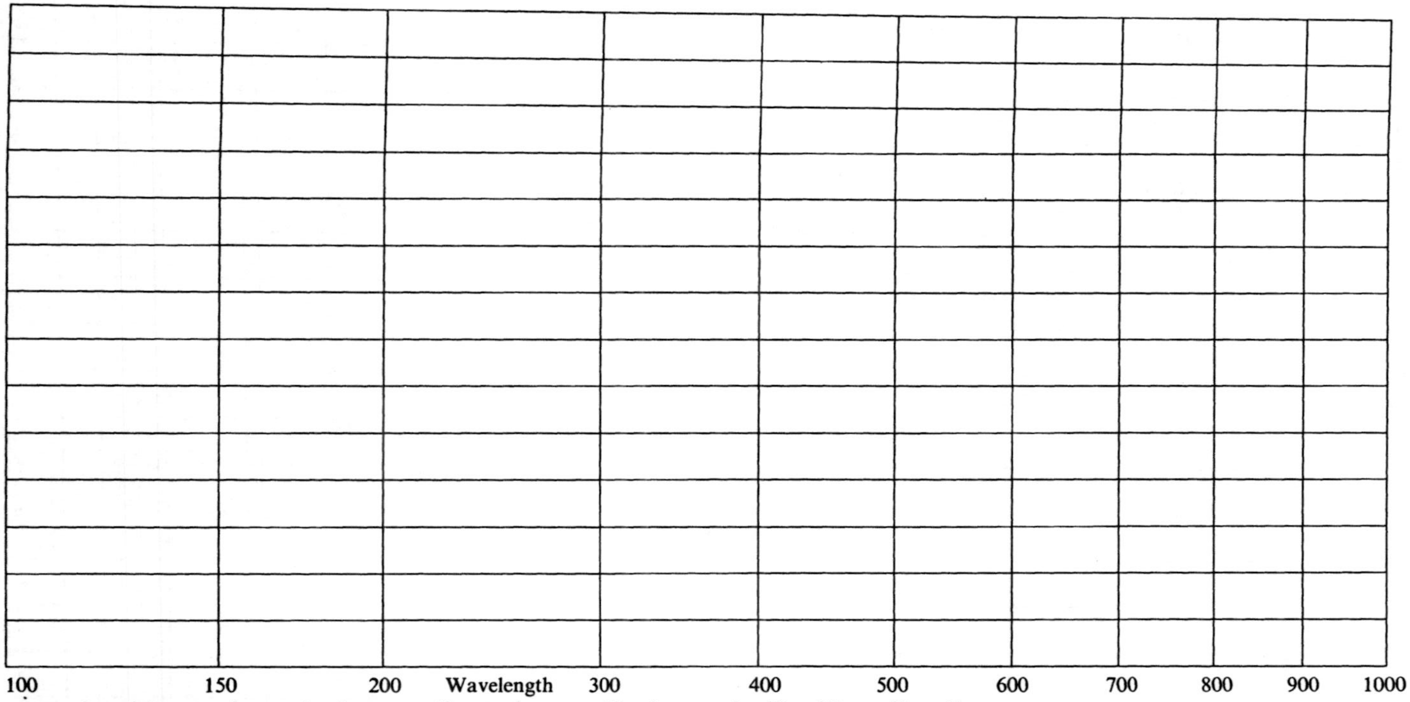

n = number of elements d = spacing between adjacent elements   Notches at: nd, $nd/2$, $nd/3, \ldots nd/(n-1)$

Peaks at: infinity, 2nd/3, 2nd/5, 2nd/7, ... d Peak at d has no attenuation

FM7

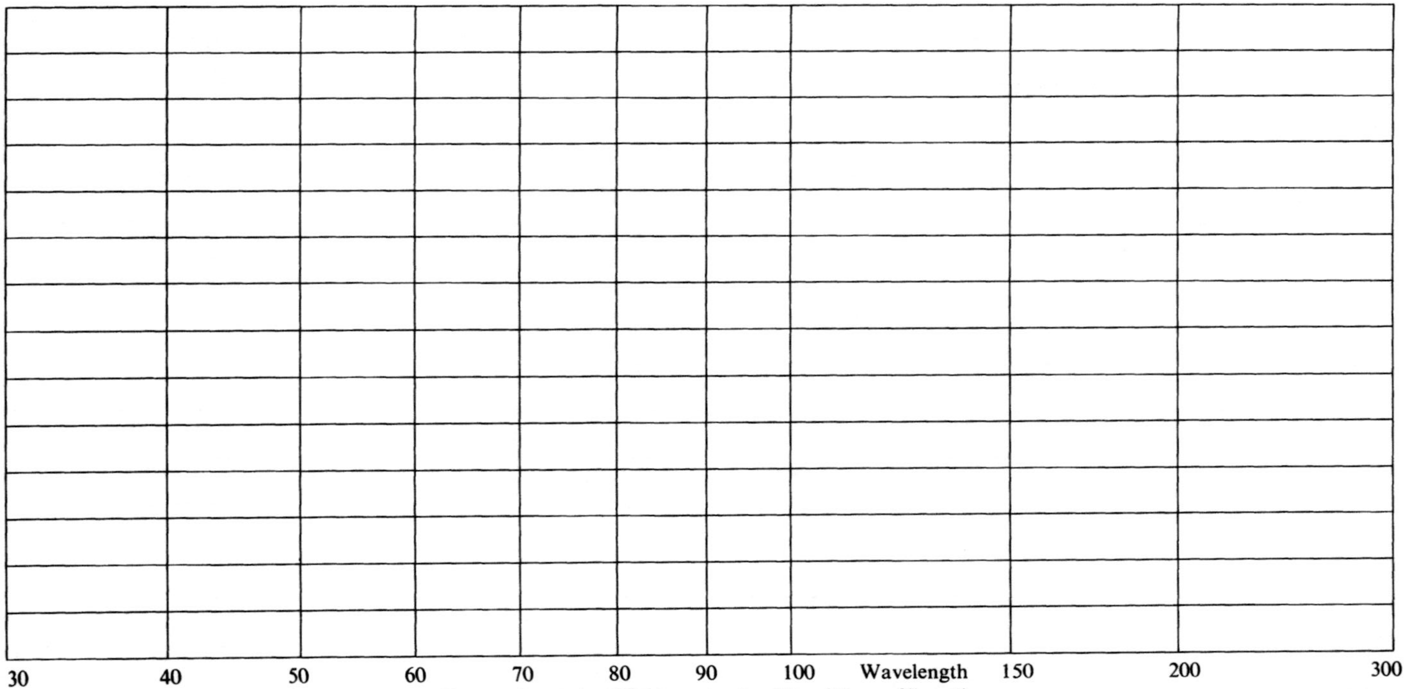

n = number of elements d = spacing between adjacent elements Notches at: nd, $nd/2$, $nd/3$, ... $nd/(n-1)$

finity, 2nd/3, 2nd/5, 2nd/7, ... d Peak at d has no attenuation

APPENDIX B—MARINE FREQUENCY BAND LIMITS

In this appendix we show how ghosts at submerged sources and ghosts at submerged detectors limit the effective frequency bandwidth of marine data.

For upward-traveling waves the reflection coefficient at a smooth water–air interface is essentially -1 (Chapter 8). Consider an air gun at a depth of d_g below the surface. When the gun is actuated, seismic energy propagates outward in all directions. The ray that travels vertically upward is reflected from the water–air interface and is reversed in polarity. It then propagates vertically downward. What is its phase relative to the wave that propagated vertically downward from the source when it was actuated? The path length of the wave reflected from the water surface is a distance $2d_g$ longer than the path length of the direct wave. The phase shift due to this extra path length is $(2d_g/\lambda)(2\pi)$ radians. The polarity reversal when the ray is reflected adds another π radians to the phase shift, making a total phase angle (delay) of:

$$\phi_g = -[\pi + (2\pi)(2d_g/\lambda)] = -[\pi + (4\pi d_g f/v)]$$

Where:

ϕ_g = phase angle (delay) of the ghost relative to the primary wave due to the gun depth.

We can represent the direct wave and the surface reflected wave by two vectors (Appendix D, Fig. D-5B) separated by the angle ϕ_g. At a far distance vertically downward the difference in the spherical divergence of the direct and reflected waves will be negligible. At depth both waves have essentially equal amplitudes which we will represent by the symbol A. Then the magnitude of the sum of the two vectors would be:

$$2A\cos(\phi_g/2) = 2A\cos[\pi/2 + (2\pi d_g f/v)]$$

at a phase angle of $\phi_g/2 = -[\pi/2 + (2\pi d_g f/v)]$

Consider the case where the amplitude of the direct wave is unity. Then the amplitude of the direct wave plus ghost would be:

$$2\cos(\phi_g/2) = 2\cos[\pi/2 + (2\pi d_g f/v)].$$

Sometimes geophysicists overlook the fact that at zero frequency the angle $\phi_g/2 = -\pi/2$, and the sum of the two downward traveling waves is zero. This ghost effect amounts to a low-cut filter which limits the effective energy at the lowest frequencies and a band rejection filter that limits the effective energy at some higher frequencies. The magnitude of the sum of the two downward travelling waves is zero whenever $2d_g f/v$ is an integer. The magnitude of the sum has a maximum value of $2A$ whenever $2d_g f/v$ is an odd number divided by 2.

At the outset we assumed a smooth water–air interface. If the sea is choppy and the height of the sea waves becomes significant compared to seismic wavelengths in water, then the upward traveling wave would be partially scattered and the magnitude of the reflection coefficient would be less than unity. (Scattering would be most severe for the higher-frequency components.) For this case let the reflection coefficient be $-R_{wa}$. The magnitude of the sum of the two downward traveling waves would be:

$$A + AR_{wa}(\cos\phi_g + j\sin\phi_g)$$

at a phase angle of

$$\text{arctangent}\,[R_{wa}\sin\phi_g/(1 + R_{wa}\cos\phi_g)]$$

where j is the imaginary square root of -1.

The relative amplitude of the primary plus the source ghost would be:

$$1 + R_{wa}(\cos\phi_g + j\sin\phi_g)$$

A rough sea does not change the frequencies at which the two waves would tend to cancel or to reinforce, but cancelation at notches and reinforcement at peaks would be less than 100 per cent particularly at the higher frequencies. For a frequency of 10 Hz the seismic wavelength in water is about 500 ft (152 m), so scattering by a rough sea surface would be minor.

Similarly, at the cable there is a direct arrival of reflections from depth and a second arrival of reflections which first reached the water–air surface and were reflected back down to the cable. The equations are identical to those for the source except the cable depth d_c replaces the gun depth d_g, and let the phase angle be ϕ_c.

For a smooth water–air interface we can write an expression for the relative amplitude resulting from the filtering effect of the receiver ghost as follows:

$$2\cos(\phi_c/2) = 2\cos[\pi/2 + (2\pi d_c f/v)]$$

with a phase angle of

$$\phi_c/2 = -[\pi/2 + (2\pi d_c f/v)]$$

The effects of the two filters are multiplicative; therefore for a smooth sea the relative response resulting from both the source and receiver ghosts would be:

Smooth sea relative amplitude due to both gun and cable depths

$$= [2\cos(\phi_g/2)][2\cos(\phi_c/2)]$$

$$= 4\{\cos[\pi/2 + (\pi)(2d_g f/v)]\}\{\cos[\pi/2 + (\pi)(2d_c f/v)]\}$$

with a phase angle of

$$(\phi_g/2) + (\phi_c/2) = -[\pi + 2\pi(f/v)(d_g + d_c)]$$

If guns and cable are operated at relatively shallow depths, the response would be improved at relatively high frequencies but would be diminished at the lower frequencies (Chapter 8). Conversely, if the guns and cable are operated at greater depths the response at lower frequencies would be improved but this improvement would be at the expense of higher frequencies.

For a rough sea the relative response resulting from both ghosts would be:

Rough sea relative amplitude due to both gun and cable depths

$$= 2 + (R_{wa})[\cos(\phi_g) + (j)\sin(\phi_g)] + (R_{wa})[\cos(\phi_c) + (j)\sin(\phi_c)]$$

at a phase angle of

$$-\arctan\{[\sin(\phi_g) + \sin(\phi_c)]/[\cos(\phi_g) + \cos(\phi_c) + 2/R_{wa}]\}$$

The velocity of sound in water depends upon the salinity and the temperature. It is approximately 5000 ft/s (1524 m/s).

APPENDIX C—NOISE FROM A DIPPING REFRACTOR

This appendix deals with noise from a dipping refractor and how this noise is suppressed by the CMP stack using both group interval arrays and double group interval arrays (Chapter 2).

Figure C-1 illustrates a dipping refractor with ray paths from a surface point to the refractor, along the refraction interface, and back up to the surface. Associated with Fig. C-1 is a derivation of equations for the apparent velocity of the refraction along the surface of the earth both updip and down dip from the source. The apparent refraction velocity updip is:

$$v_{rau} = v/[\sin(\alpha + \theta) - (\sin\theta)/(\sin\alpha)]$$

where

v = the velocity in the earth overlying the refractor.

α = the refraction angle.

θ = the angle between the surface and the refraction interface.

The apparent velocity of the refraction downdip is

$$v_{rad} = v/[\sin(\alpha - \theta) + (\sin\theta)/(\sin\alpha)]$$

We will consider a specific example with the following values for the variables:

$$v = 6000\ \text{ft/s}\ (1829\ \text{m/s});\ \alpha = 45°;\ \theta = 20°$$

By Snell's law the refraction velocity is 8485 ft/s (2586 m/s).

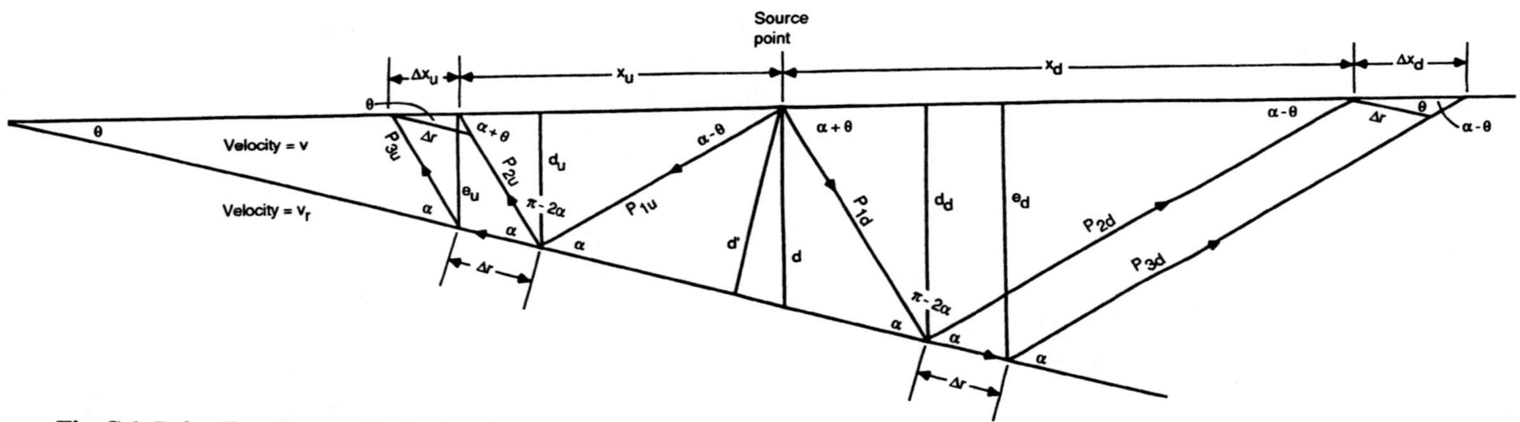

Fig. C-1 Refractions from a dipping interface

Let:

The refraction angle $\alpha = 45°$.

Then the refraction velocity $v_r = v/\sin\alpha = 2^{0.5}v$ (where v is the velocity of the first layer).

θ = the angle of dip of the refractor.

d = the depth to the refractor at the source point.

d' = the minimum distance from the source point to the refractor.

v_{rau} = the apparent velocity of the refraction in the updip direction.

v_{rad} = the apparent velocity of the refraction in the down dip direction.

$$d' = d\cos\theta$$

$P_{1u} = d'/\sin\alpha = d\cos\theta/\sin\alpha$

$d_u = P_{1u}\sin(\alpha-\theta) = d\cos\theta\sin(\alpha-\theta)/\sin\alpha$

$P_{2u} = d_u/\sin(\alpha+\theta) = d\cos\theta\sin(\alpha-\theta)/\sin\alpha\sin(\alpha+\theta)$

$P_{1d} = d'/\sin\alpha = d\cos\theta/\sin\alpha$

$d_d = P_{1d}\sin(\alpha+\theta) = d\cos\theta\sin(\alpha+\theta)/\sin\alpha$

$P_{2d} = d_d/\sin(\alpha-\theta) = d\cos\theta\sin(\alpha+\theta)/\sin\alpha\sin(\alpha-\theta)$

$P_{.} = P_{.}$

$$P_{2d}/P_{2u} = \sin^2(\alpha+\theta)/\sin^2(\alpha-\theta)$$

$$(P_{1d}+P_{2d})-(P_{1u}+P_{2u}) = (d\cos\theta/\sin\alpha)[\sin(\alpha+\theta)/\sin(\alpha-\theta)-\sin(\alpha-\theta)/\sin(\alpha+\theta)]$$

$$x_u = d_u\tan(\alpha-\theta)+\delta_u/\tan(\alpha+\theta)$$

$$x_d = d_d\tan(\alpha+\theta)+\delta_d/\tan(\alpha-\theta)$$

$$t_u = P_{1u}/v+P_{2u}/v = [d\cos\theta/v\sin\alpha][1+\sin(\alpha-\theta)/\sin(\alpha+\theta)]$$

$$t_d = P_{1d}/v+P_{2d}/v = [d\cos\theta/v\sin\alpha][1+\sin(\alpha+\theta)/\sin(\alpha-\theta)]$$

$$x_d/x_u = d_d/d_u = \sin(\alpha+\theta)/\sin(\alpha-\theta)$$

$$e_u = d_u-\Delta r\sin\theta$$

$$P_{3u} = e_u/\sin(\alpha+\theta) = (d_u-\Delta r\sin\theta)/\sin(\alpha+\theta)$$

$$\Delta P_u = \Delta r-P_{2u}+P_{3u} = \Delta r-\Delta r\sin\theta/\sin(\alpha+\theta)$$

$$\Delta t_u = \Delta r/v_r-[\Delta r\sin\theta/\sin(\alpha+\theta)]/v$$

$$\Delta t_u = (\Delta r/v)[\sin\alpha-\sin\theta/\sin(\alpha+\theta)]$$

$$e_d = d_d+\Delta_r\sin\theta$$

$$P_{3d} = e_d/\sin(\alpha-\theta) = (d_d+\Delta r\sin\theta)/\sin(\alpha-\theta)$$

$$\Delta P_d = \Delta r-P_{2d}+P_{3d} = \Delta r+\Delta r\sin\theta/\sin(\alpha-\theta)$$

$$\Delta t_d = \Delta r/v_r-[\Delta r\sin\theta/\sin(\alpha-\theta)]/v$$

$$\Delta t_d = (\Delta r/v)[\sin\alpha+\sin\theta/\sin(\alpha-\theta)]$$

$A/\sin A = B/\sin B$ (trigonometry)

$$\Delta x_u/\sin\alpha = \Delta r/\sin(\pi-\alpha-\theta) = \Delta r/\sin(\alpha+\theta)$$

$$\Delta x_u = (\Delta r\sin\alpha)/\sin(\alpha+\theta)$$

$$\Delta x_u/\Delta t_u = v_{rau} = v/[\sin(\alpha+\theta)-\sin\theta/\sin\alpha]$$

$$\Delta x_d/\sin(\pi-\alpha) = \Delta r/\sin(\alpha-\theta)$$

$$\Delta x_d = (\Delta r\sin\alpha)/\sin(\alpha-\theta)$$

$$\Delta x_d/\Delta t_d = v_{rad} = v/\sin(\alpha-\theta)+\sin\theta/\sin\alpha]$$

$$v_{rau}/v_{rad} = [\sin(\alpha-\theta)+\sin\theta/\sin\alpha]/[\sin(\alpha+\theta)-\sin\theta/\sin\alpha]$$

Using the above equations we calculate the following:

$v_{ray} = 14\,197$ ft/s (4327 m/s); $v_{rad} = 6620$ ft/s (2018 m/s); $v_r = 8485$ ft/s (2586 m/s)

The apparent refraction velocity updip is 67.3 per cent faster than the actual refraction velocity and the apparent refraction velocity down dip is 21.98 per cent slower than the actual refraction velocity. This causes the apparent wavelengths of the updip refraction to be longer than the apparent wavelengths of the downdip refraction (for any given frequency component). Obviously, the suppression of the refraction noise by a geophone array will be different for the updip cable than for the downdip cable.

For several years the author has preferred double group interval arrays over group interval arrays because of a number of concerns including changes such as those described in the Chapter 2. Intuition caused concern about group interval arrays for the case where the geophone arrays and source patterns are not equal in length and there are strong refractions from a planar interface that dips at an angle relative to a planar surface. (The refraction interface could be horizontal and the surface might slope or vice versa.)

Let the source for the example case be a single dynamite charge. Supose that the refraction has a narrow frequency bandwidth near 30 Hz. The apparent wavelength updip is $14\,197/30 = 473$ ft (144 m) and the apparent wavelength down dip is $6620/30 = 221$ ft (67 m). If the effective length of the geophone array happened to be 200 ft (61 m), the noise would be suppressed to only 10% of the maximum by the geophone array for the downdip cable. But for the updip cable the response would be 73% of maximum. Response for updip arrays $= [\sin(200\pi/473)]/[200\pi/473] = 0.73$. Response for the downdip arrays $= [\sin(200\pi/221)]/[200\pi/221] = 0.10$ (Chapter 4). So the refraction would be strong for the updip offsets but weak for the downdip offsets. Samples of the noise from the interlacing offsets in each CMP stack would be grossly different and suppression of noise by the combined stack and array would be impaired with group interval arrays. This case is more complex than it might first appear (Fig. C-2). Obviously each ray path in the CMP gather would be the same if we were to interchange source and receiver. Obviously the arrival time, and hence the phase angle, would also be the same if we interchanged source and receiver. But the angle of arrival, the apparent velocity across the geophone array and the apparent wavelength along the surface would be different.

For the example of the preceding paragraph let the group interval and the shot point interval be 200 ft (61 m). Then for an offset increase (from the source to the center of the array) of 200 ft (61 m) in the CMP gather, the offset is increased by 100 ft (30.5 m) updip and by 100 ft (30.5 m) downdip. The increased travel time is then $100/14197 + 100/6620 = 0.02215$ s. The apparent velocity for increasing offset within a CMP gather is $200/0.02215 = 9030$ fts

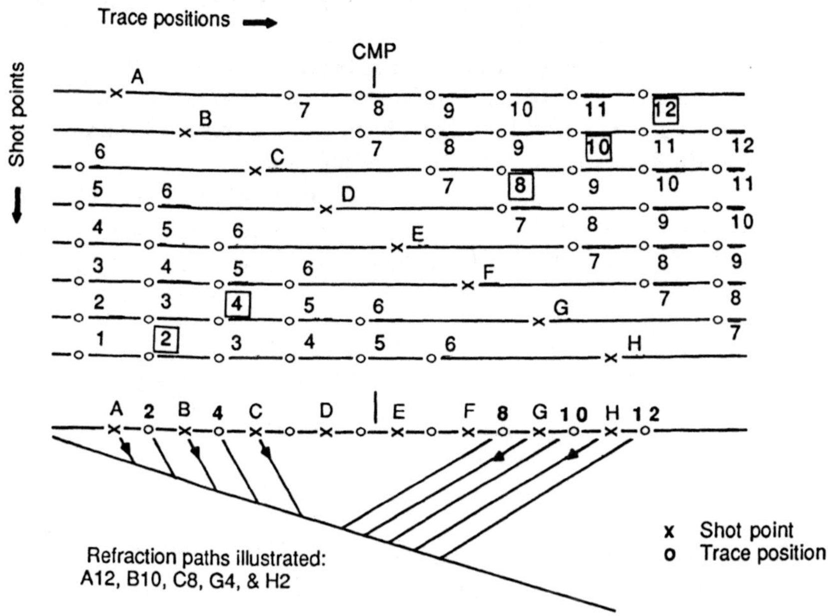

Fig. C-2 Response of a CMP gather to noise from a dipping refractor. Note that the angles of arrival of rays to updip receivers are steeper than to downdip receivers. The apparent wavelength is therefore longer for updip receivers than for downdip receivers.

(2752 m/s) which is substantially higher than the refractor velocity of 8485 fts (2586 m/s).

For the same example, double group interval arrays would probably suppress the refraction better, but not as well as supposed. (Only after reading Anstey (1986) did the author take time to analyze the case as presented here.) In each CMP gather the sum of offsets from the updip cable would not yield the desired uniform and continuous sampling of a sinusoidal component of the noise. The problem is that the phase change of the noise between traces in the CMP gather would be determined by the apparent offset velocity of 9030 ft/s (2752 m/s), but the effective length of the arrays in terms of the noise wavelengths would be determined by the apparent updip and downdip velocities of 14197 and 6620 ft/s (4327 and 2018 m/s). Consider first only traces of the updip cable in the CMP gather. The higher apparent velocity across the arrays would yield a longer apparent noise wavelength. Therefore the samples of noise seen by each such array would cover a shorter range of phase angles and there would be gaps in the sampling of the noise between one offset and the next larger or shorter offset.

These gaps in the sampling of the noise wave could result in a larger net response for the sum of the updip traces. The down-dip traces within a CMP gather would each have a weak response, but the sum of these weak responses may not be too well suppressed. The apparent low velocity across the down dip arrays would mean that the apparent wavelength would be shortened so that a given array would cover a larger range of phase angles and there would be overlaps in the sampling of the noise between one offset and the next longer or shorter offset in the CMP gather. These overlaps could result in a larger sum for the downdip traces if the sampling were uniform.

The effect of noise from a dipping refractor on geophone arrays and the CMP stack is complex, but in general the double group interval array should provide better noise suppression than the group interval array.

REFERENCES

Anstey, Nigel A. (1986). 'Whatever happened to ground roll?' *The Leading Edge* **5**, no. 3 (March), 40–45.

APPENDIX D—GEOPHONE ARRAYS, SOURCE PATTERNS, AND CMP STACKS

In this appendix we deal with the problem of suppressing unwanted noise to improve the S/N ratio. We show how geophone arrays and source patterns work with an example in which a refraction illustrates noise to be suppressed. We then discuss waves in time and in space and the representation of waves by vectors. Next we discuss a very simple method for designing effective geophone arrays and source patterns and effective combinations of the two to improve overall noise suppression. In the discussion of linear arrays and patterns we deal both with equally spaced and equally weighted configurations and with configurations with tapered weightings to further improve overall noise suppression. We discuss areal arrays and patterns including parallelograms and stars and the practical applications of areal arrays. We also discuss the suppression of noise by the CMP stack considered both by itself and in combination with the geophone array and relate the combined effect to the stack-array concept.

How geophone arrays and source patterns work

How do geophone arrays and source patterns discriminate against certain noises and at the same time allow reflected signals to be recorded? We will illustrate how geophone arrays and source patterns work with an example in which refractions are noise to be suppressed and reflections from a deeper interface are to be recorded. Consider an earth composed of three homogeneous and isotropic layers as follows:

Layer	*Velocity*	*Thickness*
1	v_1	z_1
2	$v_2 = (2)^{0.5} v_1$	z_2
3	$v_3 = 2v_1$	∞

Waves from a surface source that impinge upon the first interface will be partially reflected and partially transmitted (Fig. D-1). The angle of incidence is the angle between a normal to the interface and the incident ray path. For reflected rays the angle of reflection will be equal to the angle of incidence. For small angles of incidence, the path of the transmitted rays in the higher-velocity, second layer will be bent away from the normal so that the angle of transmission is greater than the angle of incidence. The relationship between

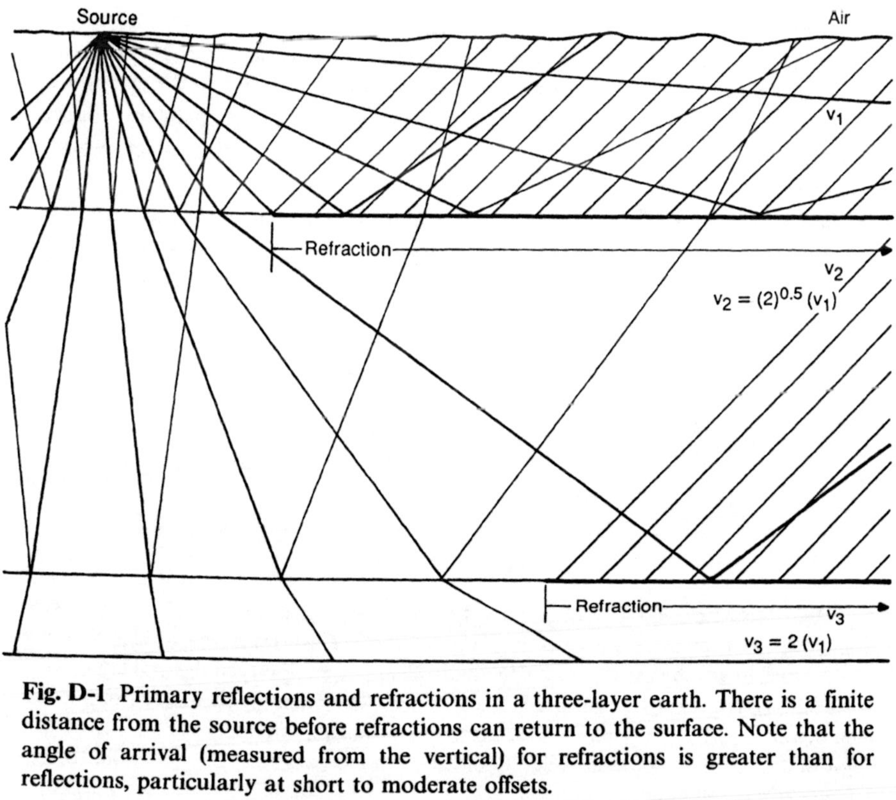

Fig. D-1 Primary reflections and refractions in a three-layer earth. There is a finite distance from the source before refractions can return to the surface. Note that the angle of arrival (measured from the vertical) for refractions is greater than for reflections, particularly at short to moderate offsets.

these two angles is defined by Snell's Law which is:

$\sin(\theta_{t2}) = (v_2/v_1)\sin(\theta_{i1})$ For this example:
$\sin(\theta_{t2}) = (2)^{0.5}\sin(\theta_{i1})$

At some angle of incidence, called the critical angle, the transmitted wave will be bent such that its path is horizontal and it becomes a refracted wave traveling in the second layer along the interface. The critical angle can be determined from Snell's law by letting the angle θ_{t2} be 90° and solving for the angle of incidence.

$\sin(\theta_{r12}) = (v_1/v_2)\sin 90°$ For this example:
$\sin(\theta_{r12}) = (0.5)^{0.5} = 0.707$
$\sin(\theta_{r12}) = v_1/v_2$ $\theta_{r12} = 45°$

where θ_{r12} is the critical angle at the interface between layer 1 and 2.

Since the refracted wave is traveling in the second layer at the faster velocity along the interface, its wavefront precedes waves traveling in the first layer. Wave energy from the refracted wave re-enters the first layer and is propagated toward the surface in an outward direction from the source. By Snell's law the angle to the vertical of these refractions is the same as the critical angle. The incident refraction ray is traveling in layer 2 and the angle of incidence at the interface is 90°. The transmitted refraction is in layer 1. We calculate: $\sin(\theta_{t1}) = (v_1/v_2)\sin(\theta_{i2}) = v_1/v_2$. There is a minimum offset from the source at which refractions reach the surface. Beyond that minimum offset, refractions can be detected.

Figure D-2 shows refracted rays arriving at the surface. Note that the time to travel from A to C at velocity v_2 is the same as the time to travel from A to E at the lower velocity v_1. The waves are in time phase at the wavefront EC and

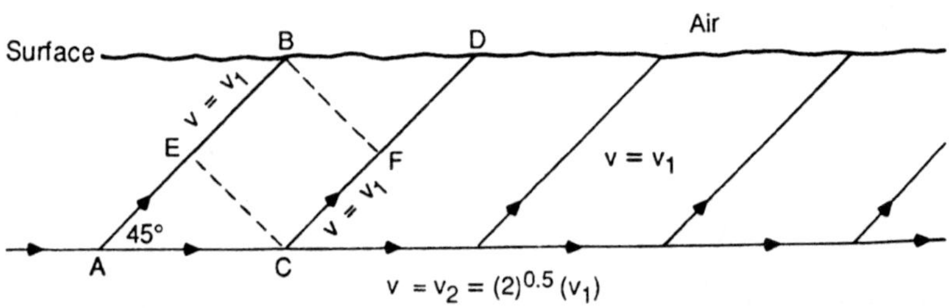

Fig. D-2 Refraction along an interface with rays returning to the surface. At any wave front such as EC or BF, the arrival times are the same. The arrival time at D is delayed relative to that at B by the distance FD divided by v_1, the velocity of the first layer. The apparent velocity along the surface is the distance BD divided by the delay time and is then equal to v_1(BD/FD) which is equal to the velocity v_1 divided by the sine of the angle of arrival. The apparent velocity of the refraction at the surface is equal to the velocity of the refractor, v_2.

again at the wavefront BF. (In isotropic media the wavefront is perpendicular to the rays. At the wavefront the rays must be in time phase.) The refracted waves travel obliquely in the first layer at the velocity v_1 as indicated, and note that there is no refracted wave traveling along the surface. But from measurements along the surface it appears that there is a refracted wave traveling horizontally along the surface. The difference in the actual arrival times at B and D is the distance FD divided by the velocity v_1. But the apparent velocity of a wave along the surface is the distance BD divided by the travel time from F to D. Therefore the apparent velocity along the surface is $BD/(FD/v_1) = v_1(BD/FD) = v_2$.

Suppose that the period (time between crests) of the refraction at point B were 0.0333 s. Then the dominant frequency would be 30 Hz. The dominant apparent wavelength would then be the apparent velocity divided by 30 Hz. Let the velocity of the first layer be 6000 ft/s (1829 m/s), then the velocity of the second layer is 8485 ft/s (2586 m/s). The velocity of the third layer is 12 000 ft/s (3658 m/s). The apparent wavelength of our 30 Hz refraction would be 8485/30 = 283 ft (86 m). Now if the distance BD were half the apparent wavelength or 141 ft (43 m), then the refraction measured at B would be half a wavelength different than the refraction measured at D. This means that the waves would be 180° different in time phase. If we added a response measured at B to a response measured at D, then the responses to the refraction would tend to cancel. (If the amplitudes of the refraction at B and D were identical and if the separation were precisely half a wavelength, then the response to the 30 Hz component of the refraction would be zero.)

By merely measuring the apparent wavelength of the refraction noise along the surface and putting two geophones one-half wavelength apart we created a very simple array to suppress the refraction noise. We did not need to know the actual velocity of the first layer nor did we need to know the refraction angle. Yet what we actually did was to make an array that discriminated against 30 Hz waves with an angle of arrival equal to the refraction angle. The preceding is summarized in the following table:

Item	*English units*	*Metric units*
Dominant frequency of refraction (f_r)	30 Hz	30 Hz
Velocity of first layer (v_1)	6000 ft/s	1829 m/s
Velocity of second layer (v_2)	8485 ft/s	2586 m/s
Velocity of third layer (v_3)	12000 ft/s	3658 m/s
Apparent velocity of refraction (v_{ar})	8485 ft/s	2586 m/s
Apparent wavelength of refraction (λ_{ar})	283 ft	86 m
One-half apparent wavelength of refraction	141 ft	43 m

Note that an array is a wavelength filter. An actual refraction wavelet would be composed of a band of frequencies rather than just a frequency of 30 Hz. Consider a component with a frequency of 35 Hz. Its wavelength would be 8485/35 = 242 ft (74 m), and with a distance between B and D of 141 ft (43 m) the responses at B and D would not be 180° apart for this 35 Hz component of the refraction. Instead the phase difference would be the separation distance divided by the wavelength times 360° (or times 2π) = (141/242)(360°) = 210° or 3.66 radians.

By relative amplitude we mean the amplitude of the sum of the signals or noises at a number of points divided by the amplitude of the signal or noise at one point times the number of measurement points. If the refraction amplitudes for the 35 Hz component were equal in amplitude at B and D (i.e. $A_B A_D$), then the relative amplitude of the sum of the two responses would be:

$$[A_B \cos(210°/2) + A_D \cos(-210°/2)]/(2A_B)$$
$$= |\cos(210°/2)| = 0.26. \text{ (Fig. D-5B, 2 vectors)}$$

Suppose our objective were to map the interface between layers 2 and 3. A reflection from this interface arrives at point G in Fig. D-1 at an angle of 15° from the vertical. The apparent velocity of this reflection is $(v_1)/\sin(15°)$ = 23 182 ft/s (7066 m/s), (Fig. 4.3). The wavelength for a 40 Hz component of this reflection at point G would then be (23 182)/40 = 580 ft (177 m). The phase difference for two geophones 141 ft (65 meters) apart would be (141/580) times (360°) = 88°. The relative amplitude of the sum of the responses for the 40 Hz component of the reflection would be:

$$|\cos(88°/2)| = 0.72$$

Relative amplitudes of reflection components at lower frequencies would be greater and would approach 1.0 as a limit (because the wavelengths are longer; therefore the angles subtended by an array are smaller). The relative amplitude of all frequency components of reflections would approach 1.0 for short offsets between the geophone array and the source position because the angles of arrival would be small and the wavelengths would be long. With more geophones in the arrays the suppression of the refraction could be effective over a wider band of frequencies of the refraction wavelet.

The previous example illustrated how a very simple geophone array could be designed to suppress a noise and yet allow a reflection to be recorded. But how does a source pattern suppress the refractions measured at the surface? Refering back to Fig. D-2, suppose that we had seismic sources at points B and D instead of geophones. Then waves would be emitted from the sources in all directions. The ray emitted from the source at B along the path BA would set up a refracted wave traveling along the interface at velocity v_2 in the second layer. Similarly the ray emitted from the source at point D along the path DC would also set up a refracted wave traveling along the interface. But for the frequency component of 30 Hz the two refracted waves would be 180° out of time phase and the two refractions would tend to cancel. Therefore two sources at the surface separated by half the apparent wavelength of a refraction will generate refractions that will tend to cancel so that there would be less refraction noise set up in the earth along a line through the two source positions. (At right angles to a line between the two source positions waves from the sources would be additive and strong refractions would result.)

What we actually did in this case was to set up a source pattern that has a null in its radiation pattern at the refraction angle (along the seismic line) for the dominant refraction frequency of 30 Hz. We did this without having to know the actual velocity of layer 1 or the refraction angle. So the source pattern that does not emit 30 Hz energy at the critical angle is identical to the geophone array that does not respond to 30 Hz energy with an angle of arrival equal to the critical angle.

Waves in time and space

Figures D-3 and D-4 may be helpful to readers who are not familiar with the representation of sinusoidal waves in time and in space. Fig. D-3A illustrates a cosine wave as a function of time at a fixed point. With time the wave goes from crest-to-trough-to-crest, etc. The period (T) of the wave is the time difference from crest-to-crest or from trough-to-trough or from any point on the wave to the next identical point of the wave ('identical' including both amplitude and its time derivative). An equation for the wave is:

$$A(t) = A_0 \cos[(2\pi/T)(t)] = A_0 \cos(2\pi ft) = A_0 \cos(\omega t)$$

where: f is the frequency in Hertz, and

ω is the angular velocity in radians per second

Figure D-3B illustrates a cosine wave as a function of position at a fixed time. With distance the wave goes from crest-to-trough-to-crest, etc. The wavelength (λ) is the distance from crest-to-crest or trough-to-trough or from any point on the wave to the next identical point on the wave. An equation for the wave is as follows:

$$A(x) = A_0 \cos[(2\pi/\lambda)(x)] = A_0 \cos(2\pi\kappa x) = A_0 \cos(\omega x)$$

where κ is the wave number.

If we had one geophone located at a point on the surface, then the peak response to this wave when the crest of the wave came by would be $K(f)A_0$, where $K(f)$ is the sensitivity of the geophone as a function of frequency. Suppose we had four geophones in series and all four were located at one point

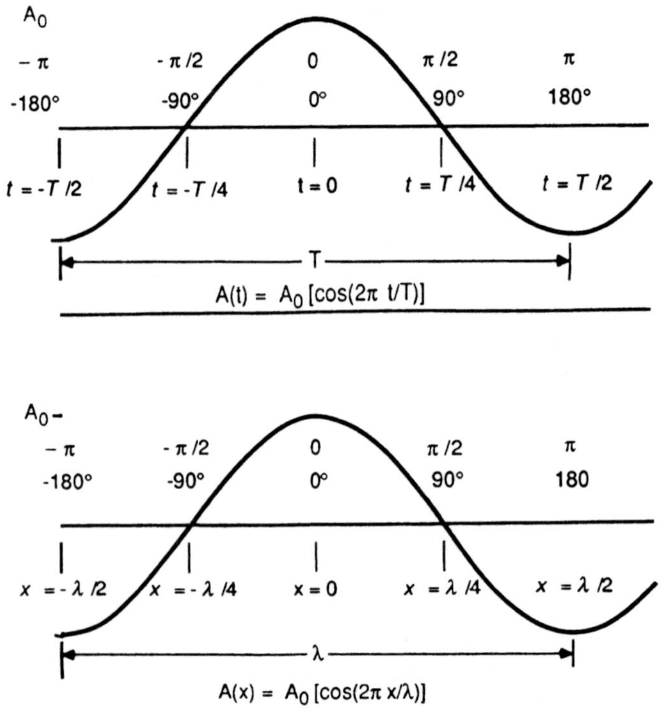

Fig. D-3 Cosine waves as functions of time and of position. Top: A cosine wave as a function of time at a fixed point. Bottom: A cosine wave as a function of position at a fixed time.

on the surface. As the crest of the wave came by the peak output would be $4K(f)A_0$. We define the relative response of an array of equally sensitive elements in series as the total response divided by the product of the response of one element times the number of elements. For the array of four elements at one point, the relative response would be 1.0.

Linear arrays and source patterns with equally spaced elements

Figure D-4 illustrates a cosine wave as a function of distance along the surface of the earth at a fixed time. Figure D-4A shows points along the surface where we have geophones to sample the wave. Suppose we had an array of four geophones in series located at ii, i, 1 and 2. The relative response as the wave reaches a crest would be less than 1.0 because when the crest of the wave was at any one geophone the other three geophones would be off the

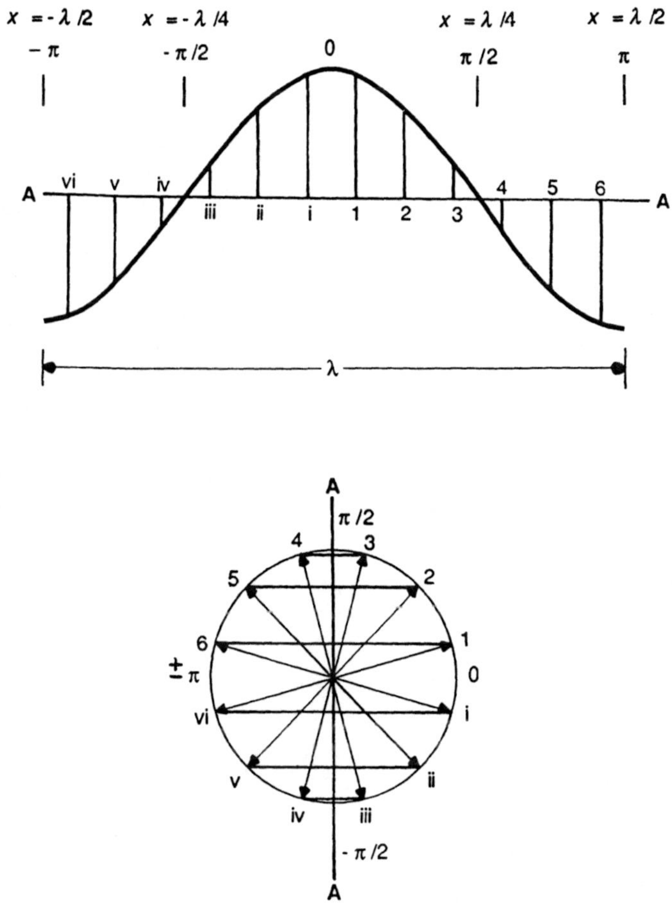

Fig. D-4 A cosine wave at a fixed time and sampled at fixed positions. Top: A cosine wave as a function of position at a fixed time that is sampled at fixed points with the amplitude of the samples relative to baseline AA. Bottom: A vector diagram with the samples of Fig. D-4A represented by vectors relative to baseline AA.

crest. The actual relative response would be the sum of the ordinates at ii, i, 1 and 2 divided by $4A_0$.

Figure D-4B is a vector diagram that represents the same wave as at the top with the same sample points. The vectors all start at the center of the circle and the radius of the circle is equal to A_0. The sum of vectors marked ii, i, 1 and 2 would be the sum of the displacements of each vector from the line AA. The components parallel to the line AA would cancel. This figure illustrates how we can use a vector diagram to represent the responses of geophone arrays to sinusoidal waves. In Fig. D-4 the geophone locations illustrated are as follows:

Sample	*Phase angle*	*Ordinate*
1	15°	$A_0 \cos(15°) = 0.966A_0$
i	−15°	$A_0 \cos(-15°) = 0.966A_0$
2	45°	$A_0 \cos(45°) = 0.707A_0$
ii	−45°	$A_0 \cos(-45°) = 0.707A_0$
3	75°	$A_0 \cos(75°) = 0.259A_0$
iii	−75°	$A_0 \cos(-75°) = 0.259A_0$
4	105°	$A_0 \cos(105°) = -0.259A_0$
iv	−105°	$A_0 \cos(-105°) = -0.259A_0$
5	135°	$A_0 \cos(135°) = -0.707A_0$
v	−135°	$A_0 \cos(-135°) = -0.707A_0$
6	165°	$A_0 \cos(165°) = -0.966A_0$
vi	−165°	$A_0 \cos(-165°) = -0.966A_0$

We could think of each sample as representing a 30° increment of the wave. For example, sample 1 would represent the increment of 0° to 30° and sample v would represent the increment from −120° to −150°. The six samples iii, ii, i, 1, 2 and 3 span only 150° but represent 180° or one-half of a wavelength. The relative response would be 0.644. If geophones were equally spaced, very close together, and spaced over a distance equal to half a wavelength, then the relative response would approach the integral of the cosine from $-\pi/2$ to $\pi/2$ divided by π which is equal to $(2/\pi)$ which is 0.637. Note how closely the sum of the six samples approximated the relative response of a continuous sample.

The integral of the $\cos(\phi)d(\phi)$ from $-\phi$ to $+\phi$ is equal to $2\sin(\phi)$. The average ordinate is this integral divided by $2(\phi)$ which is $\sin(\phi)/\phi$. We can use this relationship to evaluate the relative response of any linear geophone array where the geophones are equally and closely spaced. Note that the sum of all twelve samples in the table above represents the response over a full wavelength and is zero, which is the value of $\sin(\phi)/\phi$ when $\phi = \pi$.

For example, suppose we have an array of many geophones with an effective length of 125 ft (38 m). How would it respond to a signal or noise with an apparent wavelength along the surface of 217 ft (66 m)? The length of the array in wavelengths would be 125/217 or 0.576 wavelengths. Since one wavelength is 2π radians, the array would be $(0.576)(2\pi)$ radians long. The sampling then would be from an angle of -0.576π to $+0.576\pi$. For a many-element array we can use the integral approximation of $\sin(0.576\pi)/(0.576\pi) = 0.537$. The relative response would be 0.537.

It is important to note that the effective length of a linear geophone array or source pattern of equally spaced positions is the actual length plus one spacing. This is because each sample represents a sample width equal to the spacing between the geophones. Where there are many elements and the spacing is short, the difference between the effective and actual lengths may not be significant. But where there are few elements, the difference is quite significant. For two elements spaced one-half wavelength apart the actual length is one-half of a wavelength but the effective length is a full wavelength. The difference in the relative response is the difference between the factor of 0.637 and a factor of zero!

What works for geophone arrays also works for source patterns. Suppose we used one vibrator and vibrated every few feet over 125 ft (38 m) (or more precisely over 125 ft less one move-up). If a signal or noise had an apparent wavelength of 217 ft (66 m) along the surface the relative response to that source would be 0.537. If we combined the 125 ft source pattern with the 125 ft geophone array the relative response would be $(0.537)^2$ which is 0.288. Remember that the effective length of any linear geophone array or source pattern of equally spaced elements is one spacing more than the actual length.

Figure D-5 illustrates geophone arrays or source patterns with equally spaced elements with vector diagrams. The left diagram represents the class of cases with an odd number of elements, whereas the right diagram represents the class of cases with an even number of elements. In this figure the phase angle ϕ is the spacing between elements divided by the wavelength and multiplied by 2π for ϕ in radians (or multiplied by 360° for ϕ in degrees). Using Fig. D-5 we see that the relative amplitude of the sum of any number of elements is:

$$(\text{Sum for odd } n)/n = \frac{1}{n} + \frac{2}{n}\{\cos\phi + \cos 2\phi + \cdots \cos[(0.5)(n\text{-}1)\phi]\}, \text{ and}$$

$$(\text{Sum for even } n)/n = (2/n)\{\cos(\phi/2) + \cos(3\phi/2) + \cdots \cos[(n\text{-}1)\phi/2]\}$$

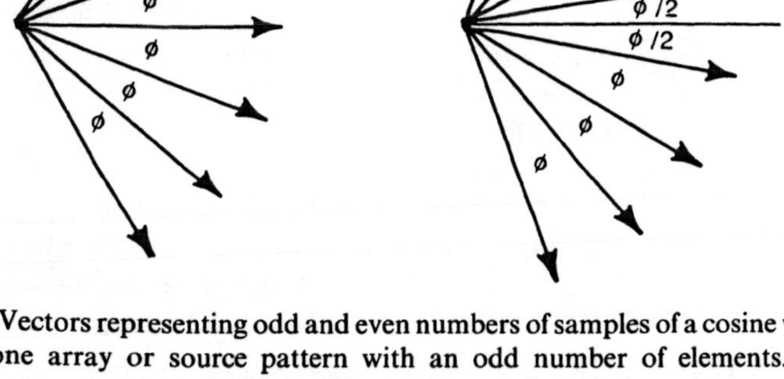

Fig. D-5 Vectors representing odd and even numbers of samples of a cosine wave. Left: A geophone array or source pattern with an odd number of elements. Right: A geophone array or source pattern with an even number of elements.

If the number of elements in an array or source pattern is less than four or if the phase angle between samples is more than 60°, then the summation equations should be used rather than the integral solution. Otherwise the integral solution is satisfactory for most purposes.

Practical design of arrays and patterns using approximations

With computers it is quite feasible to calculate noise suppression and signal loss for geophone arrays and source patterns. But we may not always have a computer handy nor do we want to thumb back and forth through a thick book of responses to design our geophone array and source pattern to yield some desired composite response. Fortunately, for linear arrays with equally spaced elements (and for some more complicated cases) there is an easy method that is precise in terms of wavelength, but is only approximate in terms of the degree of suppression. But in the field the actual responses will be altered and degraded by variations in statics, earth coupling, precise positioning, etc. Consequently the approximate suppression is adequate as a practical estimation of the field performance. Furthermore, the approximate method is something you can do with a pencil on the back of an envelope but a hand calculator would help.

The keys to the approximate method are simple expressions for the wavelengths for which suppression is theoretically infinite. These we will call notches. There are also simple expressions for the wavelengths where the suppression is poor which we will call peaks. For a few peaks there is no suppression at all. There is always a peak at an infinite wavelength and this peak has no suppression. We name this peak at an infinite wavelength the 'zero peak' we call the peak at the longest, finite wavelength the 'first peak' and the notch that precedes it the 'first notch'. For equally spaced elements the first peak will have a suppression of about 12 dB. (More precisely, the suppression for the first peak is 9.5, 11.4, and 13.4 dB for 3, 4, and 24 elements respectively and exceeds 12 dB for more than four elements.) For a wavelength that is equal to the spacing between the elements there will be a peak with no suppression (since all elements will see wave crests at the same time). We name this peak the 'first zero peak'. (Likewise there will be a second zero peak at a wavelength that is half the spacing between elements.)

The peak with the best suppression will be the middle peak between the zero peak and the first zero peak. If the number of peaks between is even, the two middle peaks will have the highest (best) suppression. A conservative estimate is that the suppression increases by about 2 dB per peak down to the middle peak or peaks and then decreases similarly. To be conservative, we limit the maximum suppression to about 24 dB. The peak that precedes the next zero peak has the same suppression as the first peak (about 12 dB). There are handy forms (one for English units and one for metric) in Appendix A for us to use which make it easy to design geophone arrays and source patterns to achieve a desirable composite response. You should duplicate these forms or make up your own. Use of these forms is illustrated later in this appendix and also in Chapter 3.

Let n be the number of element positions in a geophone array or source pattern with equal spacings of d between adjacent positions. There will be $(n-1)$ notches between the zero peak and the first zero peak. Equations for the wavelengths of these notches are given in Table D-1.

There will be $(n-2)$ peaks between the zero peak and the first zero peak. Remember that the suppression is zero at the zero peak and at the first zero

Table D-1 Notches between the zero peak and the first zero peak

Number of elements (n)	*Number of notches*	*Wavelengths of notches* (λ)
1	0	—
2	1	$\lambda = nd$
3	2	$\lambda = nd, nd/2$
4	3	$\lambda = nd, nd/2, nd/3$
5	4	$\lambda = nd, nd/2, nd/3, nd/4$
n	$n-1$	$\lambda = nd, nd/2, nd/3, nd/4, \ldots nd/(n-1)$

Table D-2 Peaks between the zero peak and the first zero peak

n	*Number of peaks*	*Wavelengths of peaks* (λ) *and the approximate suppression at each peak (dB) (limit maximum suppression to 24 dB)*
1	0	—
2	0	—
3	1	$\lambda = 2nd/3$ (9.5 dB)
4	2	$\lambda = 2nd/3, 2nd/5$ (11.4 dB) (11.4 dB)
5	3	$\lambda = 2nd/3, 2nd/5, 2nd/7$ (12 db) (14 dB) (12 dB)
6	4	$\lambda = 2nd/3, 2nd/5, 2nd/7, 2nd/9$ (12 dB) (14 dB) (14 dB) (12 dB)
7	6	$\lambda = 2nd/3, 2nd/5, 2nd/7, 2nd/9, 2nd/11$ (12 dB) (14 dB) (16 dB) (14 dB) (12 dB)
n	$n-2$	$\lambda = 2nd/3, 2nd/5, 2nd/9 \ldots 2nd/[2(n-1)-1]$

peak. The wavelengths of these peaks and the approximate suppression are given in the Table D-2.

Occasionally we will have a need for the following exact equation for the responses of linear arrays and patterns with equal spacings and weights:

$$R = 20\log_{10}\left|\frac{\sin(\pi n d/\lambda)}{(n)\sin(\pi d/\lambda)}\right| \text{ dB} \tag{D-1}$$

Weighted arrays and source patterns

The author does not advocate the use of complex arrays and source patterns that have carefully calculated weightings which theoretically can yield very high levels of suppression over a selected range of wavelengths. They just do not perform well in the field. But a case can be made for weightings which are not difficult to achieve which are designed primarily to add notches in the response at wavelengths that are not well suppressed. With the Vibroseis technique it is quite easy to create a source pattern that is weighted by spacing vibrators some distance apart in line and recording and summing the response at several locations as the group of vibrators is moved forward with some pre-selected increment. This creates a trapezoidal weighting. Figure D-6 illustrates

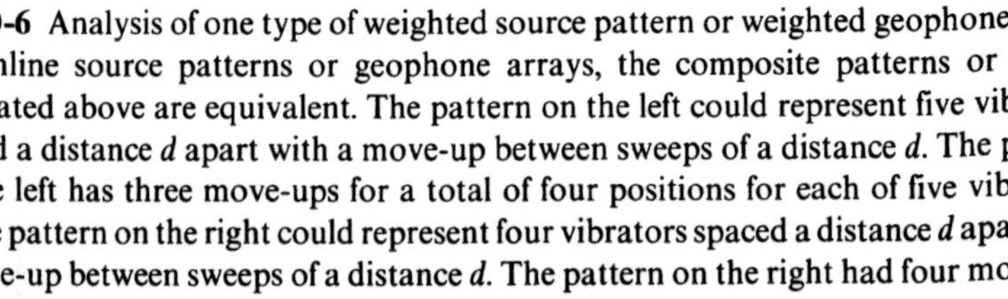

Fig. D-6 Analysis of one type of weighted source pattern or weighted geophone array. For inline source patterns or geophone arrays, the composite patterns or arrays illustrated above are equivalent. The pattern on the left could represent five vibrators spaced a distance *d* apart with a move-up between sweeps of a distance *d*. The pattern on the left has three move-ups for a total of four positions for each of five vibrators.

The pattern on the right could represent four vibrators spaced a distance *d* apart with a move-up between sweeps of a distance *d*. The pattern on the right had four move-ups for a total of five positions for each of four vibrators.

The pattern on the left must have the notches for good wavelength suppression for a five element pattern, which yields notches at wavelengths of $5d$, $5d/2$, $5d/3$, and $5d/4$. The pattern on the right must have notches for a four-element pattern, at wavelengths of $4d$, $4d/2$, and $4d/3$. Both composite patterns must have both sets of notches since they are equivalent.

We can generalize and conclude that we can expect a series of notches based upon the number of elements times the spacing and a second series of notches based upon the number of positions times the differential move-up distance.

such a weighting that can be created two different ways. The lateral displacements indicated could be zero for an in-line source pattern or for an in-line array, or the lateral displacement could be significant to form an areal, parallelogram array or pattern. The response to waves traveling parallel to the seismic line would be the same as for an in-line array. But the parallelograms would help suppress noise traveling nearly perpendicular to the line. For the following discussion consider that the lateral displacements are zero or that we are concerned with only noise traveling parallel to the seismic line.

The first pattern could be made with five vibrators with four positions with each move-up equal to the spacing between the vibrators. The equivalent source pattern could be made with only four vibrators but with the responses summed for five positions with each moveup again equal to the spacing between the vibrators. The composite response must have all the notches in the pattern suppression diagram that would result from five vibrators spaced *d* distance apart plus all of the notches for four vibrators spaced *d* distance apart. This means that we have broadened the band of wavelengths for which suppression is reasonably good.

We can generalize from Fig. D-6 and conclude that we can expect a series of rejection notches based upon the number of elements and the spacing plus a second series of notches based upon the number of positions and the incremental move-up.

Whereas it is easy to generate a trapezoidally weighted array with mobile sources, it is always troublesome to do it with geophones. But it can be done with geophone arrays and at times it is well worth doing. Also, of course, we can combine a trapezoidally weighted geophone array with a trapezoidally weighted source pattern to set up four series of notches in the composite response to achieve a very broad band of effective wavelength suppression. But we must be careful that we do not suppress also the reflected signals of interest by letting the arrays or patterns get too long.

Numerical example of a trapezoidally weighted source pattern

We illustrate the response of a trapezoidally weighted source pattern with a numerical example. We will use the source pattern shown in Fig. D-6 where the spacing *d* is 60 ft (18 m) and we use four vibrators with five positions for each vibrator. The table on the next page lists the notches and peaks for both the five positions (four move-ups) and for the four vibrators. Actually we can add the suppressions in decibels to get the total pattern suppression. (It may not be intuitively obvious, but they are independent.) We know that we will have notches at every wavelength where there is a notch either for four sources that are 60 ft (18 m) apart or for five positions that are 60 ft(18 m) apart for each vibrator.

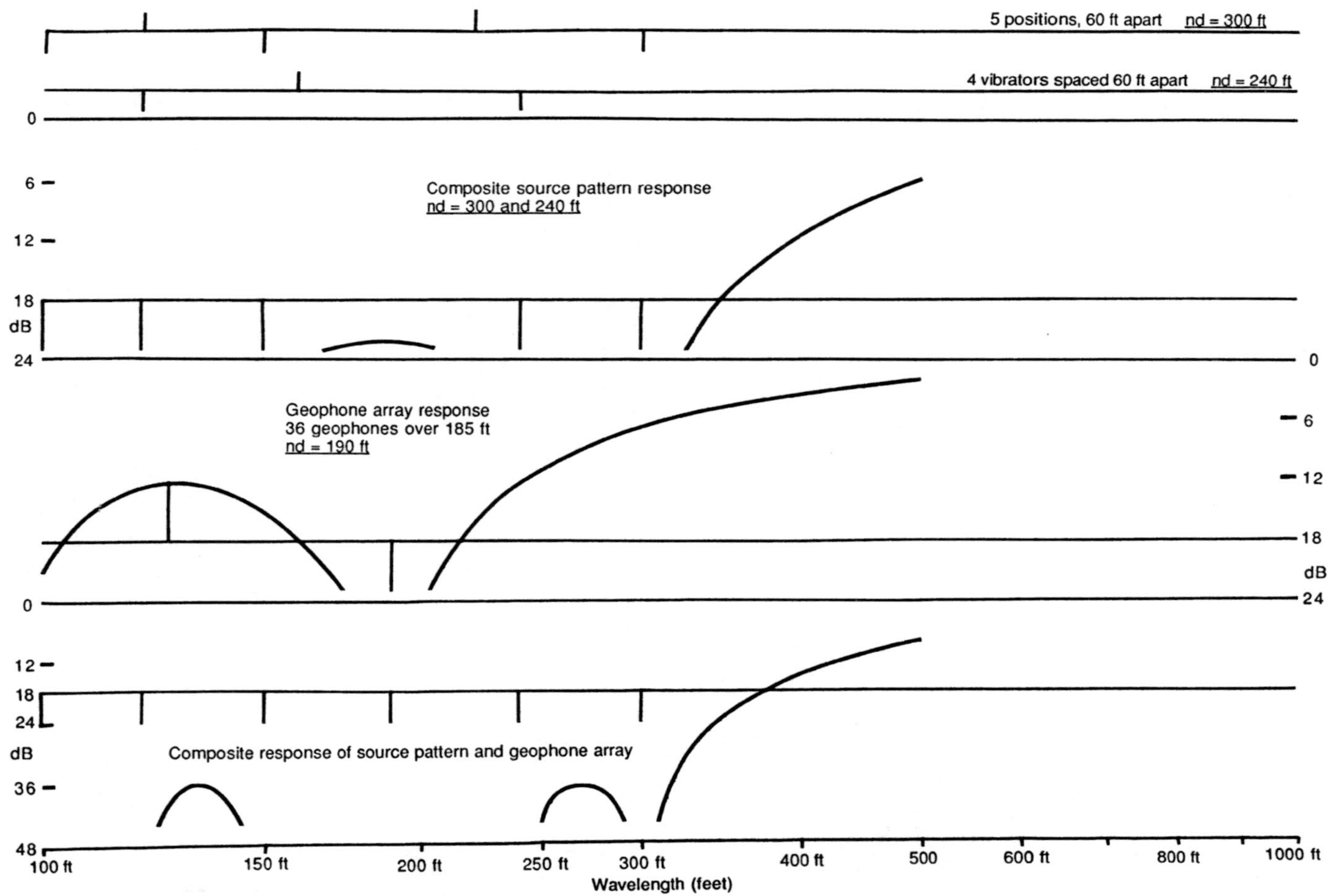

Fig. D-7A Form used for geophone array and source pattern design with an example. This form is for use primarily with distances measured in feet (but it may be used for metric units).

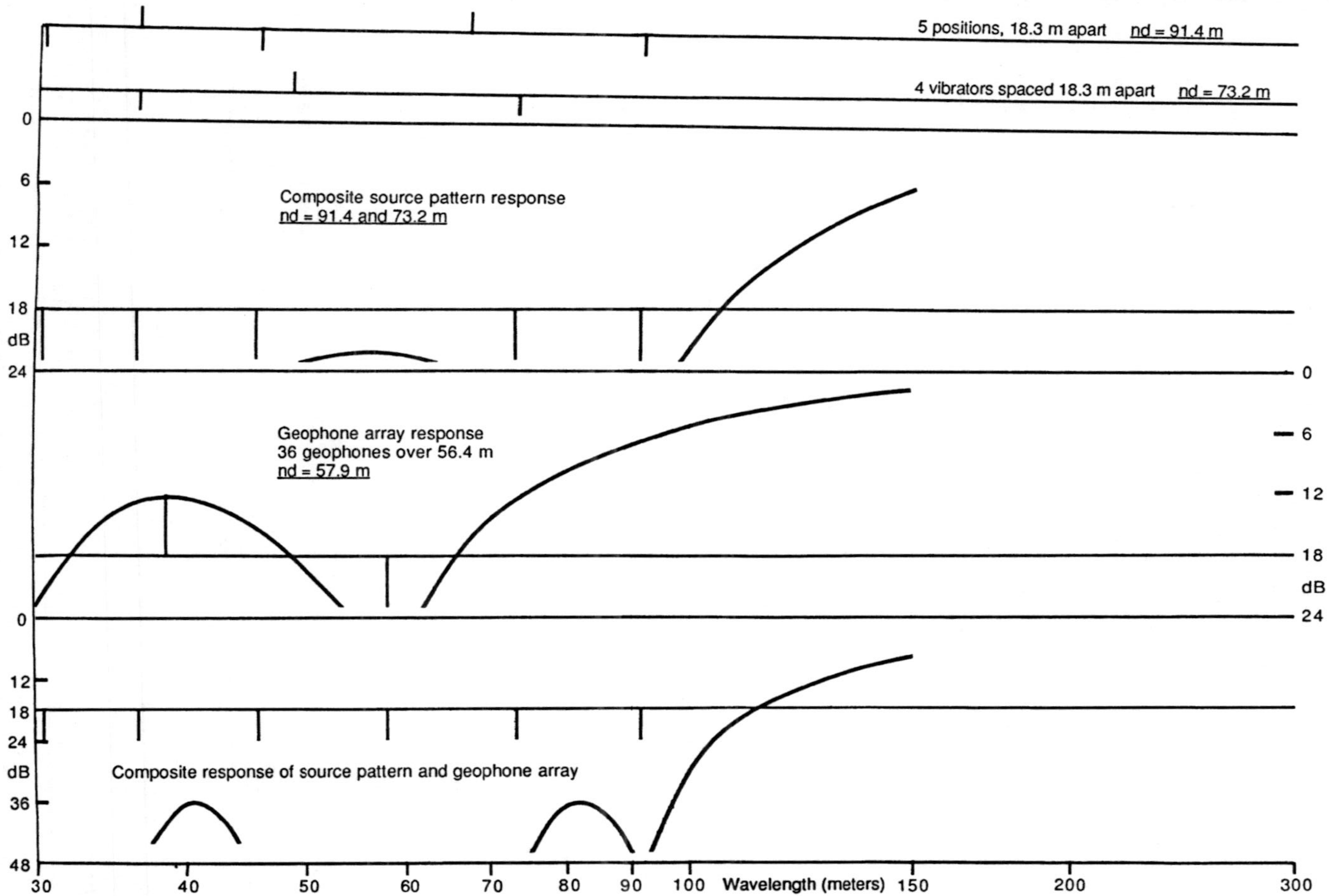

Fig. D-7B Form used for geophone array and source pattern design with an example. This form is primarily for use with distances measured in meters (but it may be used for distances in feet for short arrays).

Table D-3 Notches and peaks for a weighted source pattern and a complementary geophone array

Example is for four sources spaced 60 ft apart with five positions and four move-ups of 60 ft each

Five positions 60 ft apart nd = 300		*Four vibrators spaced 60 ft nd = 240*		*36 geophones over 185 ft nd = 190*	
Notches	*Peaks*	*Notches*	*Peaks*	*Notches*	*Peaks*
300 ft					
		240 ft			
	220 ft				
				190 ft	
			160 ft		
150 ft					
	120 ft	120 ft			127 ft
100 ft			96 ft	95 ft	
	86 ft				
		80 ft			
75 ft					76 ft
	60 ft (first zero)		60 ft (first zero)	63 ft	

Example is for four sources spaced 18 m apart with five positions and four move-ups of 18 m each

Five positions 18 m apart nd = 91		*Four vibrators spaced 18 m nd = 73*		*36 geophones over 56 m nd = 58*	
Notches	*peaks*	*Notches*	*Peaks*	*Notches*	*peaks*
91 m					
		73 m			
	67 m				
				58 m	
			49 m		
46 m					
	36 m	36 m			39 m
30.5 m			29 m	29 m	
	26 m				
		24 m			
23 m					23 m
	18 m (first zero)		18 m (first zero)	19 m	

Figure D-7 illustrates how we can use the form shown in Appendix A to sketch out the composite source response. We can be confident that wavelengths shorter than 320 ft (97.5 m) and longer than 70 ft (21 m) will be well suppressed except for wavelengths between about 170 and 205 ft (52 and 62 m) where suppression is moderate. But we can easily design a geophone array to have a notch near a wavelength of 190 ft (58 m) to yield a composite source and geophone response that is quite effective over a wide band of wavelengths (Table D-3). In Fig. D-7 vertical lines are shown at notch and peak wavelengths; downward lines for notches and upward for peaks. The source pattern suppression and the geophone array suppression are independent and the suppressions from the two in decibels can be added to get the total suppression in decibels for the composite. This is illustrated in Figs. D-7A and D-7B. Note also that we could add a notch to the source suppression at a wavelength of 60 ft (18 m) in Table D-3 by using ten positions with move-ups of 30 ft (9 m) each rather than five positions with move-ups of 60 ft (18 m) each.

Numerical example of a parallelogram geophone array

Figure D-8 illustrates a parallelogram geophone array with 24 geophones in each of two lines with spacing between adjacent geophones of 25 ft (7.6 m) and a lateral separation between the two lines of 100 ft (30.5 m). There is a longitudinal shift of 175 ft (53 m) of one line relative to the other. For noise traveling along the seismic line the response is not affected by the lateral separation of 100 ft and we can consider the array as if there were no lateral separation. The collapsed parallelogram is then equivalent to a trapezoidal array also shown in Fig. D-8. We will analyze the approximate response of the parallelogram to in-line noise. Then we will calculate the response of the array to high-frequency components of a reflection from a deep objective. For the latter calculation we will use the trapezoidal equivalent array.

From our analysis of the weighted source patterns in Fig. D-6, we know that the parallelogram array must have a series of notches for $nd = 600$ ft (183 m) [24×25 ft or 575 ft plus 25 ft (24×7.6 m or 175 m plus 7.6 m)] plus a second series of notches for $nd = 350$ ft (107 m) [2×175 ft or 175 ft plus 175 ft (2×53 m or 53 plus 53 m)]. Since n is only 2, the first zero peak is at a wavelength of 175 ft (53 m) and there is no notch at a wavelength of 175 ft (53 m). However, there will be notches at various wavelengths shorter than the first zero peak. The composite array will have notches at wavelengths of 600, 350, 300, 200, 150, 120, 100, etc. ft (183, 107, 61, 46, 30.6, 30.5, etc. m). With this information we could sketch an approximate response for the composite parallelogram array (Fig. 7.5).

Reflections from horizontal interfaces will arrive at oblique angles for all offsets except zero. The apparent velocity of a reflection along the surface is the

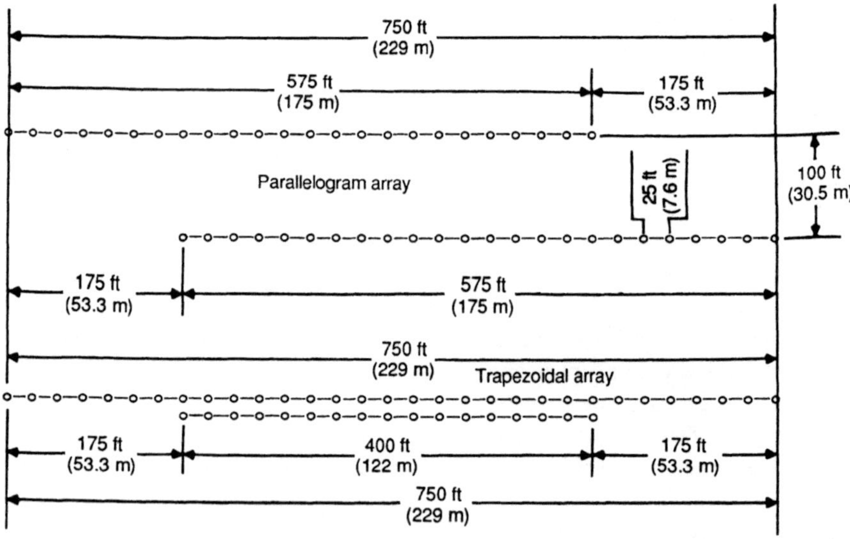

Fig. D-8 Parallelogram array and an equivalent trapezoidal array. For inline noise the trapezoidal array is equivalent to the parallelogram array and it is also equivalent for reflections from depth (in the absence of severe cross-dip).

moveout (stacking) velocity divided by the sine of the angle of arrival (Fig. 4.3). Therefore reflections have very long apparent wavelengths when the angle of arrival (from the vertical) is quite small. But as the angle of arrival increases, the wavelengths for the higher-frequency components of a reflection become short enough so that the response by a long geophone array is reduced significantly.

How can we calculate the response of the parallelogram array of Fig. D-8 to the high-frequency components of a reflection? First, we calculate the apparent velocity of the reflection at the surface location. Then for any frequency component we can calculate the apparent wavelength. We choose to use the trapezoidal equivalent array and calculate the response of the long and short lines separately. This is done by dividing the effective length of each line by the apparent wavelength of the reflection component and multiplying by 2π to get the total angle (in radians) of the component that is sampled by the array. If we define half of this angle as ϕ, then the relative response of the line is $(\sin \phi)/\phi$. This assumes that the spacing between sample points is short compared to the apparent wavelength of the component. We must weight the responses of the two lines by the number of geophones in each line. We then sum the weighted responses of the two lines and divide by the total number of geophones to get the relative response of the composite array. We illustrate these calculations with an example for the parallelogram array shown in Fig. D-8 assuming the following:

Depth of the reflector	$D = 10\,000$ ft (3048 m)
offset distance	8160 ft (2487 m)
Number of geophones in a line	n
Spacing between geophones in a line	$d = 25$ ft (7.6 m)
Stacking or moveout velocity	$v = 16\,000$ ft/s (4877 m/s)
Number of geophones in long line	$n = 31$
Number of geophones in short line	$n = 17$
Effective length of long line	$nd = 31 \times 25 = 775$ ft (236 m)
Effective length of short line	$nd = 17 \times 25 = 425$ ft (130 m)
Offset distance to nearest end of array	$L = 8160 - nd/2$ ft
Long line	$8160 - 387.5 = 7772.5$ ft (2369 m)
Short line	$8160 - 212.5 = 7947.5$ ft (2422 m)
Offset distance to farthest end of array	$L + \Delta L = 8160 + nd/2$ ft
Long line	8547.5 ft (2605 m)
Short line	8372.5 ft (2552 m)
Reflection time at zero offset	$t_0 = 2D/v = 1.25$ s

We calculate the differential time of arrival (Δt) across the long and short array lines using the exact equation from Chapter 5.

For the long line we get:

$$\Delta t = \sqrt{1.25^2 + \left(\frac{8547.5}{16000}\right)^2} - \sqrt{1.25^2 + \left(\frac{77725}{16000}\right)^2}$$

$$\Delta t = 0.018296 \text{ s}$$

For the short line we get:

$$\Delta t = \sqrt{1.25^2 + \left(\frac{8372.5}{16000}\right)^2} - \sqrt{1.25^2 + \left(\frac{7947.5}{16000}\right)^2}$$

$$\Delta t = 0.010034 \text{ s}$$

Next we calculate that angle ϕ for each line (the angle of the component subtended by the line divided by 2):

$$\phi = (\Delta t)(f)(\pi) \quad \text{radians}$$

For the long and short lines we calculate ϕ and the $(\sin\phi)/\phi$ for frequencies of 50 Hz and 40 Hz and obtain the following:

Frequency (*Hz*)	($\sin\phi$)/ϕ *for* $nd = 775$ (236 *m*)	($\sin\phi$)/ϕ *for* $nd = 425$ (129.5 *m*)
50	0.09205	0.6344
40	0.3246	0.7553

When we weight these responses by the number of geophones in each line and divide by the total number of geophones we get the composite response of the parallelogram (and of the trapezoidal equivalent) array as follows:

Response of parallelogram at 50 Hz
$= [13(0.09205) + 17(0.6344)]/48 = 0.284$

Response of parallelogram at 40 Hz
$= [31(0.3246) + 17(0.7553)]/48 = 0.477$

It may not be immediately apparent how the trapezoidal equivalent array with *nd* values of 775 and 425 ft (236 and 129.5 m) can yield the notches we know should be present at wavelengths like 600 and 350 ft (183 and 107 m). The answer is that at these wavelengths the responses of the two lines are of opposite polarity and when weighted by the number of geophones the sum should be zero. For example, we will calculate the response for wavelengths of 600 and 350 ft (183 and 107 m).

For a wavelength of 600 ft (183 m) $\phi = \pi(nd)/600$, and for a wavelength of 350 ft (107 m) $\phi = \pi(nd)/350$. For the responses of the individual lines we get:

Wavelength (*f*)	(*m*)	*Long line* $\sin\phi/\phi$	*Short line* $\sin\phi/\phi$
600	183	−0.19551	0.35652
350	107	0.08963	−0.16344

The response for the composite for a wavelength of 600 ft (183 m) is:

$$[(31)(-0.19951) + (17)(0.35652)]/48 = [-6.0608 + 6.0608]/48 = 0$$

The response for the composite for a wavelength of 350 ft (107 m) is:

$$[(31)(0.08963) + (17)(-0.16344)]/48 = [2.7785 - 2.7785]/48 = 0$$

Everything checks out and the example illustrates how the composite array response comes about.

Star geophone arrays and star source patterns

Figures D-9 and D-10 illustrate the star source pattern and star geophone array described in Chapter 12. The source pattern has fifteen shot holes with seven arms and a diameter of 200 ft (61 m). The geophone array has thirty geophones, five arms, and a diameter of 198 ft (160 m). As shown, the source pattern has one arm aligned perpendicular to the seismic line (best for suppressing in-line noise). The star geophone array as illustrated has one arm aligned parallel to the seismic line. Figure 12.6 shows the noise suppression response of this source pattern and this geophone array for noise traveling along the seismic line. The geophone array response would be somewhat improved for noise traveling along the seismic line if the array were rotated slightly such that one arm was perpendicular to the seismic line, but the response to noise arriving at steep angles to the seismic line would be degraded. This geophone array is much more effective in suppressing noise with wavelengths shorter than 100 ft (30 m).

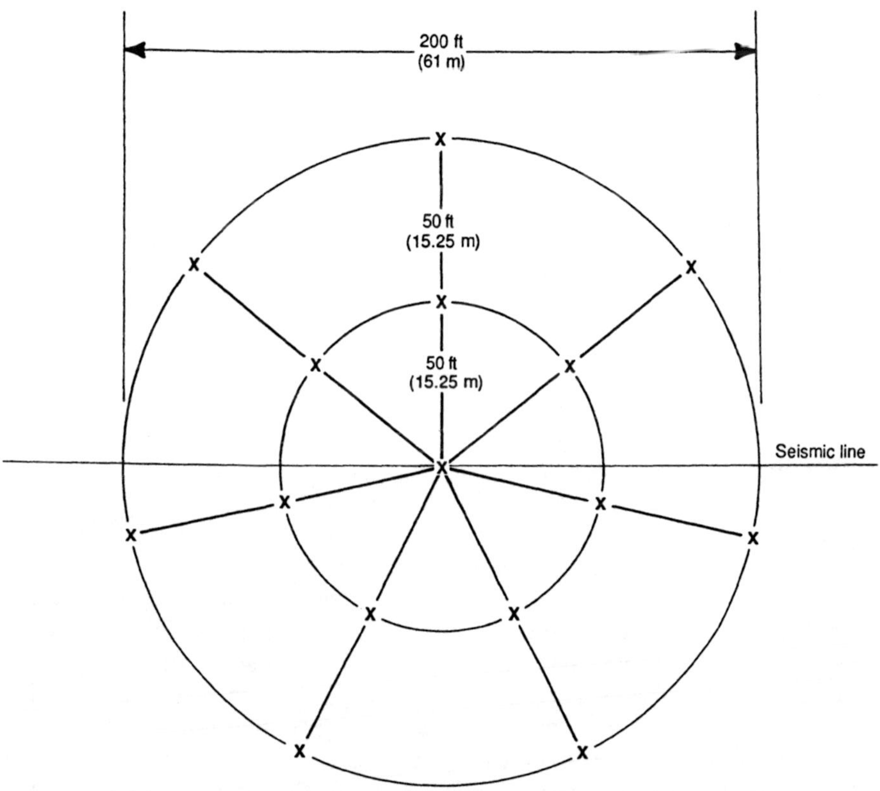

Fig. D-9 Star source pattern with seven arms and fifteen shot holes.

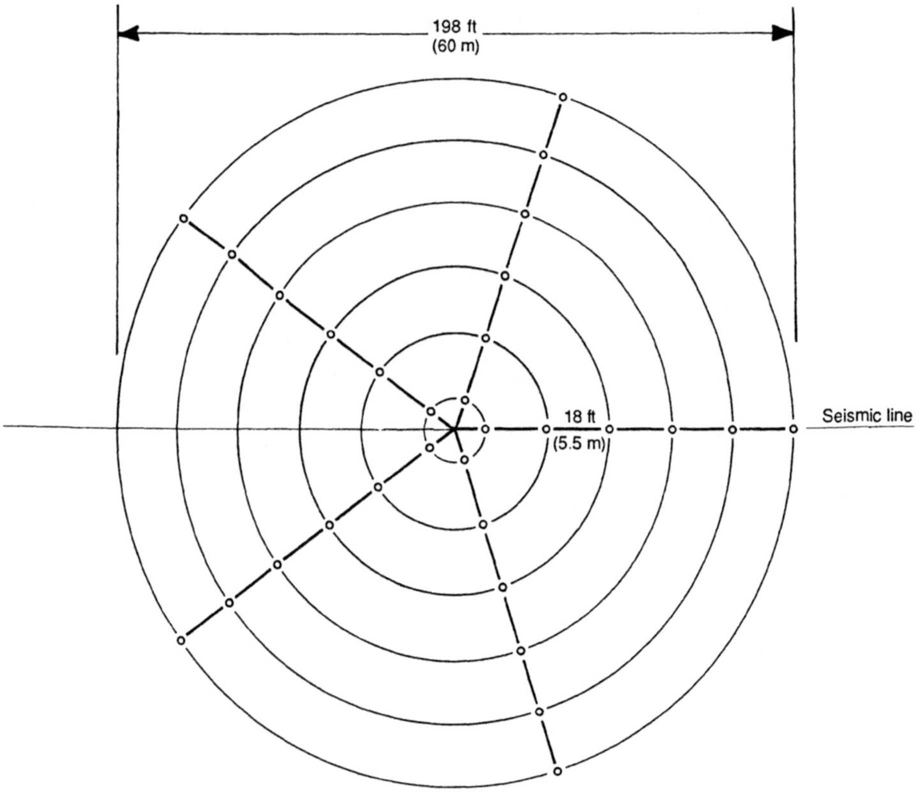

Fig. D-10 Star geophone array with five arms and thirty geophones

The preferred orientation for star geophone arrays and star source patterns depends upon the particular arrays and patterns and upon the noise problems that we are dealing with. Hopefully we will have computer programs available that we can run to evaluate possible arrays and patterns and the various orientations of each. (Star responses can be determined with a hand calculator, but it is very time consuming.) Generalities are risky, but in the absence of better information the author recommmends the use of stars with an odd number of arms and with one arm perpendicular to the expected direction of travel of the major noise. If we expect omni-directional noise, we may want to align the source pattern with one arm perpendicular to the seismic line and the receiver pattern with one arm parallel to the seismic line or vice versa.

Figure D-11 compares the response of the geophone array of Fig. D-10 with that of the same geophone array except that all of the dimensions are increased by a factor of two for one array and are decreased by a factor of two for the other array. The diamter is then 396 ft (121 m) for the largest array and 99 ft (30 m) for the smallest array. Note that this change in dimensions simply shifts

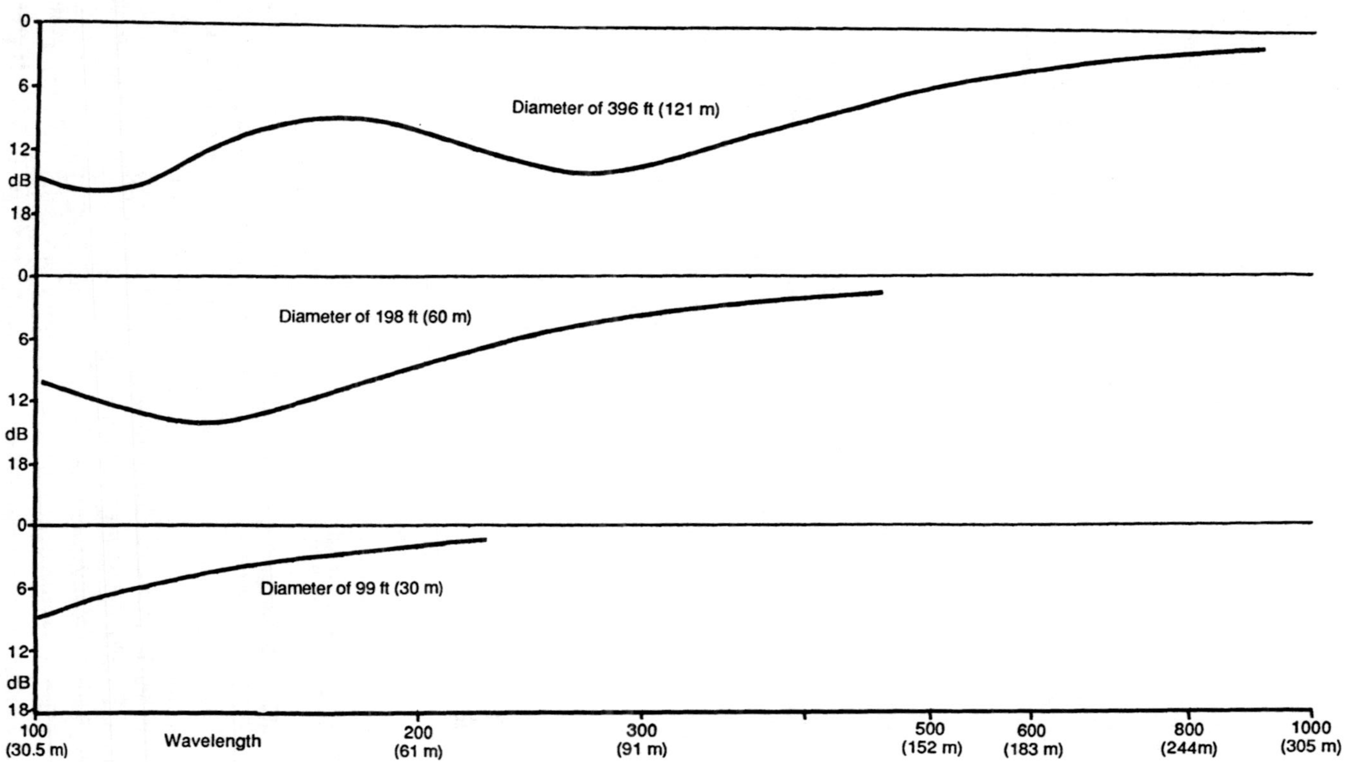

Fig. D-11 Noise suppression by six arm star arrays of various dimensions. The shape of the suppression curve is the same but it shifts with changes in the overall diameter of the array.

the response curve. The response for a wavelength of 200 ft (61 m) for the original array is the same as that for a wavelength of 400 ft (122 m) for the largest array and the same as that for 100 ft (30.5 m) for the smallest array. We can adjust the dimensions of star arrays (and other complex arrays and patterns) without having to recalculate the response. This is helpful in designing arrays and patterns in the field where we do not have access to a computer. All we need is a set of responses for a number of arrays and patterns of different types but with fixed dimensions. We can adjust the dimensions to move notches in the response to suppress noise wavelengths of our choice. Also we can adjust the source pattern to complement the receiver array or vice versa. Note that in Fig. 12.6 the first notch of the geophone array coincides with the first peak of the source pattern. This alignment was deliberate.

CMP stacks to suppress noise

The suppression of noise by the CMP stack can be considered as a separate array similar to a geophone array, but on a larger scale. Where the stack-array concept (Anstey 1986) is applicable, the CMP stack can best be considered in combination with the geophone arrays as one giant array (Chapter 2). But where the stack-array concept is inapplicable or is only marginally applicable, we need to consider the noise suppression of the CMP stack separately.

Some geophysicists erroneously assume that the CMP stack is ineffective in suppressing noise with wavelengths equal to or shorter than the offset increment in the CMP gather. For very short wavelengths the time phase for the samples from widely separated points on the surface can best be considered as being random. For very short wavelengths we can then assume that the relative noise response is the square root of the number of samples divided by the number of samples, which is the reciprocal of the square root of the number of samples (Chapter 21). For example, if the CMP fold were 48, then the relative response for noise with short wavelengths would be $1/(48)^{0.5} = 0.144$ which corresponds to 16.8 dB of suppression. This is certainly not negligible.

For moderate to long wavelengths the CMP gather is a sampling of signals and noises at various locations on the surface of the earth and is hence similar to and can be treated like a geophone array in its suppression of noise. One difference is that we try to correct the data for NMO and for static variations before the signals are stacked (summed). An NMO correction is absolutely necessary otherwise the wanted signals would be out of time phase and would be suppressed along with the noise.

We can use the same approximate techniques as we did with geophone arrays to estimate the noise suppression, or we can use equation (D-1) given previously, to calculate noise suppression:

$$R = 20\log_{10}\left|\frac{\sin(\pi nd/\lambda)}{n\sin(\pi d/\lambda)}\right| \text{ dB}$$

$$R = \left|\frac{\sin(\pi nd/\lambda)}{n\sin(\pi d/\lambda)}\right| \quad \text{as a simple ratio (not in dB)} \qquad \text{(D-2)}$$

for a CMP gather of consistent samples:

n = the CMP fold, and
d = the differential offset between consistent samples.

It is true that the CMP stack is ineffective for wavelengths approximately equal to the offset increment, but this is simply the first zero peak which is followed at progressively shorter wavelengths by additional notches and peaks (most of which have significant suppression). The theory assumes that the noise samples have approximately equal amplitudes and approximately identical amplitude spectra and that the time phase of the noise samples is proportional to the offset distance from the source point to the sample point. If the gather contains interlacing offsets from front and back cables (recorded with a split cable), then the samples from consecutive offset distances are actually from source locations on the surface which are relatively far apart and from receiver locations which are far apart. In this case variations in the acoustic properties of the surface and shallow earth section may invalidate the assumptions regarding the amplitudes and amplitude spectra.

Reciprocity of sources and receivers is not totally applicable for two reasons. First, the source patterns and receiver arrays are not identical (particularly when explosives in drilled holes are used as the seismic source). Second, there are nonlinear effects in the coupling of high-energy sources to the earth. Consequently, we may better consider a gather of just the front cable plus a separate gather of just the back cable which would double the incremental offset distance and hence double the wavelength of the first zero peak. In this case we can still use equation (D-2), but now n = half the CMP fold and d = twice the incremental offset distance in the CMP gather. Also in this case we must sum the two relative responses for the front and back cables. The responses for our reflection signals would be additive, but the responses for noise would be the sum of two vectors with some phase difference which would provide additional noise suppression.

Near zero peaks of the CMP stack it is quite desirable to provide noise suppression by geophone arrays or by source patterns, or both. Where the stack-array concept is applicable, the geophone arrays do provide noise suppression at the zero peaks of the CMP stack. We will show that the product of the suppression by the CMP stack and the suppression by the geophone array is equal to the suppression calculated by the stack-array concept. (This has to be true, but a proof may be helpful)

The suppression calculated by the stack-array concept for a group length array is given by equation (D-2), where

n_{sa} = (CMP fold) times (number of geophones per array), and
d_{sa} = the spacing between geophones in each array.

The supression for the CMP stack alone is given by equation (D-2), where

n_s = (CMP fold), and
d_s = the offset increment in the CMP gather.

Then

$n_{sa} = (n_s)$ times (number of geophones per array), and

$d_s = (d_{sa})$ times (number of geophones per array).

Therefore

$$(n_{sa}d_{sa}) = (n_s d_s).$$

The suppression of the geophone array alone is given by equation (D-2), where

n_g = (number of geophones per array), and
d_g = spacing between geophones.

The product of the relative responses of the geophone array and the CMP stack is then:

$$R = \left|\frac{\sin(\pi n_g d_g/\lambda}{n_g \sin(\pi d_g/\lambda)}\right|\left|\frac{\sin(\pi n_s d_s/\lambda)}{n_s \sin(\pi d_s/\lambda)}\right|$$

But $d_s = n_g d_g$, therefore

$$R = \left|\frac{\sin(\pi n_s d_s/\lambda)}{n_g \sin(\pi d_g/\lambda)}\right| \frac{1}{|n_s|}$$

But $n_g n_s = n_{sa}$, $n_{sa}d_{sa} = n_s d_s$, and $d_g = d_{sa}$; therefore:

$$R = \left|\frac{\sin(\pi n_{sa} d_{sa}/\lambda)}{n_{sa} \sin(\pi d_{sa}/\lambda)}\right|$$

The above equation is equal to the response under the stack-array concept and we have shown that the product of the relative response of a group length array and the relative response of a CMP gather is equal to the relative response under the stack-array concept where this concept is applicable.

For wavelengths that are long compared with the spacing between geophones, the angle in the denominator is small and the sine of the angle is approximately equal to the angle in radians. Therefore for moderate-to-long wavelengths the stack-array response is approximately equal to:

$$R_{sa} = \left| \frac{\sin(\pi n_{sa} d_{sa}/\lambda)}{\pi n_{sa} d_{sa}/\lambda} \right| \tag{D-3}$$

The above equation for the relative response of the stack-array is the approximate ($\sin \phi/\phi$) relationship used extensively throughout the book.

If the CMP samples from interlacing offsets are consistent and the effective differential offset is the group interval, then the wavelength of the first zero peak is equal to the group interval. A group interval array, double group interval array, triple group interval array, etc. (section 2.2) would all provide a notch at the wavelength of the first zero peak of the CMP stack. But if the samples from the interlacing offsets of the front and back cables are inconsistent, then the effective differential offset from just the front or back cable is twice the group interval. In this case a double group interval array would still provide a notch at the wavelength of the first zero peak of the CMP stack, but a group interval array would not.

In summary, the CMP stack is effective in suppressing noise except at wavelengths near zero peaks. It is desirable to design the geophone array or the source pattern (or both) to provide notches at wavelengths of zero peaks of the CMP response. The stack-array approach does provide these desired notches where the stack-array concept is applicable (i.e. the length of the geophone array is equal to an integer times the incremental offset between consistent samples in the CMP gathers). For very short wavelengths the stack is still reasonably effective (even near theoretical zero peaks) because of normal variations in the earth and slight imperfections in line layout (since such variations and imperfections cause the time phase of the noise to be inconsistent).

REFERENCES

Anstey, N. A. (1986) 'Whatever happened to ground roll?'. *The Leading Edge* **5**, no. 3 (March), 40–45.

APPENDIX E—ILLUSTRATIONS OF SPREADS

In this appendix we illustrate special cable layouts and operations. Noise spreads are used in testing to define the source-generated noises so that we can determine how these noises can best be suppressed. Walk-through spreads and contracting and expanding spreads are used to acquire high CMP fold data either for test purposes or for final data on lines of very limited length.

Noise spreads

Figure E-1 illustrates a noise spread. In a noise spread the cable remains in a fixed position and source points are located off one end or off of both ends. In the illustration we show only six in-line receiver positions whereas in most field cases we would likely have many more traces. At each receiver location we would have either one geophone or a string of geophones planted very close together in a 'pod'. At each source position a single source or multiple sources would be used as close together as possible. The purpose of a noise spread is to sample the noise along the surface for a range of offsets from very close to the source out to an offset about as long as we are likely to use in recording reflection data. We want a uniform sample interval that is short enough not to alias any important noise components. This requires a minimum of two samples of noise per wavelength.

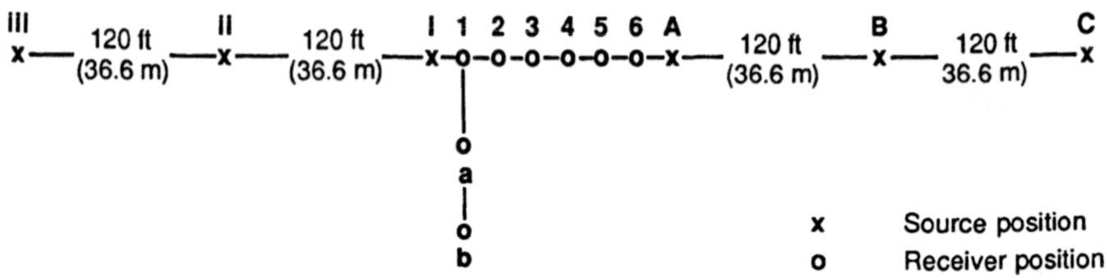

Note: The inline receiver spread uses positions 1-6.
The cross-line receiver spread uses positions 1, a & b.

Source points	Offsets in feet	Offsets in meters
I & A	20, 40, 60, 80, 100 &120	6.1, 12.2, 18.3, 24.4, 30.5 & 36.5
II & B	140, 160, 180, 200, 220 & 240	42.7, 48.8, 54.9, 61, 67.1 & 73.2
III & C	260, 280, 300, 320, 340 & 360	79.2, 85.3, 91.4, 97.5, 103.6 & 109.7

Fig. E-1 Illustration of a noise spread

1. The illustration is for six inline receiver positions (1–6), three cross-line receiver positions (1, a & b), and six source points (i, ii, iii, A, B, C);
2. The illustrated inline receivers are 20 ft (6.1 m) apart and source points A and i are 20 ft (6.1 m) from either end;
3. Noise is sampled in both, in-line directions for offsets ranging from 20ft (6.1 m) to 360 ft (110 m) with a uniform, in-line sample interval of 20 ft (6.1 m);
4. In an actual field measurement we would probably use many more traces and measure noises for inline offsets ranging out to about the maximum offset we planned to use in recording data;
5. The cross-line receivers are used to measure noises scattered from off line. The cross-line receivers may or may not be spaced farther apart than the inline receivers. We may or may not choose to use cross-line receivers in a noise spread measurement.

Often, instead of a linear noise spread cable we layout a cable in the shape of an 'L' so as to be able to define the azimuth of arrival of noises and to measure the wavelength and velocity of noises that arrive at azimuths that are nearly perpendicular to the seismic line. Generally we can use a longer trace spacing in the arm that is perpendicular to the seismic line.

Walk-throughs

Figure E-2 illustrates a walk-through spread which is used to acquire high CMP fold reflection data over a limited length of line. In a walk-through, the cable remains fixed and data are recorded for a series of source points as illustrated. The maximum CMP fold is equal to the number of traces used. The fold tapers down to one at each end of the CMP coverage. Walk-throughs are frequently used as the final step in a test where there is a need to acquire a limited amount of high-fold, CMP data which can be processed and stacked to

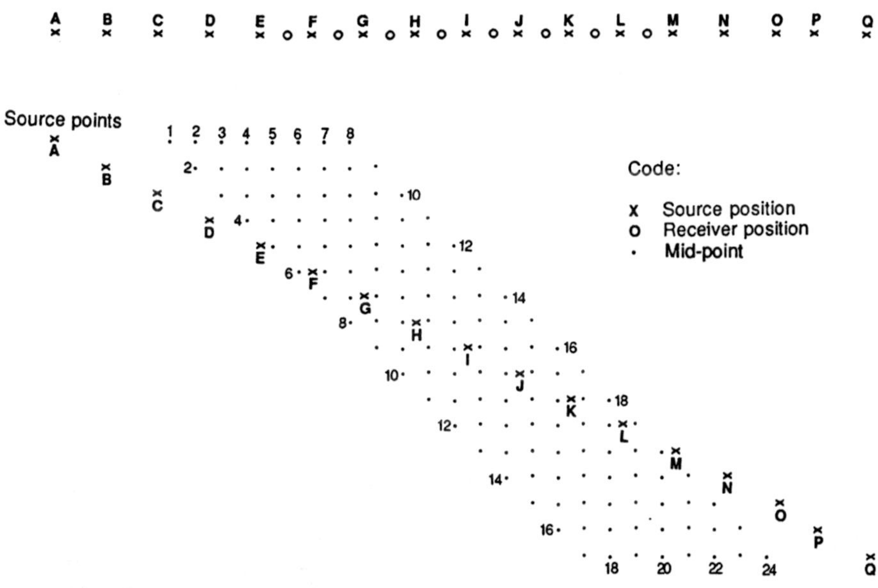

Fig. E-2 Walk-through spread. Note:

1. The cable remains in one location. The illustration is for eight receiver positions and seventeen source points;
2. The maximum CMP fold is equal to the number of traces recorded per record. In this illustration the maximum fold of eight is achieved for depth points 8–17. The fold tapers down to one at both ends;
3. In an actual field measurement we would probably use many more traces and record data for many more source points. Some offsets may be too long to be effective.

see what quality of data can be achieved for the final sections in the area. (A walk-through test is not usually done unless we are in a bad record area or beginning a very important and large program.) Frequently when walk-through spreads are used for a test, the cable is folded back so that we have two or more receiver take-outs at each receiver position. This allows us to acquire stackable data with two or more different geophone arrays. We can then compare the stacked data to decide which geophone array to use in the main recording program in the area.

In actual practice we would probably disconnect (or at least not record) data from receiver positions that are very close to a given source position. Therefore the number of live traces would be decreased whenever the source position for a record is located near or within the zone occupied by receivers. This slightly reduces the maximum CMP fold.

As illustrated in Fig. E-2, the range of offsets changes for each record and the offsets in each CMP gather also change. The offsets (in group intervals) for

Table E-1 Offsets in each mid-point gather for example of Fig. E-2

Mid-point	*Range of offsets in group intervals*							
1								4.5
2							5.5	3.5
3						6.5	4.5	2.5
4					7.5	5.5	3.5	1.5
5				8.5	6.5	4.5	2.5	0.5
6			9.5	7.5	5.5	3.5	1.5	0.5
7		10.5	8.5	6.5	4.5	2.5	0.5	1.5
8	11.5	9.5	7.5	5.5	3.5	1.5	0.5	2.5
9	10.5	8.5	6.5	4.5	2.5	0.5	1.5	3.5
10	9.5	7.5	5.5	3.5	1.5	0.5	2.5	4.5
11	8.5	6.5	4.5	2.5	0.5	1.5	3.5	5.5
12	7.5	5.5	3.5	1.5	0.5	2.5	4.5	6.5
13	6.5	4.5	2.5	0.5	1.5	3.5	5.5	7.5
14	5.5	3.5	1.5	0.5	2.5	4.5	6.5	8.5
15	4.5	2.5	0.5	1.5	3.5	5.5	7.5	9.5
16	3.5	1.5	0.5	2.5	4.5	6.5	8.5	10.5
17	2.5	0.5	1.5	3.5	5.5	7.5	9.5	11.5
18	1.5	0.5	2.5	4.5	6.5	8.5	10.5	
19	0.5	1.5	3.5	5.5	7.5	9.5		
20	0.5	2.5	4.5	6.5	8.5			
21	1.5	3.5	5.5	7.5				
22	2.5	4.5	6.5					
23	3.5	5.5						
24	4.5							

each gather are given in Table E-1, starting with the mid-point near source point C which we will call mid-point 1.

We note from the table that in each CMP gather (horizontal lines) that the offsets from front and back (or ahead and behind) cables do interlace. This is because we chose to locate source positions halfway between receiver positions. But the number from front and back differ; consequently the interlacing is not complete. (In a few cases all of the offsets are front and for

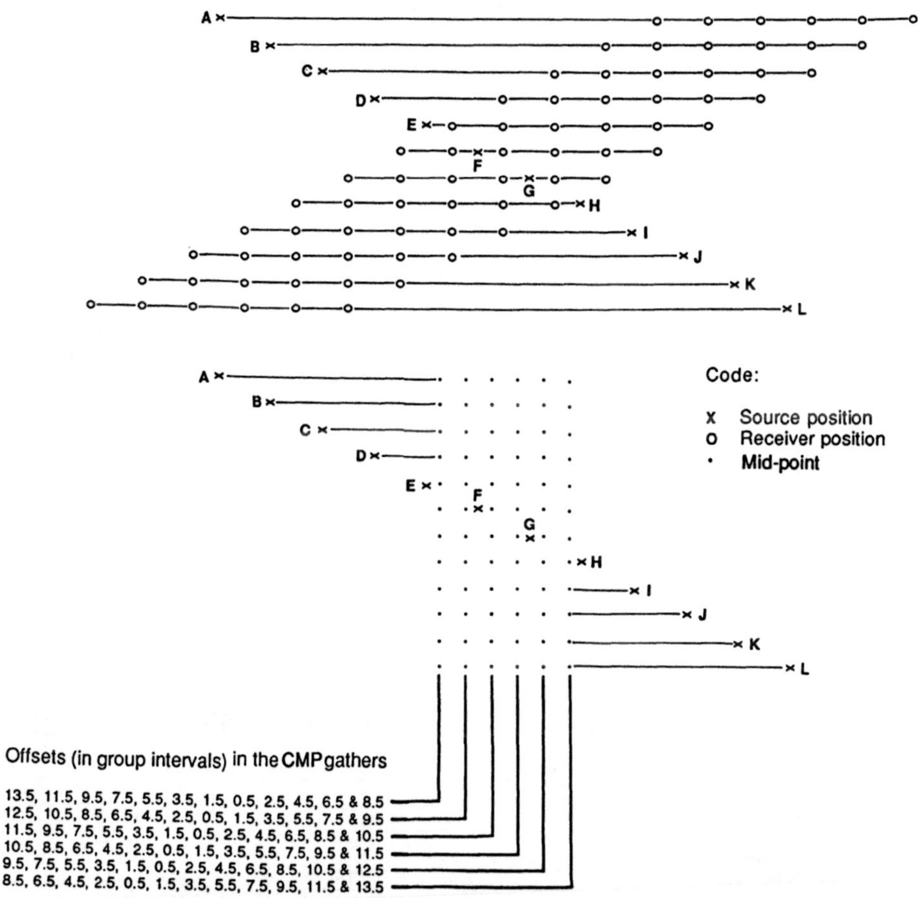

Fig. E-3 Contracting and expanding spread. Note:

1. The illustration is for twelve source points, with six receiver positions per record;
2. The CMP fold is equal to the number of source points;
3. In an actual field measurement we would probably use many more traces and record data for many more source points. Some offsets may be too long to be effective.

others all of the offsets are back.) With a double group length array (Chapter 2) we would have the uniform sampling desired in the CMP stack. But with group length arrays the lack of consistent interlacing of offsets from the front and back would not yield the desired uniform sampling in most of the gathers. Even so, a walk-through test with group length arrays is usually satisfactory for determining the attainable data quality and for evaluating group length arrays.

Contracting and expanding spreads

Figure E-3 illustrates a method to achieve a very high CMP fold for a very limited length of line. We refer to this spread as a 'contracting and expanding spread', which is similar to the walk-through except both the source and the cable are moved one group interval after each record. The receiver positions and the source positions lie on a single line. (The lateral displacements shown in Fig. E-3 are used to separate the successive layouts for successive records.) For the first record the source is rather far off one end of the cable. After each record is taken, both the source point and the whole cable are both moved one group interval. The two movements are in opposite directions such that initially the separation between the source and the cable is reduced. With successive recordings and moves the source point moves through cable and then on beyond the cable.

The CMP points remain fixed with this system and the CMP fold is then equal to the total number of recordings. In practice we would probably not record data from receivers too close to the source and the CMP fold would then be slightly reduced. The line length of CMP data is half the length of the cable. Since we chose to locate source positions half way between receiver positions, the front and back offsets interlace within the CMP gathers. But the number front and back are generally not equal; so the interlacing is not complete. Double group length arrays would provide the desired uniform sampling, but group length arrays would not be as satisfactory in this respect.

APPENDIX F—SHORT PROGRAMS FOR PROGRAMMABLE CALCULATORS

This appendix contains seven programs to calculate:

1. apparent horizontal velocity from the normal moveout,

2. differential NMO across an array and $f_{0.64}$, $f_{0.3}$, and f_0,
3. half of the Fresnel Zone radius,
4. uphole velocity survey corrections,
5. NMO velocity from reflection time and offset pairs,
6. sweep for a Pelton nonlinear sweep generator, and
7. sweep for a Texas Instruments nonlinear sweep generator.

These seven programs can be used separately or they can all be entered and retained for use as needed in a Texas Instruments model TI-66 programmable calculator (available in Europe). If the combined program is entered, each program is labeled and is called by its label when needed. These programs are listed at the end of this appendix. These programs can be adapted for use with other programmable calculators. Similar programs have been used with the TI-65 calculator, but only one program can be stored at a time because of its limited memory.

Since the overall program contains 323 steps, it is necessary to partition the memory of the TI-66 to allow that number of steps before the combined program is entered. This can be done as follows:

Key stroke	Display
2nd	
Part	
23	327.22

This allows up to 328 program steps (000 to 327) and up to 23 memories (00 to 22).

Apparent horizontal velocity from the normal moveout

We need to know the apparent horizontal velocity of our reflections so that we can calculate the apparent horizontal wavelengths by dividing by the frequency. Then we can see how much some geophone array or a source pattern would suppress the reflections.

The first program calculates the apparent horizontal velocity of reflections from horizontal interfaces.

Label	A
Input	Offset, NMO velocity, and the two-way reflection time
Output	Apparent horizontal velocity

Procedure and key strokes

Key Stroke	Procedure
2nd CMs	
	Enter NMO velocity
STO	
00	
	Enter two way reflection time
STO	
01	
	Enter offset
A	
	Program runs. Read horizontal velocity.

Example with English units:

NMO velocity	10 000 ft/s
Two-way reflection time	1 s
Offset	5000 ft
Apparent horizontal velocity	22 361 ft/s

Second example with metric units:

NMO velocity	4000 m/s
Two-way reflection time	2 s
Offset	3000 m
Apparent horizontal velocity	11 392 m/s

Differential NMO across an array and $f_{0.64}$, $f_{0.3}$ and f_0

This program calculates the differential normal moveout across an array or source pattern for a reflection from a horizontal interface. This information is then used to calculate the frequencies for which the response decreases to 64, 30 and 0 per cent of the maximum response.

Label	A′
Input	Offset, NMO velocity, two-way reflection time, and the effective length of an array or pattern.
Output	Differential NMO across the array or pattern, and the frequencies where the response drops to 64, 30 and 0 per cent of the maximum response.

Procedure and key strokes

Key stroke	Procedure
2nd CMs	
	Enter NMO velocity
STO	
00	
	Enter two-way reflection time
STO	
01	
	Enter effective length of an array or pattern
X↔T	
	Enter offset
2nd, A′	
	Program runs. Read differential travel time.
R/S	
	Program runs. Read $f_{0.64}$.
R/S	
	Program runs. Read $f_{0.30}$.
R/S	
	Program runs. Read f_0.

Example with English units:

NMO velocity	10 000 ft/s
Two way reflection time	1 s
effective length of array	330 ft
Offset	5000 ft
Differential travel time	0.01476 s
$f_{0.64}$, $f_{0.3}$, and f_0	33.9, 50.8 and 67.8 Hz

Second example with metric units:

NMO velocity	4000 m/s
Two-way reflection time	2 s
Effective length of array	150 m
Offset	3000 m
Differential travel time	0.00878 s
$f_{0.64}$, $f_{0.3}$, and f_0	57.0, 85.4 and 113.9 Hz

Half of the Fresnel Zone radius

Half of the Fresnel Zone radius is a measure of the attainable horizontal resolution for reflections from an objective interface.

Label	B
Input	Frequency, two-way reflection time, average velocity to the objective and velocity of the interval *overlying* the objective. (The interval considered should be that which determines the wavelength of the wave that impinges upon the objective interface.)
Output	One-half the Fresnel Zone radius

Procedure and key strokes

Key stroke	Procedure
2nd CMs	
	Enter average velocity to the objective
STO	
00	
	Enter velocity of interval *overlying* the objective
STO	
01	
	Enter two-way reflection time to the objective
X↔T	
	Enter frequency
B	
	Program runs. Read 1/2 Fresnel Zone radius.

Example with English units:

Average velocity	15 000 ft/s
Velocity of *overlying* interval	14 000 ft/s
Two-way reflection time	2.1 s
Frequency	35 Hz
1/2 Fresnel Zone radius	889 ft

Second example with metric units:

Average velocity	4000 m/s
Velocity of *overlying* interval	4500 m/s
Two-way reflection time	2 s
Frequency	30 Hz
1/2 Fresnel Zone radius	274 m

Uphole velocity survey corrections

This is a program for use in correcting uphole velocity survey data so as to yield corrected depths, corrected vertical uphole times, and corrected average vertical velocities.

Label	B′
Input	Shot depth, time break correction, shot offset, receiver depth, and measured uphole time.
Output	Corrected depth, corrected uphole time, and the average vertical velocity.

Procedure and key strokes

Key strokes	Procedure
2nd CMs	
	Enter shot depth
STO	
00	
	Enter the time break correction
STO	
01	
	Enter the shot offset distance
STO	
02	
	Enter the receiver depth
X↔T	
	Enter the measured uphole time
2nd B′	
	Program runs. Read corrected depth.
R/S	
	Program runs. Read corrected uphole time.
R/S	
	Program runs. Read average velocity.

Example with English units:

Shot depth	25 ft
Time break correction	0.001 s
Offset	30 ft
Receiver depth	293 ft
Uphole time	0.032 s
Corrected depth	268 ft
Corrected uphole time	0.031 s
Average velocity	8699 ft/s

Second example with metric units:

Shot depth	10 m
Time break correction	0.002 s
Offset	15 m
Receiver depth	103 m
Uphole time	0.029 s
Corrected depth	93 m
Corrected uphole time	0.027 s
Average velocity	3489 m/s

NMO velocity from reflection time and offset pairs

Reflection time and offset pairs are picked for a single reflection. The data are analyzed statistically for the best estimate of the actual NMO velocity and the zero offset time. Then reflection times can be calculated for any number of offsets.

These two programs can also be used rather easily to generate any number of reflection time and offset pairs given a zero offset reflection time (t_0) and an NMO velocity (v). (This is useful for making overlays of moveouts as described in Chapter 4.) The procedure is to enter t_0 and 0.0 as the first time and offset pair. Then use any convenient offset (L) and calculate a precise reflection time $[(t_0)^2 + (L/v)^2]^{0.5}$, and enter these as a second time and offset pair. The program will run and display a correlation coefficient of 1.0, a zero offset time of t_0, and a NMO velocity of v. You can then enter offset values and calculate reflection times for as many offsets as you want.

Label	C and C′
Input	Any number of pairs of reflection time and offset.
Output	Number of data pairs, correlation coefficient, NMO velocity, and zero offset time.
Later inputs	Offset.
Later outputs	Corresponding reflection time for offset given.

Procedure and key strokes

Key stroke	Procedure
2nd CSR	(Note change. This clears the statistical registers)
	Enter reflection time
X↔T	
	Enter corresponding offset
C	
	Data is input. Number of pairs entered is displayed.
	Enter next reflection time.
X↔T	
	Enter corresponding offset
C	
	Data is input. Number of pairs entered is displayed.

Continue until all data pairs have been entered.

2nd C′	
	Program runs. Correlation coefficient of data is displayed.
R/S	
	Zero offset reflection time is displayed.
R/S	
	NMO velocity is displayed.
	Enter any offset desired.
R/S	
	Corresponding reflection time is displayed.

You can continue for as many offsets as desired.

Example with English units:

Data pairs	(2.000 s, 770 ft), (2.002 s, 1210 ft), (2.006 s, 2090 ft), (2.009 s, 2970 ft), (2.026 s, 4730 ft)
Correlation coefficient	0.997
Zero offset time	2.000 s
NMO velocity	14 649 ft/s
Offset of 5000 ft	Reflection time of 2.029 s
Offset of 7000 ft	Reflection time of 2.056 s

Second example with metric units:

Data pairs	(2.500 s, 300 m), (2,505 s, 700 m), (2.515 s, 1000 m), (2.530 s, 1500 m), (2.550 s, 2000 m)

Correlation coefficient	0.997
Zero offset time	2.500 s
NMO velocity	3937 m/s
Offset of 2500 m	Reflection time of 2.579 s
Offset of 3000 m	Reflection time of 2.613 s

Third example for given t_0 and v: for either English or metric units

Zero offset reflection time	2.0 s
NMO velocity	6000 ft/s or m/s
First data pair	
Time	2.0
Offset	0.0 ft or m
Assumed second offset	3000 ft or m
Calculated second time	$[2^2 + (3000/6000)^2]^{0.5} = [4.25]^{10.5}$ = 2.061552813
Correlation coefficient	1
Calculated zero offset time	2
Calculated NMO velocity	6000 ft/s or m/s
Enter offset of 2000	
Reflection time	2.0276 s
Enter offset of 4000	
Reflection time	2.1082 s

Program for Pelton nonlinear sweep function

This program calculates the frequency and rate of change of frequency at any time during a sweep using a Pelton, nonlinear sweep generator.

Label	D
Input	Start frequency, end frequency, pre-emphasis, sweep length, and the time during the sweep.
Output	Frequency at the input time and the rate of change of frequency at that time.

Procedure and key strokes

Key stroke	Procedure
2nd CMS	
	Enter start frequency
STO	
00	
	Enter end frequency

(*Continued*)

Key stroke	Procedure
STO	
01	
	Enter pre-emphasis
STO	
02	
	Enter sweep length
STO	
03	
	Enter time
D	
	Program runs. Read frequency.
X↔T	
	Read rate of change of frequency.

First example:		
Start frequency	10 Hz	10 Hz
End frequency	100 Hz	100 Hz
Pre-emphasis	6 dB	6 dB
Sweep length	10 s	10 s
Time *t*	1 s	9 s
Frequency at time *t*	33.015 Hz	94.921 Hz
Rate of change of frequency	14.993 Hz/s	5.214 Hz/s
Second example:		
Start frequency	12 Hz	12 Hz
End frequency	90 Hz	90 Hz
Pre-emphasis	8 dB	8 dB
Sweep length	12 s	12 s
Time *t*	1 s	9 s
Frequency at time *t*	32.319 Hz	79.664 Hz
Rate of change of frequency	12.479 Hz/s	3.748 Hz/s

Program for Texas Instruments nonlinear sweep function

This program calculates the frequency and rate of change of frequency at any time during a sweep using a Texas Instruments, logarithmic sweep generator.

Label	D′
Input	Start frequency, end frequency, sweep length, and the time during the sweep.

Output	Frequency at the input time and the rate of change of frequency at that time.

Procedure and key strokes

Key stroke	Procedure
2nd CMS	
	Enter start frequency
STO	
00	
	Enter end frequency
STO	
01	
	Enter sweep length
STO	
03	
	Enter time
2nd D′	
	Program runs. Read frequency.
X↔T	
	Read rate of change of frequency.

Note that memory 02 is not used in this program.

First example:

Start frequency	10 Hz	10 Hz
End frequency	100 Hz	100 Hz
Sweep length	10 s	10 s
Time *t*	1 s	9 s
Frequency at time *t*	36.016 Hz	94.423 Hz
Rate of change of frequency	18.766 Hz/s	3.753 Hz/s

Second example:

Start frequency	12 Hz	12 Hz
End frequency	90 Hz	90 Hz
Sweep length	12 s	12 s
Time *t*	1 s	9 s
Frequency at time *t*	32.079 Hz	82.022 Hz
Rate of change of frequency	15.205 Hz/s	3.041 Hz/s

Program listing

Step no.	Key	Comments
000	LBL	Program for apparent horizontal velocity of reflections
001	A	
002	1/X	
003	STO	
004	02	1/offset
005	RCL	
006	01	t
007	PRD	
008	02	t/offset
009	RCL	
010	00	
011	PRD	
012	02	vt/offset
013	RCL	
014	02	
015	X^2	$[vt/\text{offset}]^2$
016	+	
017	1	
018	=	
019	$X^{0.5}$	$\{1 + [vt/\text{offset}]^2\}^{0.5}$
020	x	
021	RCL	
022	00	
023	=	$v\{1 + [vt/\text{offset}]^2\}^{0.5}$
024	R/S	
025	LBL	Program to calculate differential NMO across an array and the frequencies where the array response drops to 64, 30, and 0 percent of maximum.
026	A′	
027	–	
028	X↔T	offset – array length in T register
029	STO	
030	02	array length
031	/	
032	2	
033	=	
034	STO	
035	03	offset – (array length)/2
036	/	

Program listing (*Continued*)

Step no.	Key	Comments
037	RCL	
038	00	
039	=	
040	X↔T	offset − array length in display [offset − (array length)/2]/v in T register
041	RCL	
042	01	t
043	INV	
044	P⇒R	
045	X↔T	$\{t^2 + [(\text{offset} - \text{array length}/2)/v]^2\}^{0.5}$
046	EXC	
047	03	$\{t^2 + (1/v)^2[\text{offset} - (\text{array length}/2)]^2\}^{0.5} = (t_a)$ in 03
048	+	
049	RCL	
050	02	
051	=	offset + (array length)/2
052	/	
053	RCL	
054	00	
055	=	[offset + (array length)/2]/v
056	X↔T	[offset + (array length)/2]/v in T register
057	RCL	
058	01	t
059	INV	
060	P⇒R	
061	RCL	
062	02	offset + (array length)/2
063	X↔T	$\{t^2 + (1/v)^2[\text{offset} + (\text{array length})/2]^2\}^{0.5} = t_b$
064	−	
065	RCL	
066	03	
067	=	$t_b - t_a = \Delta t$
068--------	R/S-------	differential arrival time across array = Δt
069	x	
070	2	
071	=	$2\Delta t$
072	1/X	$1/2\Delta t$
073---------	R/S----------	frequency where response is 64% = $f_{0.64}$
074	x	

Program listing (*Continued*)

Step no.	Key	Comments
075	1	
076	.	
077	5	
078	=	$3/4\Delta t$
079--------	R/S---------	frequency where response is 30% = $f_{0.30}$
080	x	
081	4	
082	/	
083	3	
084	=	$1/\Delta t$
085---------	R/S---------	frequency where response drops to zero = f_0
086	LBL	Program for 1/2 the Fresnel zone radius (horizontal resolution)
087	B	
088	STO	
089	03	f
090	X↔T	f in T register
091	STO	
092	02	t
093	x	
094	RCL	
095	00	tv
096	x	
097	RCL	
098	01	tvv_i
099	/	
100	4	
101	/	
102	RCL	
103	03	
104	=	
105	STO	
106	04	$tvv_i/4f$
107	RCL	
108	00	v
109	/	
110	4	
111	/	
112	RCL	

Program listing (*Continued*)

Step no.	Key	Comments
113	03	
114	=	$v/4f$
115	X^2	$(v/4f)^2$
116	SUM	
117	04	$(v/4f)^2 + tvv_i/4f$
118	RCL	
119	02	t
120	X↔T	t in T register
121	RCL	
122	04	$(v/4f)^2 + tvv_i/4f$
123	$X^{0.5}$	$[(v/4f)^2 + tvv_i/4f]^{0.5}$
124	/	
125	2	
126	=	$(1/2)[(v/4f)^2 + tvv_i/4f]^{0.5}$
127----------	R/S---------	1/2 the Fresnel radius
128	LBL	Uphole velocity survey analysis
129	B′	
130	STO	
131	03	uht (uphole time)
132	CLR	
133	X↔T	receiver depth
134	−	
135	RCL	
136	00	shot depth
137	=	
138	STO	
139	05	receiver depth − shot depth
140	R/S	
141	RCL	
142	02	offset
143	X^2	
144	+	
145	RCL	
146	05	
147	X^2	
148	=	$(\text{receiver depth} - \text{shot depth})^2 + (\text{offset})^2$
149	$X^{0.5}$	slant path
150	1/X	1/slant path

Program listing (*Continued*)

Step no.	Key	Comments
151	x	
152	RCL	
153	05	(rec. depth – shot depth)
154	x	
155	(	
156	RCL	
157	03	uht
158	–	
159	RCL	
160	01	time break = tb
161	)	
162	=	(rec. depth – shot depth)(uht-tb)/slant path
163	STO	
164	06	
165----------	R/S----------	corrected uphole time
166--------	RCL	
167	05	
168	/	
169	RCL	
170	06	
171	=	(rec. depth – shot depth)/(corrected uht)
172---------	R/S--------	average velocity
173	LBL	Data entry of pairs of reflection times and offsets for calculations of NMO velocity, zero offset reflection time, and reflection times for other offsets
174	C	
175	X^2	
176	$X \leftrightarrow T$	
177	X^2	
178	$\Sigma +$	enter reflection time and offset pair
179	R/S	
180	LBL	Calculations of NMO velocity, zero offset reflection time, and reflection times for other offsets
181	C′	
182	OP	
183	13	
184----------	R/S----------	correlation coefficient of the data set
185	OP	
186	12	
187	$X^{0.5}$	
188------------	R/S----------	zero offset reflection time
189	$X \leftrightarrow T$	

Program listing (*Continued*)

Step no.	Key	Comments
190	$X^{0.5}$	
191	1/X	
192----------	-R/S---------	NMO velocity Input offset value into display
193	LBL	
194	NOP	Return point in program
195	X^2	
196	OP	
197	14	
198	$X^{0.5}$	
199-----------	- R/S-----------	-reflection time for offset input above
200	GTO	
201	NOP	Return to NOP point in program

202	LBL	Program to calculate frequency and rate of
203	D	change of frequency for a Pelton nonlinear sweep
204	STO	
205	07	time
206	RCL	
207	02	pre-emphasis
208	/	
209	6	
210	+	
211	1	
212	=	
213	STO	
214	06	1 + pre-emphasis/6 = C
215	RCL	
216	00	start frequency = f_1
217	Y^x	
218	RCL	
219	06	
220	=	
221	STO	
222	05	$f_1^C = B$
223	+/−	
224	STO	
225	04	$-f_1^C$
226	RCL	
227	01	end frequency = f_2

Program listing (*Continued*)

Step no.	Key	Comments
228	Y^x	
229	RCL	
230	06	
231	=	f_2^C
232	SUM	
233	04	$f_2^C - f_1^C$
234	RCL	
235	03	sweep length = SL
236	INV	
237	PRD	
238	04	$(f_2^C - f_1^C)/(\mathrm{SL}) = A$
239	RCL	
240	07	t
241	x	
242	RCL	
243	04	
244	+	
245	RCL	
246	05	
247	=	$At + B$
248	Y^x	
249	(	
250	RCL	
251	06	
252	1/X	
253	−	
254	1	$1/C - 1$
255	)	$(At + B)^{1/c-1}$
256	x	
257	RCL	
258	04	
259	/	
260	RCL	
261	06	
262	=	$[A/C](At + B)^{1/c-1} = \mathrm{d}F/\mathrm{d}t$
263	PAU	$\mathrm{d}F/\mathrm{d}t$ = rate of change of frequency in Hz/s
264	X↔T	
265	RCL	
266	07	t
267	x	
268	RCL	

Program listing (*Continued*)

Step no.	Key	Comments
269	04	*At*
270	+	
271	RCL	
272	05	
273	=	$At + B$
274	Y^x	
275	RCL	
276	06	
277	1/X	
278	=	$(At + B)^{1/c}$
279------------	R/S -----------	Frequency

280	LBL	Program to calculate frequency and rate
281	D′	of change of frequency for a Texas
282	STO	Instruments Nonlinear Sweep
283	05	time t in register 05
284	RCL	
285	01	f_2 ending frequency
286	−	
287	RCL	
288	00	f_1 starting frequency
289	=	$f_2 - f_1$
290	/	
291	(	
292	RCL	
293	03	sweep length = SL
294	+	
295	1	
296	)	
297	LNx	
298	=	
299	STO	
300	04	$(f_2 - f_1)$/Ln(1 + SL)
301	1	
302	SUM	
303	05	$1 + t$ in 05
304	RCL	
305	05	$1 + t$
306	1/X	
307	x	

Program listing (*Continued*)

Step no.	Key	Comments
308	RCL	
309	04	
310	=	$(f_2 - f_1)/(1+t)\mathrm{Ln}(1+\mathrm{SL}) = \mathrm{d}F/\mathrm{d}t$
311	X↔T	$(f_2 - f_1)/(1+t)\mathrm{Ln}(1+\mathrm{SL})$ in the T register
312	RCL	
313	05	$1+t$
314	LNx	
315	x	
316	RCL	
317	04	
318	+	
319	RCL	
320	00	
321	=	$f_1 + \mathrm{Ln}(1+t)\{(f_2 - f_1)/[\mathrm{Ln}(1+\mathrm{SL})]\}$
322------------	R/S---------	frequency

APPENDIX G—WAVELETS, PHASE AND DEGREE

In this appendix we discuss wavelets that are minimum-phase, zero-phase, and mixed-phase. For the zero-phase wavelets we differentiate between wavelets that are zero degree and ninety degree or some degree between these two limits.

Minimum-phase wavelets

Before the development of the Vibroseis technique, geophysics was simpler in that we dealt with only minimum-phase wavelets. For an impulsive seismic source, such as a dynamite explosion, there is no source-generated signal prior to detonation. The resulting signal wavelets in the earth are said to be causal wavelets since they are zero prior to detonation and remain zero until the source-generated wavelet propagates from the source to the point of observation. The propagation time is equal to the path length from the source to a point of observation divided by the group velocity (Chapter 22) of the fastest component of the received wavelet that is significant. The wavelets are said to be causal in that there must be an input signal to cause an output signal. For a minimum-phase wavelet the phase of every frequency component of an output wavelet is the minimum phase that will allow the composite wavelet to be zero for all times prior to the excitation time plus the propagation time. Hence the name 'minimum-phase wavelet.'

The equation for a minimum-phase wavelet is a Fourier Integral of the form:

$$W(t) = A_1 \cos(2\pi f_1 t + \phi_1) + A_2 \cos(2\pi f_2 t + \phi_2) + A_3 \cos(2\pi f_3 t + \phi_3) + \cdots + A_n \cos(2\pi f_n t + \phi_n)$$

where f_n takes on all values from zero to infinity including all non-integer values. The constants $A_1, A_2 \cdots A_n$ and $\phi_1, \phi_2 \cdots \phi_n$ determine the shape and amplitude of the minimum-phase wavelet and the constants $\phi_1, \phi_2 \cdots \phi_n$ have the minimum values that will result in a causal wavelet. (It is beyond the scope of this book to determine the values of these phase angles.)

The earth acts like a low-pass filter which attenuates all frequencies and the higher the frequency the greater the attenuation. The earth, like all physically realizable filters, is a minimum-phase filter. Hence a minimum-phase wavelet propagating through the earth remains a minimum-phase wavelet. The geophone or hydrophone acts like a band-pass filter and, being a physically realizable filter, it is minimum-phase. The amplifiers and the galvanometers are minimum-phase if they are not overdriven. The output wavelets on the old analog records would have been minimum-phase except for the automatic gain control (AGC). In analog recording AGC was absolutely essential, but since it is nonlinear it affected the phase in some undetermined manner. Even so, the signal remained causal. One might argue that the dynamite source is nonlinear and hence the input signal might not be minimum-phase. But the nonlinear effects occur within a fraction of a millisecond and within just a few feet of the explosion. The response to the detonation is causal and the assumption of minimum-phase is justified beyond the nonlinear zone.

With the advent of digital processing of seismic data there was an opportunity to alter the phase of the wavelets. But early digital filters for seismic data processing were designed to be minimum-phase and the AGC used in processing was probably no worse that the AGC used in the field. In modern digital recording systems there is no field AGC (a credit), but unfortunately the digital filters used in processing minimum-phase data are frequently zero-phase (a debit). In processing, gain compensation can be handled in a variety of ways (a credit) making the use of AGC optional.

It is often assumed by seismologists that dispersion is zero and that all seismic frequency components have precisely the same velocity. Dispersion may be quite low, but it cannot be zero. If dispersion were zero, then the Q of the earth would be infinite and nobody makes that claim. Therefore, the propagation time is the travel time for the frequency component with the highest group velocity (Chapter 22) that is present in the measured wavelet at a significant level. In seismology, the fastest frequency component for body waves (as opposed to ground roll, Love waves, Stoneley waves, etc.) will be the highest frequency component that is significant in the received wavelet.

What is meant by 'significant'? Consider the case of a velocity survey of a well where we shoot explosives in shallow holes and record the first arrivals at various depths in the well. What we are trying to do is to measure the onset

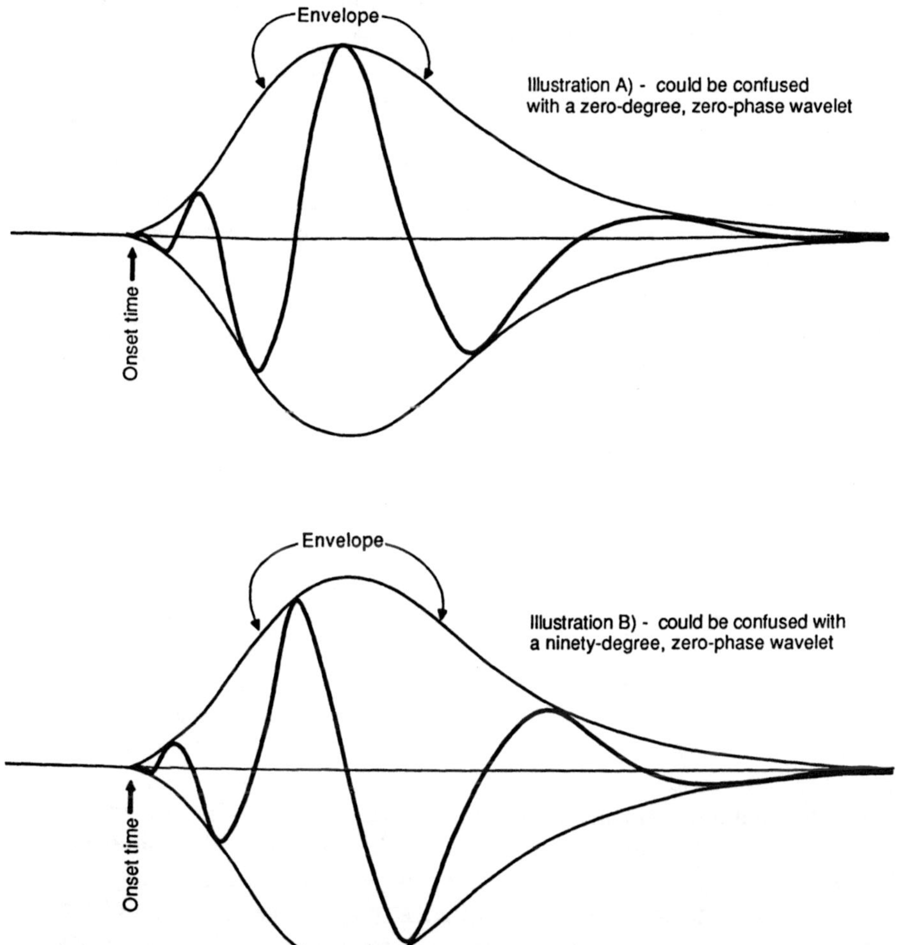

Fig. G-1 Minimum-phase wavelets. Minimum-phase wavelets (as opposed to zero-phase wavelets) are 'causal'. That is, they result from a disturbance and hence cannot precede the disturbance in time. The onset time is delayed after the disturbance time by the propagation distance divided by the fastest group velocity of the significant frequency components of the received wavelet. In a typical earth the velocity increases with frequency, so that the leading part of a propagated wavelet is dominated by the higher-frequency components.

times of the minimum-phase wavelets. Under ideal conditions of very low noise we can use a low gain for the amplifiers and measure an onset time. But if we turn up the gain 10-fold or 100-fold and measure the onset time again, we are very likely to measure a slightly shorter onset time. By turning up the gain we are making some higher-frequency components in the received wavelet become significant. In other words, to some degree, significance, like beauty, is in the eye of the beholder. Not very scientific, but true. (Shooting heavier charges may or may not shorten the onset time because heavier charges may actually decrease the relative amplitude of the higher frequencies due to the finite time of the explosion and due to nonlinear effects near the shot.) If the records are noisy, then it is more difficult to pick the onset time and it is unlikely that we could note any change in this onset time with gain, since the noise would be amplified along with the signal. Usually at seismic frequencies the dispersion is small enough so that the difference in measured onset times is relatively small.

Figure G-1 illustrates wavelets that are approximately minimum-phase. (The wavelets shown are sketches for illustrative purposes and are not precisely calculated, minimum-phase wavelets.) In these diagrams the envelope of the wavelets is also indicated. The envelopes of minimum-phase wavelets are not symmetrical about a center time but rather the envelope expands rapidly from the onset time and then decays more slowly with an elongated tail (Fig. 22.4b).

Zero-phase wavelets

With the advent of Vibroseis came the practical opportunity to produce seismic records with zero-phase wavelets. Several calculated and precisely displayed zero-phase wavelets are shown in Fig. H-3 (Appendix H). Figure G-2 illustrates three wavelets. The unit impulse is the strange wavelet that is both minimum-phase and zero-phase (of zero degree). The unit impulse has components at all frequencies from zero to infinity and can be represented by a Fourier integral (previous section) of cosine terms with the phase angle of every term being zero degrees. This makes it a zero-phase wavelet of zero degree, but it is also causal in that it has a discrete onset time. If we pass the unit impulse through a minimum-phase filter it remains a minimum-phase

Fig. G-2 Sketches illustrating minimum-phase wavelets and zero-phase wavelets. The unit impulse is both a minimum-phase and a zero-phase wavelet (of zero degree). Both the unit impulse and the rectangular pulses of finite widths have infinite bandwidths. If these pulses are filtered so that they are then band-limited, they become pulses with theoretically infinite or semi-infinite widths.

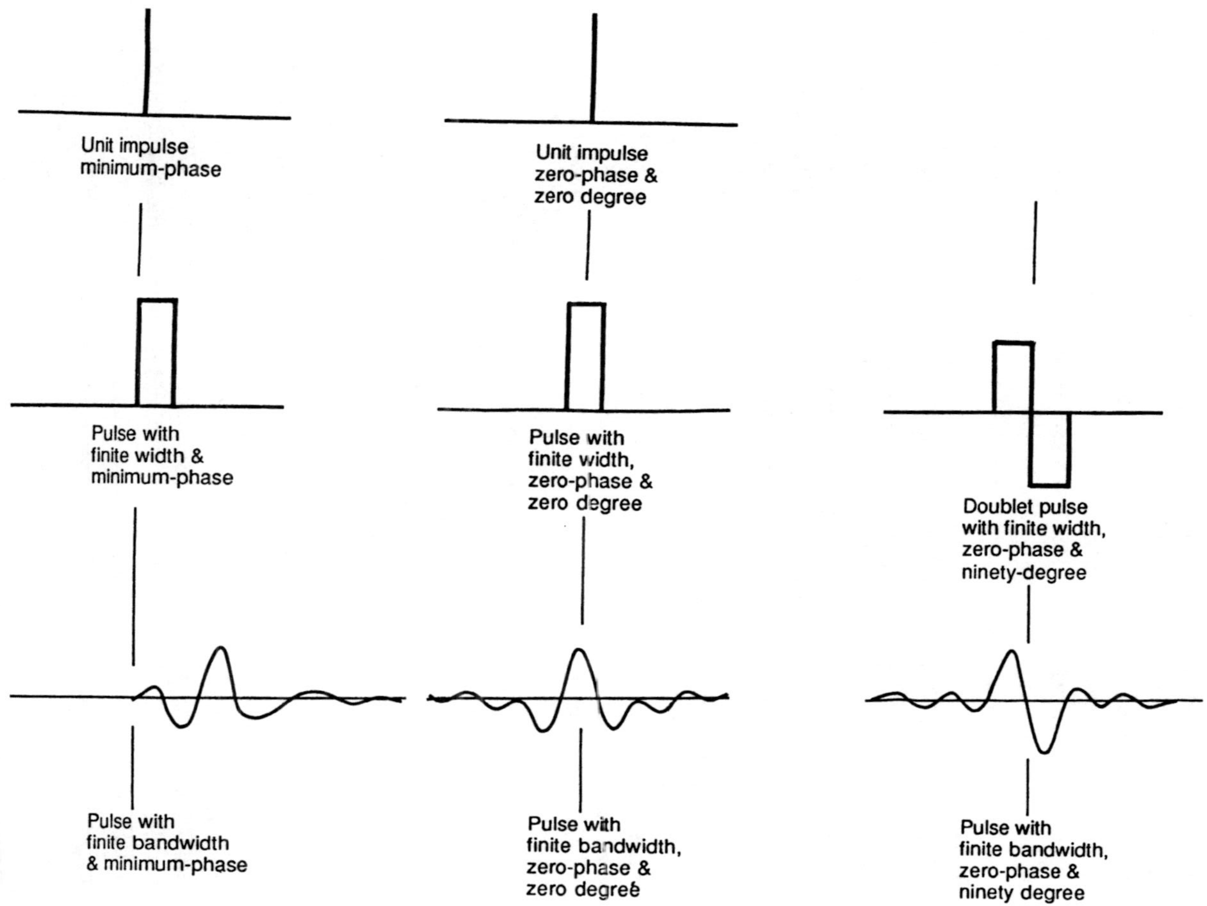

Unit impulse
minimum-phase
Pulse with
finite width &
minimum-phase
Pulse with
finite bandwidth
& minimum-phase
Unit impulse
zero-phase &
zero degree
Pulse with
finite width,
zero-phase &
zero degree
Pulse with
finite bandwidth,
zero-phase &
zero degree
Doublet pulse
with finite width,
zero-phase &
ninety-degree
Pulse with
finite bandwidth,
zero-phase &
ninety degree

wavelet with a discrete onset time; however, it is no longer a zero-phase wavelet. This is illustrated at the bottom of the first column of Fig. G-2. If we use a computer and apply a zero-phase filter to a unit impulse, it remains a zero-phase wavelet of zero degree, but it is no longer causal in that there are ripples that precede the previous onset time. Hence it is no longer a minimum-phase wavelet. This is illustrated at the bottom of the second column of Fig. G-2. A zero-phase wavelet of ninety degree is illustrated in the third column of Fig. G-2.

Zero-phase wavelets have a theoretical onset time at minus infinity. But here comes that word 'significant' again. As we move farther in time either way from the center of a zero-phase wavelet, the side lobes become ripples that diminish until they are no longer significant. But there is not even an approximate, finite time at which the wavelets can be said to begin, even if noise is quite low and even if there are no other events to overlap in time. Whatever time we may pick for the beginning time of a zero-phase wavelet would be extremely sensitive to the noise level; therefore there is no point in trying to pick a beginning or onset time for zero-phase wavelets.

Zero-phase wavelets of zero degree can be represented by a Fourier integral of cosine terms (previous section) identical in form to the Fourier integral that represents a unit impulse with all phase angles being zero, hence the designation of zero degree. The values of the constants $A_1, A_2, A_3, \ldots A_n$, etc. will be different from those for a unit impulse. For very high frequency components the constants will be essentially zero, and the component for zero frequency will be zero. Since the Fourier integral that represents a zero-phase wavelet of zero degree has only cosine terms with phase angles of zero degree it is not surprising that it has cosine symmetry about the central time of the wavelet.

All zero-phase wavelets of ninety degree can be represented by the Fourier integral but all the phase angles are ninety degrees, hence the designation of ninety degree. Since the cosine of ninety degrees is the sine of zero degrees, it is not surprising that zero-phase wavelets of ninety degree have a symmetry about the central time similar to the symmetry of a sine wave. Zero-phase wavelets may have degrees other than zero and ninety. But the phase angles will all be identical. For example, a zero-phase wavelet may have a degree of forty-five which means that all of the constants $\phi_1, \phi_2 \cdots \phi_n$ are 45°.

Mixed-phase wavelets

Wavelets may be neither minimum-phase nor zero-phase, in which case they are called mixed-phase wavelets. In this case the constants $\phi_1, \phi_2 \cdots \phi_n$ are not identical. Therefore mixed-phase wavelets are not the same as zero-phase wavelets of some degree that is intermediate between zero and ninety. Mixed-phase wavelets can be of almost any shape (if multi-valued at any one time these values must be continuous).

If we passed a zero-phase wavelet through a minimum-phase filter (such as propagation through the earth) it would become mixed-phase. It may remain dominantly zero-phase for those frequency components where the filter response was flat (the phase response of the filter would be nearly linear for those frequencies), but for frequencies where the filter response was not flat, the phase of the output wavelet would definitely be mixed. Similarly if we passed a minimum-phase wavelet through a zero-phase filter the output wavelet would be mixed-phase. The phase of the output wavelet may remain dominantly minimum-phase for those frequency components where the filter response was flat.

Determining the phase of wavelets in seismic data

Note that the minimum-phase wavelet illustrated in Fig. G-1 A looks something like a zero-phase wavelet of zero degree whereas the minimum-phase wavelet illustrated in Fig. G-1 B looks more like a zero-phase wavelet of ninety degree. In the presence of noise or in the presence of other wavelets not well-separated in time, we may have difficulty, by visual inspection, in deciding whether our wavelets are zero-phase or minimum-phase. Even with computers it is difficult if wavelets overlap in time (for example, because of reflection interfaces that are closely spaced), or if the S/N ratio is low. Usually it is difficult to determine whether wavelets are minimum-phase or zero-phase and, if zero, of what degree. There are computer programs designed to extract this information, but they may or may not succeed. Some programs assume that the wavelets are zero-phase and simply try to solve for the degree. Since the earth filter is minimum-phase, this is not a good assumption and often it is very inappropriate.

In one case the author's team ran a line directly over a well location where the sonic log showed an ideal interface with an abrupt velocity increase of 60 per cent. The sonic log showed few, minor changes in velocity within 600 ft above and below this abrupt change in velocity. The two-way travel time from this interface to the nearest significant velocity change was more than 0.1 s. On this line we recorded beautiful data with a high S/N ratio, wide bandwidth, and excellent resolution (because of the use of nonlinear sweeps that put more energy into the earth at the higher frequencies and because of the spectral shaping done in processing). The effective frequency band was 20–120 Hz; therefore the periods of the lowest, average and highest frequency components of the reflected wavelet were 0.05, 0.0143, and 0.00833 s respectively. We used a minimum-phase correlation. The reflection from this interface on our seismic data was a beautiful example of a classic minimum-phase wavelet not only near the well but for quite some distance to either side. Also the polarity was correct. (Often we worry much about phase when polarity is questionable, or we only think we know polarity.)

Until that time we were less than sure what kind of wavelets we were actually getting back from the earth and displaying on our sections. They were supposed to be minimum-phase and should have been minimum-phase, but our interpreter was 'sure' they were zero-phase and zero-degree. The point is that it is difficult and often impossible to determine from field data the phase of wavelets and, for zero-phase wavelets, to determine the degree. To find stratigraphic traps we need to model the earth and the earth changes we seek, but we need to use the same wavelets for our model as we obtain from the earth (after processing).

If we do not use the same wavelets for the model as we measure from the earth, then a match between an earth response and a model response means that the model does *not* precisely represent the actual earth. But the difference may or may not be significant, depending upon the phase difference. A phase difference of 45° represents a time difference of only 0.833 ms at a frequency of 150 Hz but it represents a time difference of 8.33 ms at 15 Hz. Small phase deviations at high frequencies may not be too critical.

It is helpful if we are lucky enough to have a known, ideal, isolated interface to provide an answer which, hopefully, is consistent with expectations. We should look for such opportunities and take advantage of them.

APPENDIX H—VIBROSEIS

In this appendix we discuss correlation of Vibroseis wavelets to achieve either zero-phase of minimum-phase wavelets. For zero-phase correlated wavelets we discuss wavelets of zero and ninety degrees and the shape of the resulting wavelets. We also discuss mixed-phase wavelets. Other topics discussed include: harmonic ghosts, reasonable standards for vibrator phase, and vibrator design for performance at low frequencies.

Correlation

Correlation is a digital procedure by which wavelets are compared for the degree of correspondence between them. In the Vibroseis technique the long sweep signal is correlated with the earth response to produce a usable record with relatively compact wavelets for reflections from interfaces at depth. In seismology we deal with continuous waves with continuous derivatives, but it is easier to illustrate correlation with square waves of unit amplitude. Figure H-1 illustrates the procedure for correlating a variable period, unit

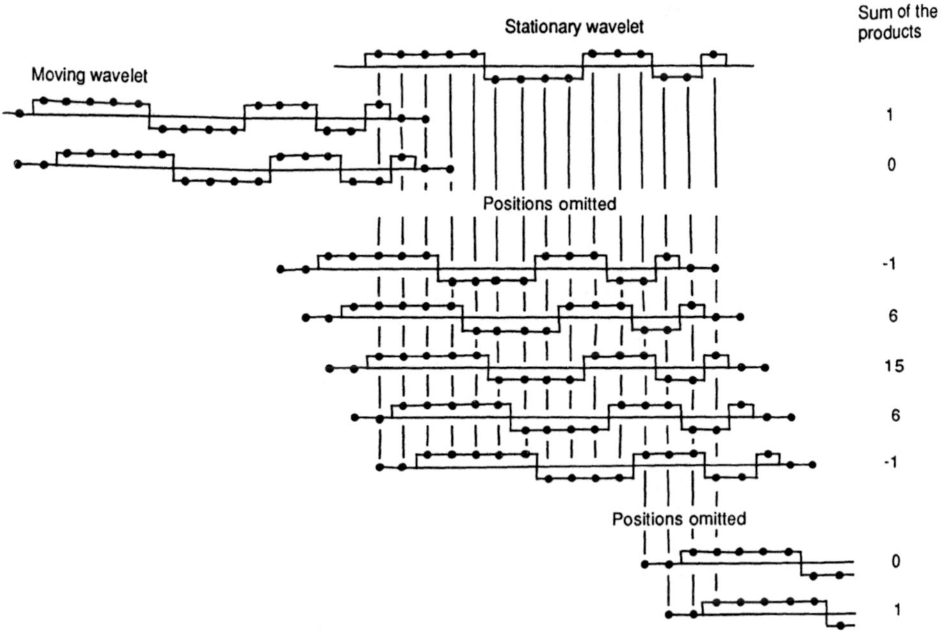

Fig. H-1 Example of an autocorrelation using square waves of unit amplitudes. Note:
1. The digitized sampling points are indicated by dots;
2. A complete autocorrelation begins and ends with both wavelets at zero with zero derivatives. The shift increment is uniform and continuous without any breaks. The breaks in the figure are to simplify the illustration.

square wave with itself—an autocorrelation. To correlate two wavelets you multiply their values at the sampled points and sum these products to produce one value of the correlated wavelet. Then you shift one wavelet with respect to the other by one sample increment and repeat. This procedure of multiplying and summing the products and then shifting one sample increment is carried out from a position of one wavelet well ahead of the second until this first wavelet is well behind the second. If the two wavelets are not identical then the result is a crosscorrelation.

Figure H-1 illustrates a few shift positions and the sums of the products for these shift positions. The wave train at the top is held stationary and the calculations to obtain the autocorrelation are shown for several positions of the moving wave train. The wiggly lines indicate gaps in the illustrated shift sequence. Table H-1 lists these sums of products for this unit square wave example and these sums of the products are plotted in Fig. H-2. The result is a correlated wavelet with a relatively compact central peak with high amplitude with broad side lobes of relatively small amplitude. This particular correlated

Table H-1 Autocorrelation values for Fig. H-1

Times	*Value*
$t-17$ and $t+17$	0
$t-16$ and $t+16$	0
$t-15$ and $t+15$	0
$t-14$ and $t+14$	+1
$t-13$ and $t+13$	0
$t-12$ and t+12	−1
$t-11$ and $t+11$	0
$t-10$ and $t+10$	+1
$t-9$ and $t+9$	0
$t-8$ and $t+8$	+1
$t-7$ and $t+7$	+2
$t-6$ and $t+6$	−1
$t-5$ and $t+5$	−2
$t-4$ and $t+4$	−5
$t-3$ and $t+3$	−4
$t-2$ and $t+2$	−1
$t-1$ and $t+1$	+6
t	+15

wavelet is zero-phase and zero degree (Appendix G and Chapter 3). It has cosine symmetry which means that it can be represented by a sum of cosine waves with phase angles of zero degrees. This type of symmetry is like that of an object with its mirror image.

After a long Vibroseis sweep signal is reflected from an interface at depth, the reflected sweep signal arrives at a surface geophone. Correlation of the input sweep signal with the response from the geophone results in a correlated wavelet centered at the time delay for that reflection.

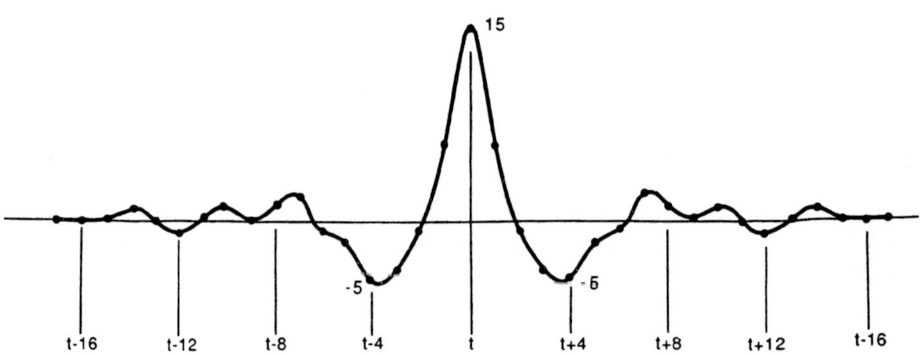

Fig. H-2 Autocorrelation for the unit square waves example. Note: The digitized sample points are indicated by dots

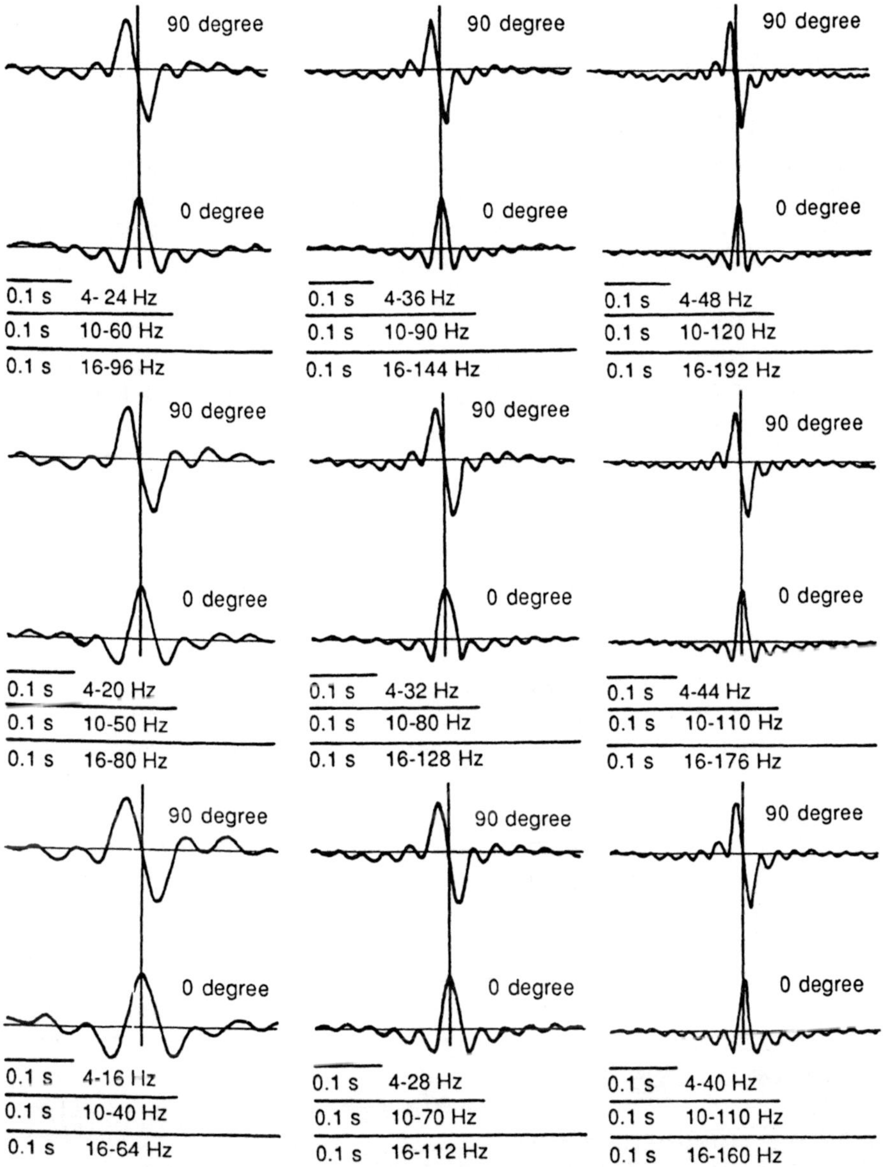

Fig. H-3 Correlated Vibroseis wavelets for various sweeps. Note:

1. The frequency bands of the wavelets are determined by the time scales that are used;
2. The larger the ratio of the upper frequency limit to the lower frequency limit the better the pulse shape in terms of the amplitude ratio of the main lobe, or lobes, to the side lobes.

Zero-phase Vibroseis wavelets

Figure H-3 shows some vibroseis wavelets generated by autocorrelations of various sweeps. The time scale chosen determines the frequency band for each illustrated wavelet. Other frequency bands can be obtained by using different time scales. The low-frequency limit is 4 Hz for a time scale of two large divisions per 100 ms, but it is 6 Hz for three large divisions per 100 ms, 8 Hz for four large divisions per 100 ms, etc. All of the wavelets shown in Fig. H-3 are zero-phase wavelets as opposed to being minimum-phase or mixed-phase wavelets. To specify a zero-phase wavelet we also must specify the degree. Both zero and ninety-degree, zero-phase wavelets are shown here. The bandwidth ratio (the maximum frequency divided by the minimum frequency) for the wavelets shown ranges from 4(2 octaves) to 12(3.585 octaves). The greater the ratio bandwidth the better the wavelet in terms of the ratio of the amplitude of the main lobe (zero degree wavelet) or lobes (ninety-degree wavelet) to the amplitudes of the side lobes.

The average of the minimum and the maximum frequencies is the average frequency which has a period that is very close to the period of the main lobes of the ninety-degree wavelet as measured between zero crossings. The period of the main lobe of the zero degree wavelet could be determined by doubling the time between the first zero crossings to either side of the main lobe. In this case, this period is very close to the period of the average frequency. Zero crossing are not easy to determine visually on field records (but are easy with the computer) and may not be that easy on final stacked sections depending upon the type of display. For zero-degree wavelets we can take the time period between the maximum amplitudes of the first side lobes to either side of the central peak as a measure of the period. This will yield a shorter period and hence indicate a somewhat higher frequency. Specifically the period will be about 88 per cent of the period of the average frequency and the reciprocal of that period will be about 14 per cent higher than the average frequency. For the ninety-degree wavelets, if the period is estimated by doubling the time between the main crest and trough, then the period will be somewhat less than the period of the average frequency, and the reciprocal of this estimate will be slightly greater than the average frequency.

Ninety-degree wavelets have a symmetry like the symmetry of a sine wave. These wavelets can be represented by a Fourier integral (Appendix G) containing only cosine waves with phase angles of ninety degrees, hence the name.

For a zero-phase seismic record the author's preference is for a zero degree wavelet with a polarity such that a reflection from an isolated interface with a positive reflection coefficient results in a positive central crest. Thus a refraction from an isolated refractor should also display a positive central crest.

Minimum-phase correlation for minimum-phase wavelets

As was pointed out in Chapter 5 and Appendix G (and mentioned in Chapter 17), one can use a minimum-phase correlation of Vibroseis records and have minimum-phase reflection wavelets instead of zero-phase wavelets like those shown in Fig. H-3. This is particularly advantageous if we plan to record some portions of lines with impulsive sources and other portions with vibrators. Since the earth, geophone and analog filters in the instruments are all minimum-phase, we should use a minimum-phase correlation and minimum-phase processing to get minimum-phase reflection wavelets. It is highly advantageous for stratigraphic work to know the phase of the wavelets; therefore the use of a minimum-phase correlation and minimum-phase processing should be advantageous over methods where the phase is less well known. A filtered sweep should not be used for a minimum-phase correlation (next section).

Mixed-phase wavelets

Suppose that we do a perfect job of injecting a Vibroseis sweep signal into the earth, we correlate our reflection data with a zero-phase correlation, and process the data with zero-phase programs. The reflection wavelets will be mixed-phase rather than zero-phase because of the minimum-phase filtering by the earth. They may approach zero-phase if we limit the upper frequency of the data to a frequency much lower than the effective cut-off frequency that is imposed by the earth filter. But ordinarily we want to use frequencies up to (or very close to) the effective cut-off frequency of the earth in order to achieve as much vertical resolution as possible. In high-resolution work, particularly where we use nonlinear sweeps to offset severe earth attenuation of the higher frequencies, the minimum-phase filter of the earth will result in mixed-phase data. In fact, for the higher frequencies the phase may become more nearly minimum-phase than zero-phase except for very shallow reflections.

However, if the problem were confined to just the highest frequencies, it might not be too significant to explorationists. For example, a 45° phase shift for a 100 Hz frequency component is a time shift of only 1.25 ms. Is that really going to be very serious? It would change a wavelet shape somewhat, but not much. But the deviation of phase will not be limited to just the higher frequencies. Studies need to be made to determine the severity of the problem from a practical standpoint. Results from such studies of mixed-phase wavelets would have an impact on field geophysicists in their use of filters and geophones.

For a near zero-phase wavelet the processor needs from the field a sweep that is filtered just like the data are filtered. The phase effects of the filters are offset when the data are correlated with a filtered sweep. But if a minimum-

Table H-2 Harmonic ghost example

Sweep: Linear downsweep from 52 Hz to 12 Hz over 10 seconds

Time (s)	*Sweep frequency* (Hz)		*Second harmonic* (Hz)
−10	52		104
−9	48		96
−8	44		88
−7	40		80
−6	36		73
−5	32	5 s delay	64
−4	28		56
−3	24	4 s delay	48
−2	20		40
−1	16	3 s delay	32
0	12		24

phase correlation is to be used to get a near minimum-phase wavelet, a sweep that has not been filtered should be provided and used for the correlation (so that the minimum-phase effects of physically realizable filters will not be canceled in correlation). The safest thing for the field geophysicist to do is to provide both a filtered and an unfiltered sweep so that if someone changes his mind later, the door is still open to process the data to best meet the needs of the interpreter. In the past the author was guilty of negligence and oversight in this matter and calls it to your attention to help you to do better.

As regards the geophones, if zero-phase wavelets are to be used, the data should be recorded with geophones with resonant frequencies that are well below the lower-frequency limit of the sweep. As an alternative, or in addition, the phase effects of the geophones can approximately be corrected for in processing. No geophone phase correction should be made with minimum-phase correlation and processing.

Harmonic ghosts

Harmonic ghosts are described briefly in Chapter 5. Vibrators do not couple perfect sinewaves into the earth and harmonic frequencies are generated throughout each sweep. For compressional-wave work the main problem is with the second harmonics. (In shear-wave work the main problem is usually with the third harmonic. The reason is that the push to one side by a horizontal vibrator to generate shear waves is likely to be very much the same as the push to the other side and if distortion occurs it affects wave peaks and troughs identically. Waves that are distorted, but have symmetry about zero amplitude do not contain even harmonics; therefore with horizontal vibrators the first significant harmonic is usually the third.)

Table H-2 illustrates harmonic ghosts with the example of a downsweep from 52 to 12 Hz over a period of 10 s. The formula for the minimum ghost delay time for a linear downsweep is the lowest frequency divided by the rate of change of frequency during the sweep. In this case the lowest frequency is 12 Hz and the rate of change of frequency is $(52 - 12)/10 = 4$ Hz/s. The minimum ghost delay time is then $12/4 = 3$ s. Note that this is consistent with Table H-1. There is a 3 s delay between the time the sweep signal is 24 Hz and the time the second harmonic of the sweep is 24 Hz. Then every event on the record that has a component with a frequency of 24 Hz will have a harmonic ghost 3 s later. Similarly, there is a 4 s time delay between a 32 Hz component in the sweep and the time the second harmonic of the sweep is 32 Hz. Therefore the 32 Hz component of the harmonic ghosts will lag 4 s behind every event with a 32 Hz frequency component.

If we use a long sweep, we can make the minimum ghost delay long enough to avoid any problem. In the example used, if we lengthened the sweep to 20 s, the minimum harmonic ghost delay would be 6 s which should be long enough in most cases. Another way to avoid the harmonic ghost problem is to use upsweeps. Then the harmonic ghosts precede the primary events. The harmonic ghost of a deep reflection then competes with primary data from more shallow events which have a much higher signal level, so the ghost is overwhelmed. Also one can use a long sweep and this will put the ghosts from even the deepest event before time zero on our records where it does not have a 'ghost of a chance' of competing with any primary reflection signal.

Reasonable standards for vibrator phase

Table H-3 shows the error in milliseconds for various vibrator phase angle errors in degrees for frequencies ranging from 8 Hz to 150 Hz. At one time we specified that the vibrator phase had to be within $\pm 10°$ at all frequencies within the sweep. This specification is too loose at very low frequencies but at the very high seismic frequencies used today in high resolution work it is hard to meet and maintain in the field. We decided to specify instead that at all but the lowest frequencies the phase error could not be more than that which would result in a time error of ± 0.35 ms. Table H-4 shows the error in degrees that corresponds to time errors of ± 0.35 ms. This specification allows a greater phase error of $\pm 18.9°$ at 150 Hz. The new specification is about the same as the old $\pm 10°$ at a frequency of 80 Hz. At frequencies below 80 Hz the new standard is tougher than the previous standard, in fact, at the

Table H-3 Error times in milliseconds for various phase errors om degrees

Frequency (Hz)	*Error times (ms) for phase angle errors of:* 10°	15°	20°	25°	30°
8	3.47	5.21	6.94	8.68	10.42
10	2.78	4.17	5.55	6.94	8.33
12	2.31	3.47	4.63	5.79	6.94
20	1.39	2.08	2.78	3.47	4.17
30	0.93	1.39	1.85	2.31	2.78
40	0.69	1.04	1.39	1.74	2.08
60	0.46	0.69	0.93	1.16	1.39
80	0.35	0.52	0.69	0.87	1.04
100	0.28	0.42	0.56	0.69	0.83
120	0.23	0.35	0.46	0.58	0.70
130	0.21	0.32	0.43	0.53	0.64
140	0.20	0.30	0.40	0.50	0.60
150	0.19	0.28	0.37	0.46	0.56

very lowest frequencies it is too tough and is not necessary. Instead, below 20 Hz we relaxed the allowable time error to ± 1 ms which is what the old standard was at about 28 Hz. Below 28 Hz the ± 1 ms new standard is tougher than the old $\pm 10°$, but it needed to be tougher. Our good crews had little trouble and no complaint about meeting or exceeding these standards.

Table H-4 Phase error in degrees for a time error of 0.35 ms

Frequency (Hz)	*Phase error* (degrees)
8	1.01
10	1.26
12	1.51
20	2.52
30	3.78
40	5.04
60	7.56
80	10.08
100	12.6
110	13.9
120	15.1
130	16.4
140	17.6
150	18.9

Are the new standards tight enough for better seismic data? How bad will it be for a worst case where at each VP exactly half of the sweeps are at the positive limit and half are at the negative limit? If the maximum frequency we need to consider is 150 Hz, then the relative amplitude at the highest frequency would be equal to $\cos(18.9°) = 0.946$ instead of 1. For a worst case that does not look bad. The worst case time error would be if all of the sweeps from all of the vibrators were either ahead or behind by the limit of 0.35 ms. It is difficult to see how a worst case time error of 0.35 ms could hurt us very much.

At the low end the biggest deviation will almost certainly occur at the lowest frequency that is swept. Consider a frequency of 8 Hz. The period is 125 ms and we would allow an error of up to ± 1 ms which is $\pm 3.86°$. This is much tighter than the old standard of $\pm 10°$. The possible amplitude loss is trivial. The worst possible effect is likely to be a slight upset in the symmetry of the reflection wavelets, but that is not likely to cause much of a problem.

Our experience with these new standards was good and they are recommended as being both practical and adequate over the frequency range of 8–150 Hz.

Vibrator performance at low frequencies

In vibrator specifications there is a quantity called the displacement limit which has the units of seconds^{-1} and represents the lowest frequency for which the maximum displacement of the reaction mass could not exceed the piston stroke and hence 'hit the stops'.

Assume that the displacement of the reaction mass is sinusoidal including negligible delay in the operation of the control valves.

Let:

f = the displacement limit sought in Hertz (with units of seconds^{-1}),
S = usable stroke length in feet,
s = usable stroke length in inches = $12S$ therefore $S = s/12$,
F = hydraulic force in pounds available to accelerate and move the reaction mass (the reaction mass is also the piston),
t = time in seconds,
W = reaction mass in pounds
g = acceleration due to gravity and is approximately 32.2 ft/s/s or in metric units 980 cm/s/s, and
$X(t)$ = the sinusoidal displacement = $(S/2)\sin(2\pi f t)$. (1)

The displacement varies sinusoidally between the limits of $-S/2$ and $+S/2$. The force in poundals is equal to the mass times the acceleration or:

$$Fg = Wa_0$$

and

$$a_0 = FgW \qquad (2)$$

Where a_0 = maximum acceleration of the reaction mass in ft/s/s.
For sinusoidal motion the acceleration as a function of time is:

$$a(t) = -a_0 \sin(2\pi f t) \qquad (3)$$

(The negative sign was arbitrarily chosen.)
Since the double integral of acceleration is the displacement we can write:

$$X(t) = [a_0/(2\pi f)^2] \sin(2\pi f t) \qquad (4)$$

Combining equations (1) and (4) we can write:

$$S/2 = a_0/(2\pi f)^2$$

Solving for f we get:

$$f = (a_0/[(S/2)(2\pi)^2])^{0.5}$$

Substituting for a_0 from equation (2) we get:

$$f = \sqrt{\frac{Fg}{W\frac{S}{2}(2\pi)^2}}$$

Simplifying we get:

$$f = \sqrt{\frac{Fg}{2WS\pi^2}} \qquad (5)$$

This equation is valid for any consistent set of units.
If we want to use the stroke length in inches rather than feet we get:

$$f = \sqrt{\frac{6gF}{WS\pi^2}}$$

Substituting the approximate value of 32.2 ft/s/s for g we get:

$$f = \sqrt{\frac{19.56F}{sW}} \qquad (6)$$

with *s* in inches, *F* and *W* in pounds (or any consistent unit such as tons), and *g* in ft/s/s—the mixed bag we have chosen to use in the United States. In metric units the equation is:

$$f = \sqrt{\frac{49.65F}{sW}} \tag{7}$$

with *s* in cm, *F* and *W* in any consistent units, and *g* in cm/s/s.

The preceding also assumed that the baseplate was constrained by the earth such that the baseplate motion is negligible in comparison with the motion of the reaction mass. At low frequencies the wavelength in the earth is long; therefore the earth mass is very large as compared with the reaction mass which makes this assumption appropriate. The assumption fails if the earth surface is too soft such that the baseplate is not well constrained.

APPENDIX I—GROUND FORCE PHASE LOCK AND DRIVE LEVEL CONTROL

Ground force phase lock and drive level control is an improved and advantageous method for controlling vibrators. We couple vibrator signals into the deep earth through shallow earth layers which can be represented by masses and springs with various spring constants and various resonances. Our objective is to couple a Vibroseis sweep signal to the deep earth such that the signal from a perfect geophone at depth when correlated with the sweep would yield the desired wavelet. The desired wavelet would be either minimum-phase, zero-phase and zero-degree, or zero-phase and ninety-degree, depending upon the choice of the geophysicist. (In this discussion we will assume an unreal earth that is not dispersive and has an infinite *Q*, and consequently does not attenuate seismic waves.) We would prefer that the power level of the signal at depth be constant throughout the sweep. But this is secondary in that the amplitude spectrum can be dealt with in processing if we can achieve a satisfactory S/N over the required bandwidth.

For many years the phase of a vibrator was controlled by a signal from an accelerometer attached to the baseplate of the vibrator through a feedback loop. The feedback signal was used to make the baseplate motion in phase with the sweep signal. Then if the baseplate is tightly coupled to the earth, the motion of the shallow earth is the motion desired at depth. But the shallow earth motion is not a good representation of the motion of the earth at depth (even neglecting the time lag due to propagation). This difference in the motion between the shallow and deep earth can be illustrated by attaching five number 18 rubber bands together and attaching a magnet (like those we use to put seismic sections on a magnetic wall) to one end. (Dick Pelton used this illustration in his schools.) The magnet is only a suitable mass that a geophysicist is likely to have handy; it does not have to be a magnet. Loop the other end of the string of rubber bands around a finger and let the magnet dangle on the rubber bands until it comes to rest.

Your hand represents the baseplate of a vibrator, the magnet a mass of the deep earth, and the rubber band the spring between the shallow earth and the deep earth. If you slowly move your hand up and down, the magnet will move up and down in phase with it. So far so good. Now slowly increase the frequency of the hand motion like an increase in the frequency during a vibrator upsweep. At some point the phase of the motion of the magnet will start to deviate from the phase of the motion of the hand. Now you are approaching a resonant frequency of the rubber band spring and the mass of the magnet. Indeed at some frequency the two motions will be 180° out of phase and the magnet will be moving up when your hand is going down. Likewise as a vibrator approaches and goes through a shallow earth resonance, the phase of the motion of the baseplate and the motion of the earth deviate.

Empirical measurements have confirmed a theory that provides a way to control the vibrator so as to couple a signal into the deep earth with the phase that we want (Sallas 1984). The theory is that the force exerted by the baseplate on the earth is in phase with the deep earth displacement except for the time lag due to propagation. It is not feasible to measure this force directly but it can be measured by combining signals from a baseplate accelerometer and an accelerometer attached to the reaction mass of the vibrator. The force on the earth is a combination of the force resulting from the acceleration of the reaction mass and the force resulting from the acceleration of the baseplate. An equation for this ground force is:

$$F_g = (a_{rm})(m_{rm}) + (a_{bp})(m_{bp})$$

where

F_g is the ground force,
a_{rm} is the acceleration of the reaction mass,
m_{rm} is the mass of the reaction mass,
a_{bp} is the acceleration of the baseplate, and
m_{bp} is the mass of the baseplate.

Electrical signals from the two accelerometers are scaled to account for the difference in the two masses and then added to provide the ground force feedback signal to control the vibrator. This is called ground force phase lock on ground force phase control. The use of the ground force signal for phase control of vibrators is a significant step forward in a long history of improvements in Vibroseis equipment and technique. Now that it is generally available we should always use ground force phase lock. It goes a long way toward meeting the objective stated in the first paragraph to achieve the desired phase for the deep earth signal.

We stated a secondary objective of wanting the power level of the deep earth signal to be constant throughout the sweep. Ground force amplitude or drive level control can help us approach this goal. Ground force control of the vibrator drive level is particularly advantageous for vibrators that have peak force capability that is much greater than the hold-down weight of the vibrator. On an earth surface with a high Q resonance these vibrators can literally jump off the ground unless the drive level is reduced. With ground force amplitude control the drive level is automatically varied so as to try to maintain a constant ground force amplitude that is somewhat less than the hold-down weight of the vibrator.

Assume that a high peak force vibrator starts an upsweep at a frequency well below any shallow earth resonance. A maximum ground force level has been set that is somewhat below the hold-down weight of the vibrator. At low frequencies near the start of the sweep the drive level may be maximum and yet the ground force may be below the set maximum level. In this case the drive level remains maximum until at some higher frequency the ground force signal reaches the set maximum level. The drive level will then be reduced as required to maintain the ground force signal at the set maximum level as the sweep frequency continues to increase. As an earth resonance is approached the ground force signal will start to rise but the drive level will automatically be reduced by the feedback control to maintain the ground force signal at the set maximum level.

At resonance the drive level may be much lower than maximum with the ground force signal still near the desired level somewhat below the hold-down weight, so that the baseplate is still coupled to the earth. As the sweep frequency rises above the resonance, the ground force will start to decrease, but the feedback control will raise the drive level to maintain the set maximum ground force level. At high frequencies the maximum drive level may be reached and then the full peak force is applied to try to maintain the set maximum ground force level. However, the peak force may be insufficient: consequently the signal level may drop off somewhat. But the ground force amplitude control keeps the vibrator from decoupling at earth resonances and uses the full capability of the vibrator to inject as much energy as possible at the higher frequencies and at some of the low frequencies. Ground force amplitude control is desirable and is probably worth the additional cost in most cases. It should always be used with vibrators with high ratios of peak force to holddown weight.

It may be worthwhile to record the ground force signal (or the drive level) throughout each sweep. Without amplitude control it is the ground force signal that we would want. With amplitude control the ground force signal is stabilized and it is the drive level variations that we want. These signals respond to shallow earth resonances which tell us something about the shallow earth and about changes in the shallow earth from one VP to the next. Changes in the shallow earth cause static errors that we seek to correct. Now we have a signal we can record that might help us in making static corrections For example, suppose we are recording a line where the shallow earth has a thin soil layer underlain by a moderately high velocity layer with a rather high Q resonance. If the line crosses a buried channel just below the surface (now filled with soil and other poorly consolidated material) then the resonance will be affected. The Q will diminish and the ground force signal (or the drive level signal) near resonance will change.

The static correction needed will also change. Could this change in shallow earth resonance reflected by the recorded signal be used to improve static corrections (by helping us model the shallow earth change that caused the static problem)? It is worth trying in areas where static corrections are a big problem not well handled by presently used techniques. But there is no use in recording the extra signal unless we get someone interested who can and will try to do something useful with the data. The best progress requires a team effort.

REFERENCES

Sallas, J. J. (1984). 'Seismic vibrator control and the downgoing P-wave.' *Geophysics* **49**, 732–40.

APPENDIX J—NONLINEAR SWEEP THEORY

This appendix discusses nonlinear sweeps and what sweep function we might choose to partially offset earth attentuation and limitations of the effectiveness of our vibrator sources as a function of frequency.

Nonlinear sweeps to partially offset earth attention

One reason for using nonlinear sweeps is to boost the higher frequencies relative to the lower frequencies because the earth filter attenuates the higher frequencies more severely. We approach the problem of designing a good sweep for this purpose as if we would flatten the spectrum in the field for reflections from an objective horizon. Actually, we do not want to go that far in boosting the higher frequencies, but we want to know the type of nonlinear sweep that could flatten the spectrum. We can then use this equation but moderate the boost of the higher frequencies to whatever degree we decide is necessary and practical. We define the high frequency boost as the sweep time per Hertz at the highest frequency divided by the sweep time per Hertz at the lowest frequency.

We start by assuming earth attenuation (Chapter 22) is at least approximately defined by the following equation (neglecting divergence):

$$A(d) = A_0 e^{-cfd}$$

where

$A(d)$ is the amplitude of some frequency component as a function of distance traveled through the earth,
A_0 is the initial amplitude of the frequency component at the source,
c is a constant of the earth filter,
f is the frequency, and
d is the distance traveled in the earth.

The above equation is based upon the expectation that the earth attenuation exponent is proportional to frequency to the power of 1.

The energy in any narrow frequency band of the sweep is proportional to the time spent sweeping the narrow frequency band and is inversely proportional to the rate of change of frequency. To offset attenuation that increase with frequency we need to satisfy the equation:

$$\frac{dt}{df} e^{-Cf} = c_1$$

where

dt/df is the reciprocal of the rate of change of frequency,
C 2(cd), and
$c_1$is a constant.

Then

$$e^{cf}\, df = dt/c_1$$

we then integrate from $t=0$ to $t=t$ and get:

$$\frac{1}{C}[e^{Cf(t)} - e^{Cf(0)}] = \frac{t}{c_1} = \frac{0}{c_1}$$

where

$f(0)$ is the starting frequency at $t=0$
$f(t)$ is the frequency at $t=t$, and
L_s is the length of the sweep.

then

$$f(t) = \frac{1}{C}\ln\left(\frac{tC}{c_1} + e^{Cf(0)}\right)$$

Let

$(fs) = \text{starting frequency} = f(t = 0)$
$(fe) = \text{ending frequency} = f(t = L_s)$

then

$$(fe) = \frac{1}{C}\ln\left(\frac{CL_s}{c_1} + e^{Cfs}\right)$$

$$e^{C(fe)} = \frac{CL_s}{c_1} + e^{C(fs)}$$

$$c_1 = \frac{L_s C}{e^{C(fe)} - e^{C(fs)}}$$

$$f(t) = \frac{1}{C}\ln\left[\frac{t}{L_s}(e^{C(fe)} - e^{C(fs)}) + e^{C(fs)}\right] \qquad (1)$$

Equation (1) can be rearranged and written in a neater form by putting more of the mess in the constants.

$$f(t) = (1/C)\ln(tC_2 + C_3)$$

$$f(t) = (1/C)\ln[C_2(t + C_3/C_2)]$$

$$f(t) = (1/C)\ln[C_2] + (1/C)\ln[t + C_3/C_2)]$$

$$f(t) = C_4 + C_5\ln(t + C_6) \qquad (2)$$

where

$$C_4 = \frac{1}{C}\ln([e^{C(fe)} - e^{C(fs)}]/L_s)$$

$$C_5 = 1/C, \quad \text{and}$$

$$C_6 = \frac{L_s}{e^{C(fe) - C(fs)} - 1}$$

The amount of boost of the high frequencies is determined largely by the constant C_6. Plots of this equation for reasonable sweep lengths and frequencies indicated that the value for the constant C_6 should probably be between 0.5 for a strong boost and 2 for a moderate boost of the higher frequencies. If we choose to let $C_6 = 1$, we can solve for the other constants as follows:

$$f(t) = C_4 + C_5 \ln(t+1), \tag{3}$$

$$C_4 = (fs), \quad \text{and}$$

$$C_5 = \frac{fe - fs}{\ln(1 + L_s)}$$

We called this our 'log of $(t + 1)$' sweep and plotted sweeps for various sweep lengths and for various frequencies ranges of interest to us. We also had a 'log of $(t + 2)$' sweep, a 'log of $(t + 0.5)$' sweep and others that we considered.

We had erasable, programmable, read-only memory (EPROM) chips coded to provide specific logarithmic sweeps that we used successfully, but not without some problems with the EPROMS from time to time. We bought a lot of EPROMS in order to have relatively few sweeps available to use for our tests. When good, nonlinear sweep generators became available that did not require us to code and carry a box full of EPROMS, we looked to see what nonlinear sweeps were available with these units. The Texas Instruments sweep generator had one, and only one, logarithmic sweep available which turned out to be our 'log of $(t + 1)$' sweep that we had already adopted as our main sweep to use. But now we had the flexibility of changing sweep length, starting frequency and ending frequency, which formerly took separate sets of EPROMS for each change. Multiply by sets of EPROMS for each vibrator and for the recording truck and you can better appreciate the advantages of the adjustable nonlinear sweep generators with no EPROMS to code and manually interchange for each change in a single sweep parameter.

However, with the Texas Instruments 'log of $(t + 1)$' sweep we had limited capability of increasing or decreasing the amount of boost of the higher frequencies. Previously we said that the constant C_6 largely controlled the boost and with the Texas Instruments nonlinear sweep generator this is fixed at a value of 1. However, the sweep length also affects the high frequency boost with an increase in the boost for an increase in sweep length. The reason for this difference is that it is really the ratio of the maximum value of t (which is the sweep length L_s) to the constant C_6 that determines the boost. So we can increase the boost by either making C_6 smaller or making L_s longer. The equation for the Texas Instruments logarithmic sweep is:

$$F(t) = F_I + \frac{(F_F - F_I) \ln(t+1)}{\ln(SL+1)}$$

Table J-1 Example of the effect of sweep length on the high-frequency boost of the Texas Instruments logarithmic sweep

Frequency band 20–120 Hz *Sweep length = 8 s* Time(s)	F(t)	$\Delta f/\frac{1}{2}s$	Avg. f	Relative normalized boost 2(ms/Hz)	*Sweep length = 16 s* Time(s)	F(t)	$\Delta f/s$	Avg. f	Relative normalized boost (ms/Hz)	*Sweep length = 24 s* Time(s)	F(t)	$\Delta f/1\frac{1}{2}s$	Avg. f	Relative normalized boost (ms/Hz)/1.5
0	20				0	20				0	20			
		18.454	29.22	54.19			24.465	32.2	40.88			28.466	34.23	35.1
0.5	38.454				1	44.465				1.5	48.466			
		13.092	45.00	76.38			14.311	51.6	69.88			14.602	55.77	68.5
1	51.546				2	58.786				3	63.068			
		9.844	56.62	101.58			9.856	63.7	101.46			9.893	68.01	101.1
1.5	61.702				3	68.930				4.5	72.961			
		8.298	65.85	120.51			7.876	72.9	127.0			7.492	76.71	133.5
2	70.000				4	76.806				6	80.453			
		7.016	73.51	142.53			6.435	80.0	155.4			6.032	83.47	165.8
2.5	77.016				5	83.241				7.5	86.485			
		6.077	80.05	164.55			5.441	86.0	183.8			5.049	89.01	198.1
3	83.093				6	88.682				9	91.534			
		5.361	85.77	186.53			4.713	91.0	212.2			4.342	93.71	230.3

3.5	88.454				7	93.395				10.5	95.876			
		4.795	90.85	208.55			4.157	95.5	240.6			3.809	97.78	262.5
4	93.249				8	97.552				12	99.685			
		4.337	95.42	230.57			3.719	99.4	268.9			3.392	101.4	294.8
4.5	97.586				9	101.271				13.5	103.077			
		3.960	99.57	252.5			3.364	103.0	297.3			3.058	104.6	327.0
5	101.546				10	104.635				15	106.135			
		3.643	103.37	274.5			3.071	106.0	325.6			2.784	107.5	359.2
5.5	105.189				11	107.706				16.5	108.919			
		3.373	106.88	296.5			2.825	109.0	354			2.555	110.2	391.4
6	108.562				12	110.531				18	111.474			
		3.140	110.13	318.5			2.616	111.8	382			2.361	112.7	423.5
6.5	111.702				13	113.147				19.5	113.835			
		2.937	113.17	340.5			2.435	114.4	411			2.194	114.9	455.8
7	114.639				14	115.582				21	116.029			
		2.760	116.02	362.3			2.278	116.7	439			2.049	117.1	488.0
7.5	117.399				15	117.860				22.5	118.078			
		2.601	118.70	384.5			2.140	118.9	467.3			1.922	119.0	520.3
8	120				16	120				24	120			

384.5/54.19 = 7.1 = high-frequency boost

467.3/40.88 = 11.4 = high-frequency boost

520.3/35.1 = 14.8 = high-frequency boost

Table J-2 Example comparison between a Texas Instruments logarithmic sweep and Pelton sweeps of 9, 12, and 15 dB/octave

Sweep length — *16 s*
Sweep band — *20 Hz to 120 Hz (2.585 octaves)*

	Texas Instruments log sweep		*Pelton nonlinear sweeps*					
			9 dB/octave		*12 dB/octave*		*15 dB/octave*	
Time (s)	*Frequency (Hz)*	$\Delta f/s$	*Frequency (Hz)*	$\Delta f/s$	*Frequency (Hz)*	$\Delta f/s$	*Frequency (Hz)*	$\Delta f/s$
0	20.00		20.00		20.00		20.00	
		24.47		22.15		28.70		34.78
1	44.47		42.15		48.70		54.78	
		14.31		11.70		11.94		11.72
2	58.79		53.85		60.64		66.50	
		9.86		8.77		8.50		8.06
3	68.93		62.62		69.14		74.56	
		7.88		7.23		6.81		6.33
4	76.81		69.85		75.94		80.89	
		6.44		6.25		5.76		5.29
5	83.24		76.10		81.71		86.17	
		5.44		5.56		5.05		4.58

6	88.68		81.67		86.76		90.75	
		4.71		5.05		4.52		4.07
7	93.40		86.71		91.28		94.82	
		4.16		4.64		4.11		3.67
8	97.55		91.34		95.39		98.49	
		3.72		4.32		3.79		3.36
9	101.27		95.67		99.18		101.85	
		3.36		4.03		3.52		3.10
10	104.64		99.70		102.69		104.96	
		3.07		3.81		3.29		2.89
11	107.71		103.51		105.99		107.84	
		2.83		3.61		3.10		2.71
12	110.53		107.12		109.08		110.55	
		2.62		3.43		2.93		2.55
13	113.15		110.55		112.02		113.10	
		2.44		3.28		2.79		2.42
14	115.58		113.83		114.80		115.52	
		2.28		3.15		2.66		2.30
15	117.86		116.98		117.46		117.81	
		2.14		3.02		2.54		2.19
16	120.00		120.00		120.00		120.00	

where

F_I = initial frequency,
F_F = final frequency,
SL = sweep length, and
t = time.

Table J-1 shows the effect of sweep length on the boost. To compare sweeps of different lengths we normalized to the 16 s sweep. Each increment of time in this table is then 6.25% of the sweep length. The table gives the sweep frequency as a function of time. But what we are most interested in is the relative time spent in sweeping each frequency increment. Therefore for each time increment we show the average frequency and the normalized milliseconds per 1 Hz change in frequency. (The absolute milliseconds per 1 Hz change in frequency is half that shown for the 8 s sweep and is 1.5 times as much as is shown for the 24 s sweep.) From Table J-1 the high-frequency boost is 7.1, 11.4, and 14.8 for sweep lengths of 8, 16, and 24 seconds respectively. (This was calculated as the normalized milliseconds per Hertz change for the last time 6.25 per cent time increment, divided by the normalized milliseconds per Hertz

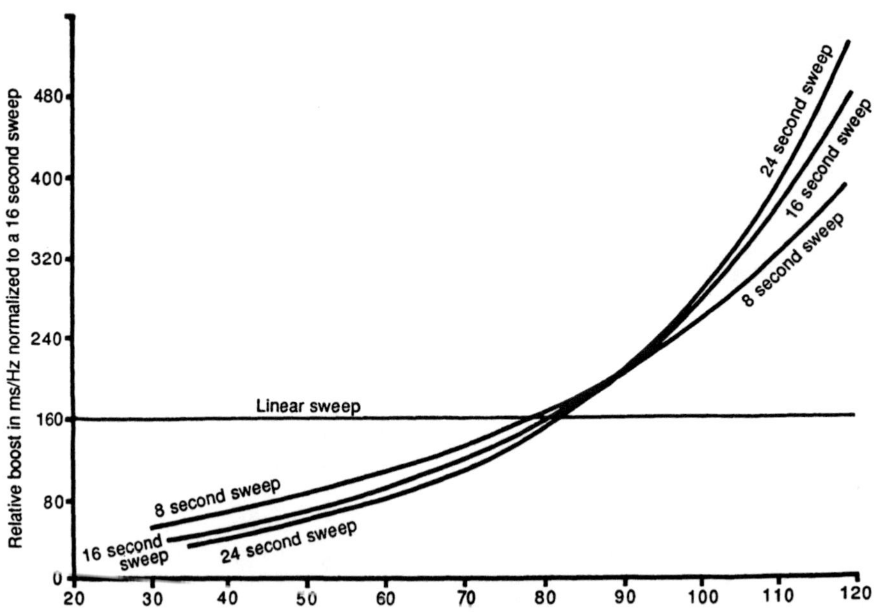

Fig. J-1 Normalized relative boost with a Texas Instruments logarithmic sweep. The relative boost has been normalized relative to a 16 s sweep. The longer the sweep length the greater the boost of the high frequencies relative to the low frequencies.

change for the first 6.25 per cent time increment of the sweep.) This proved to be an adequate range of boosts for our purposes. (We could have increased the range of boost more by using still shorter and till longer sweeps if we had wanted more range. Also, Texas Instruments provides, as an option, a sweep that is intermediate between linear and logarithmic, but we never chose to use it.)

Figure J-1 plots data from Table J-1 to compare the normalized, relative high-frequency boost for the Texas Instruments logarithmic sweep for sweep lengths of 8, 16, and 24 s for a sweep band of 20–120 Hz. It is believed that at least one other manufacturer of nonlinear sweep generators uses a similar or possibly identical equation for a sweep to boost the higher frequencies, and probably their units also would be perfectly satisfactory.

Other nonlinear sweeps to boost high frequencies

The Pelton sweep generator had a different types of nonlinear sweep to boost the higher frequencies. Their boost function range is 1–15 dB/octave in steps of

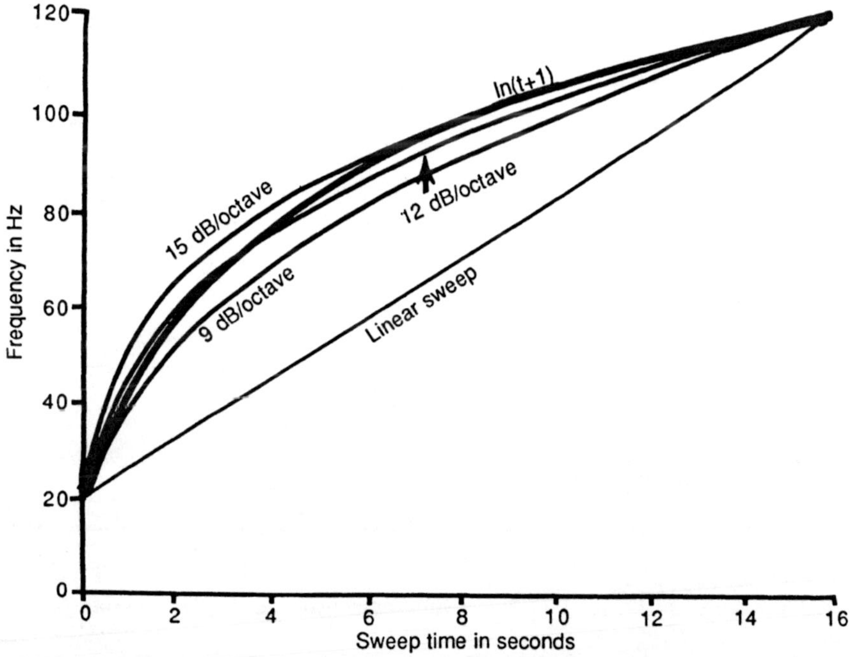

Fig. J-2 Frequency as a function of time for Texas Instruments and Pelton sweeps. The illustrated Texas Instruments and Pelton sweeps rapidly sweep through the lower frequencies and spend more time sweeping through the higher frequencies.

1 dB/octave. The equation is quite different and is the equation we would derive if we wanted to compensate for a loss of effectiveness of the vibrators at the higher frequencies due to a limited peak force. If that were the only consideration all we would need to offset the acceleration limit would be a 6 dB octave boost rather than the range of 1–15 dB/octave. But a vibrator system has elasticity including elasticity of the hydraulic fluid and this also contributes to the loss of efectiveness at the higher frequencies. Table J-2 compares 16 s Pelton sweeps of 9, 12, and 15 dB/octave with a 16 s Texas Instruments logarithmic sweep. Figure J-2 compares the frequency versus time curve for these Pelton sweeps with the Texas Instruments logarithmic sweep. Figure J-3 compares the relative high-frequency boost as a function of frequency for the Pelton 12 and 15 dB/octave sweeps with the Texas Instruments logarithmic sweep. (All sweeps plotted in Figure J-3 are 16 s in length.)

We had good success with both sweep generators since we used the boost simply to improve the S/N ratio enough at the higher frequencies so that we would be able in processing to flatten the spectrum and have a good S/N ratio over the bandwidth of our sweep.

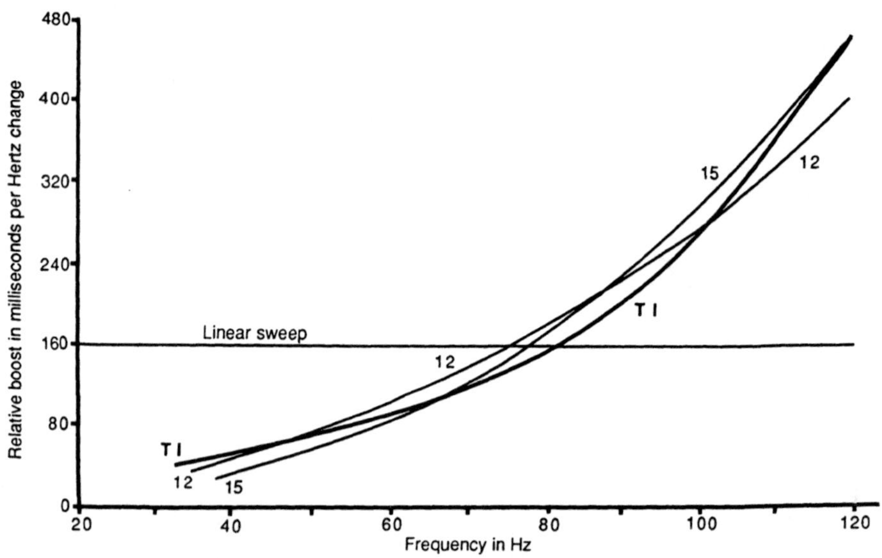

Fig. J-3 Relative high-frequency boost for Texas Instruments and Pelton sweeps. Note:

1. A Texas Instruments logarithmic sweep of 16 s duration is indicated by the letters 'TI';
2. Pelton sweeps of 16 s duration are indicated by the numbers '12' and '15' which also indicate the selected boost in dB per octave.

The equation used by Pelton for their nonlinear sweep is

$$f = (At + B)^{1/c}$$

where

f is the frequency as a function of time.
$A = [(F_2)^c - (F_1)^c]/SL$,
$B = (F_1)^c$,
$c = 1 + (D/6)$,
D = pre-emphasis in dB/octave ($-15 \leqslant D \leqslant 15$),
SL = sweep length in seconds,
F_1 = start frequency, and
F_2 = end frequency

There are several other manufacturers that offer nonlinear sweep generators that are regarded as excellent and have quite satisfactory sweeps to use to boost the higher frequencies. Some manufacturers also offer a T^2 sweep that is not recommended because it results in a tremendous increase in the boost right at the very end of the sweep. This would be useful if we just had to get up to a frequency of 80 Hz and 79 Hz required a modest boost but we had to spend a lot of sweep time to increase the boost to get 80 Hz. This is just not the way the earth filter works nor is it the way other factors work that affect high-frequency response.

Special purpose nonlinear sweeps

Particularly for shear-wave seismology we need good energy output at very low frequencies where vibrator performance may be limited by the pump capacity or by the displacement limit of the reaction mass. In this case we can apply a boost to the low frequencies instead of boosting the high frequencies. This can be done with Texas Instruments and Pelton sweep generators and with sweep generators offered by other manufacturers. But it is likely that we would want to apply this boost of the low frequencies only up to a frequency where the vibrators were more effective and then would want either a linear sweep or a sweep to boost the higher frequencies. Otherwise, we would probably end up with records with extremely little energy at the higher frequencies for reflections from depth. For applications such as this, EPROMS may offer the best solution. We can code the EPROMS to provide whatever boost we may want at any and all frequencies. For example, we might use a log sweep to boost the lower frequencies, a different log sweep to boost the higher frequencies, and a linear sweep through the middle frequencies.

The author's experience with EPROMS was with Pelton sweep generators. We had good results with EPROMS coded for us by Pelton to fit equations specified by us. We also coded EPROMS ourselves. There were commercially available equipment and software for coding EPROMS that did some pretty crazy things particularly at the lowest frequencies. To try to fit some specified frequency versus time curve, one system would let the rate of change of frequency gyrate and even be negative. The instructions for this equipment did caution the user about limiting the required rate of change of frequency, but the limit they recommended was not sufficient to prevent gyrations in the frequency as a function of time. Even so, our results with EPROMS coded in this crazy manner were not too bad. But after we realized what this system did, we chose not to use it anymore.

Comments on nonlinear sweeps

There was much reticence on the part of some developers of the Vibroseis technique and Vibroseis equipment to the use of nonlinear sweeps. This was based upon a strongly held conviction that what was needed was a flat spectrum for the wavelet injected into the earth and hence, a flat spectrum for the correlation of that wavelet. But it is not the wavelet injected into the earth that is important. What is important are the wavelets of the reflections from depth. We want these to have a flat spectrum after processing. To achieve this goal we need an adequate S/N ratio for all of the frequency components within our sweep band (or at least all that we can feasibly get and we should not be sweeping much higher than that). This requires a lot of sweep time at the higher frequencies to partially offset earth attenuation. If we chose to stick with linear sweeps, we would then be choosing to waste unneeded sweep time at the lower frequencies in order to get enough sweep time at the higher frequencies. We chose to boost the higher frequencies with nonlinear sweeps. This is the practical solution notwithstanding the reservations of some prominent people who have contributed much to the development of the Vibroseis technique and Vibroseis equipment. We all need to be careful not to become so attached to some idea or conviction that we cannot see the potential value of a contrary idea. Likewise you should question ideas and convictions expressed in this book, taking what helps and makes sense to you, but then move on to acquire still better seismic data.

Index

Where several pages are referenced the order of importance is indicated by bold face numbers for most important and by enclosure in parentheses for the least important.

Printed in the United Kingdom
by Lightning Source UK Ltd.
132655UK00001B/91/A

9 780412 350009